Data Engineering for Machine Learning Pipelines

From Python Libraries to ML Pipelines and Cloud Platforms

Pavan Kumar Narayanan

Apress®

Data Engineering for Machine Learning Pipelines: From Python Libraries to ML Pipelines and Cloud Platforms

Pavan Kumar Narayanan
Sheridan, USA

ISBN-13 (pbk): 979-8-8688-0601-8	ISBN-13 (electronic): 979-8-8688-0602-5
https://doi.org/10.1007/979-8-8688-0602-5

Copyright © 2024 by Pavan Kumar Narayanan

This work is subject to copyright. All rights are reserved by the Publisher, whether the whole or part of the material is concerned, specifically the rights of translation, reprinting, reuse of illustrations, recitation, broadcasting, reproduction on microfilms or in any other physical way, and transmission or information storage and retrieval, electronic adaptation, computer software, or by similar or dissimilar methodology now known or hereafter developed.

Trademarked names, logos, and images may appear in this book. Rather than use a trademark symbol with every occurrence of a trademarked name, logo, or image we use the names, logos, and images only in an editorial fashion and to the benefit of the trademark owner, with no intention of infringement of the trademark.

The use in this publication of trade names, trademarks, service marks, and similar terms, even if they are not identified as such, is not to be taken as an expression of opinion as to whether or not they are subject to proprietary rights.

While the advice and information in this book are believed to be true and accurate at the date of publication, neither the authors nor the editors nor the publisher can accept any legal responsibility for any errors or omissions that may be made. The publisher makes no warranty, express or implied, with respect to the material contained herein.

Managing Director, Apress Media LLC: Welmoed Spahr
Acquisitions Editor: Celestin Suresh John
Development Editor: Laura Berendson
Editorial Project Manager: Gryffin Winkler

Cover designed by eStudioCalamar

Cover photo by Bradley Dunn on Unsplash

Distributed to the book trade worldwide by Springer Science+Business Media New York, 1 New York Plaza, Suite 4600, New York, NY 10004-1562, USA. Phone 1-800-SPRINGER, fax (201) 348-4505, e-mail orders-ny@springer-sbm.com, or visit www.springeronline.com. Apress Media, LLC is a California LLC and the sole member (owner) is Springer Science + Business Media Finance Inc (SSBM Finance Inc). SSBM Finance Inc is a **Delaware** corporation.

For information on translations, please e-mail booktranslations@springernature.com; for reprint, paperback, or audio rights, please e-mail bookpermissions@springernature.com.

Apress titles may be purchased in bulk for academic, corporate, or promotional use. eBook versions and licenses are also available for most titles. For more information, reference our Print and eBook Bulk Sales web page at http://www.apress.com/bulk-sales.

Any source code or other supplementary material referenced by the author in this book is available to readers on GitHub. For more detailed information, please visit https://www.apress.com/gp/services/source-code.

If disposing of this product, please recycle the paper

To my beloved father

Table of Contents

About the Author xix

About the Technical Reviewer xxi

Introduction xxiii

Chapter 1: Core Technologies in Data Engineering 1

    Introduction 1

    Python Programming 2

        F Strings 2

        Python Functions 3

        Advanced Function Arguments 4

        Lambda Functions 5

        Decorators in Python 6

        Type Hinting 7

        Typing Module 7

        Generators in Python 8

        Enumerate Functions 8

        List Comprehension 11

        Random Module 13

    Git Source Code Management 16

        Foundations of Git 16

        GitHub 17

        Core Concepts of Git 23

    SQL Programming 31

        Essential SQL Queries 31

        Conditional Data Filtering 32

        Joining SQL Tables 32

TABLE OF CONTENTS

    Common Table Expressions 33

    Views in SQL 34

    Temporary Tables in SQL 36

    Window Functions in SQL 37

Conclusion 39

Chapter 2: Data Wrangling using Pandas 41

Introduction 41

Data Structures 42

    Series 42

    Data Frame 43

Indexing 44

    Essential Indexing Methods 45

    Multi-indexing 49

    Time Delta Index 52

Data Extraction and Loading 56

    CSV 56

    JSON 57

    HDF5 57

    Feather 58

    Parquet 58

    ORC 59

    Avro 59

    Pickle 60

    Chunk Loading 60

Missing Values 61

    Background 61

    Missing Values in Data Pipelines 62

    Handling Missing Values 63

Data Transformation 65

    Data Exploration 66

    Combining Multiple Pandas Objects 69

Data Reshaping 75
    pivot() 75
    pivot_table() 76
    stack() 78
    unstack() 80
    melt() 81
    crosstab() 82
    factorize() 85
    compare() 87
    groupby() 89
Conclusion 91

Chapter 3: Data Wrangling using Rust's Polars 93
Introduction 93
Introduction to Polars 94
    Lazy vs. Eager Evaluation 95
Data Structures in Polars 96
    Polars Series 97
    Polars Data Frame 98
    Polars Lazy Frame 100
Data Extraction and Loading 102
    CSV 102
    JSON 102
    Parquet 103
Data Transformation in Polars 103
    Polars Context 104
    Basic Operations 109
    String Operations 113
    Aggregation and Group-By 113
Combining Multiple Polars Objects 114
    Left Join 114
    Outer Join 116

TABLE OF CONTENTS

    Inner Join 117

    Semi Join 118

    Anti Join 119

    Cross Join 119

  Advanced Operations 120

    Identifying Missing Values 121

    Identifying Unique Values 122

    Pivot Melt Examples 122

  Polars/SQL Interaction 123

  Polars CLI 128

  Conclusion 130

Chapter 4: GPU Driven Data Wrangling Using CuDF 133

  Introduction 133

  CPU vs. GPU 134

    Introduction to CUDA 135

  Concepts of GPU Programming 135

    Kernels 135

    Memory Management 136

  Introduction to CuDF 136

    CuDF vs. Pandas 136

    Setup 137

    File IO Operations 143

  Basic Operations 144

    Column Filtering 145

    Row Filtering 146

    Sorting the Dataset 147

  Combining Multiple CuDF Objects 148

    Left Join 148

    Outer Join 150

    Inner Join 150

Left Semi Join	151
Left Anti Join	151
Advanced Operations	152
Group-By Function	152
Transform Function	153
apply()	154
Cross Tabulation	156
Feature Engineering Using cut()	157
Factorize Function	158
Window Functions	159
CuDF Pandas	160
Conclusion	161

Chapter 5: Getting Started with Data Validation using Pydantic and Pandera 163

Introduction	163
Introduction to Data Validation	164
Need for Good Data	164
Definition	165
Principles of Data Validation	166
Introduction to Pydantic	168
Type Annotations Refresher	168
Setup and Installation	170
Pydantic Models	171
Fields	173
JSON Schemas	174
Constrained Types	176
Validators in Pydantic	176
Introduction to Pandera	178
Setup and Installation	179
DataFrame Schema in Pandera	180
Data Coercion in Pandera	182
Checks in Pandera	183

Table of Contents

Statistical Validation in Pandera 186

Lazy Validation 189

Pandera Decorators 192

Conclusion 196

Chapter 6: Data Validation using Great Expectations 197

Introduction 197

Introduction to Great Expectations 198

Components of Great Expectations 198

Setup and Installation 200

Getting Started with Writing Expectations 203

Data Validation Workflow in Great Expectations 207

Creating a Checkpoint 216

Data Documentation 220

Expectation Store 222

Conclusion 223

Chapter 7: Introduction to Concurrency Programming and Dask 225

Introduction 225

Introduction to Parallel and Concurrent Processing 226

History 226

Python and the Global Interpreter Lock 226

Concepts of Parallel Processing 227

Identifying CPU Cores 227

Concurrent Processing 228

Introduction to Dask 230

Setup and Installation 230

Features of Dask 231

Tasks and Graphs 231

Lazy Evaluation 232

Partitioning and Chunking 232

Serialization and Pickling	232
Dask-CuDF	233
Dask Architecture	233
Core Library	234
Schedulers	234
Client	234
Workers	236
Task Graphs	236
Dask Data Structures and Concepts	236
Dask Arrays	236
Dask Bags	239
Dask DataFrames	241
Dask Delayed	244
Dask Futures	246
Optimizing Dask Computations	249
Data Locality	250
Prioritizing Work	251
Work Stealing	251
Conclusion	252

Chapter 8: Engineering Machine Learning Pipelines using DaskML 253

Introduction	253
Machine Learning Data Pipeline Workflow	254
Data Sourcing	255
Data Exploration	255
Data Cleaning	255
Data Wrangling	256
Data Integration	256
Feature Engineering	256
Feature Selection	257
Data Splitting	257
Model Selection	258

TABLE OF CONTENTS

- Model Training 258
- Model Evaluation 259
- Hyperparameter Tuning 259
- Final Testing 259
- Model Deployment 260
- Model Monitoring 260
- Model Retraining 260
- Dask-ML Integration with Other ML Libraries 260
  - scikit-learn 261
  - XGBoost 261
  - PyTorch 261
  - Other Libraries 262
- Dask-ML Setup and Installation 262
- Dask-ML Data Preprocessing 263
  - RobustScaler( ) 263
  - MinMaxScaler( ) 264
  - One Hot Encoding 265
  - Cross Validation 266
- Hyperparameter Tuning Using Dask-ML 269
  - Grid Search 269
  - Random Search 270
  - Incremental Search 270
- Statistical Imputation with Dask-ML 272
- Conclusion 276

Chapter 9: Engineering Real-time Data Pipelines using Apache Kafka 277

- Introduction 277
- Introduction to Distributed Computing 278
- Introduction to Kafka 279
- Kafka Architecture 281
  - Events 281
  - Topics 282

Table of Contents

- Partitions 282
- Broker 282
- Replication 282
- Producers 283
- Consumers 283
- Schema Registry 284
- Kafka Connect 284
- Kafka Streams and ksqlDB 284
- Kafka Admin Client 286
- Setup and Development 287
- Kafka Application with the Schema Registry 296
  - Protobuf Serializer 301
- Stream Processing 307
  - Stateful vs. Stateless Processing 308
- Kafka Connect 320
  - Best Practices 321
- Conclusion 322

Chapter 10: Engineering Machine Learning and Data REST APIs using FastAPI ... 323

- Introduction 323
- Introduction to Web Services and APIs 324
  - OpenWeather API 324
  - Types of APIs 324
  - Typical Process of APIs 326
  - Endpoints 326
  - API Development Process 327
  - REST API 328
  - HTTP Status Codes 329
- FastAPI 330
  - Setup and Installation 331
  - Core Concepts 332
  - Path Parameters and Query Parameters 332

xiii

TABLE OF CONTENTS

    Pydantic Integration 337

    Dependency Injection in FastAPI 342

    Database Integration with FastAPI 343

    Object Relational Mapping 347

    Building a REST Data API 350

    Middleware in FastAPI 354

    ML API Endpoint Using FastAPI 355

Conclusion 359

Chapter 11: Getting Started with Workflow Management and Orchestration 361

Introduction 361

Introduction to Workflow Orchestration 362

    Workflow 362

    ETL and ELT Data Pipeline Workflow 365

    Workflow Configuration 368

    Workflow Orchestration 369

Introduction to Cron Job Scheduler 371

    Concepts 371

    Crontab File 372

    Cron Logging 374

    Cron Job Usage 375

    Cron Scheduler Applications 380

    Cron Alternatives 381

Conclusion 381

Chapter 12: Orchestrating Data Engineering Pipelines using Apache Airflow 383

Introduction 383

Introduction to Apache Airflow 384

    Setup and Installation 384

Airflow Architecture 389

    Web Server 390

    Database 390

- Executor 390
- Scheduler 391
- Configuration Files... 391
- A Simple Example.. 393
- Airflow DAGs 395
  - Tasks 397
  - Operators.. 397
  - Sensors... 399
  - Task Flow.. 400
  - Xcom... 401
  - Hooks.. 402
  - Variables... 403
  - Params 406
  - Templates 407
  - Macros.. 407
- Controlling the DAG Workflow 408
  - Triggers... 411
- Conclusion 413

Chapter 13: Orchestrating Data Engineering Pipelines using Prefect 415

- Introduction.. 415
- Introduction to Prefect 416
  - Setup and Installation.. 417
- Prefect Server 418
- Prefect Development... 421
  - Flows 421
  - Flow Runs 421
  - Interface 424
  - Tasks 424
  - Results.. 427
  - Persisting Results.. 428
  - Artifacts in Prefect... 432

TABLE OF CONTENTS

    States in Prefect 438

    State Change Hooks 441

    Blocks 442

    Prefect Variables 444

    Variables in .yaml Files 445

    Task Runners 446

Conclusion 448

Chapter 14: Getting Started with Big Data and Cloud Computing 451

Introduction 451

Background of Cloud Computing 452

    Networking Concepts for Cloud Computing 453

Introduction to Big Data 454

    Hadoop 454

    Spark 455

Introduction to Cloud Computing 456

    Cloud Computing Deployment Models 456

    Cloud Architecture Concepts 458

    Cloud Computing Vendors 461

    Cloud Service Models 461

    Cloud Computing Services 463

Conclusion 471

Chapter 15: Engineering Data Pipelines Using Amazon Web Services 473

Introduction 473

AWS Console Overview 474

    Setting Up an AWS Account 474

    Installing the AWS CLI 488

AWS S3 489

    Uploading Files 492

AWS Data Systems 493

    Amazon RDS 493

Amazon Redshift	497
Amazon Athena	504
Amazon Glue	506
AWS Lake Formation	506
AWS SageMaker	519
Conclusion	530

Chapter 16: Engineering Data Pipelines Using Google Cloud Platform 531

Introduction	531
Google Cloud Platform	532
Set Up a GCP Account	533
Google Cloud Storage	534
Google Cloud CLI	535
Google Compute Engine	537
Cloud SQL	541
Google Bigtable	547
Google BigQuery	552
Google Dataproc	554
Google Vertex AI Workbench	559
Google Vertex AI	562
Conclusion	569

Chapter 17: Engineering Data Pipelines Using Microsoft Azure 571

Introduction	571
Introduction to Azure	572
Azure Blob Storage	574
Azure SQL	577
Azure Cosmos DB	584
Azure Synapse Analytics	591
Azure Data Factory	593
Azure Functions	602

TABLE OF CONTENTS

Azure Machine Learning 606

Azure ML Data Assets 609

Azure ML Job 611

Conclusion 616

Index 617

About the Author

Pavan Kumar Narayanan has an extensive and diverse career in the information technology industry, with a primary focus on the data engineering and machine learning domains. Throughout his professional journey, he has consistently delivered solutions in environments characterized by heterogeneity and complexity. His experience spans a broad spectrum, encompassing traditional data warehousing projects following waterfall methodologies and extending to contemporary integrations that involve application programming interfaces (APIs) and message-based systems. Pavan has made substantial contributions to large-scale data integrations for applications in data science and machine learning. At the forefront of these endeavors, he has played a key role in delivering sophisticated data products and solutions, employing a versatile mix of both traditional and agile approaches. Pavan completed his Master of Science in Computational Mathematics at Buffalo State University, Operations Research at the University of Edinburgh, and Information Technology at Northern Arizona University. He can be contacted at pavan.narayanan@gmail.com.

About the Technical Reviewer

Suvoraj Biswas is a leading expert and emerging thought leader in enterprise generative AI, specializing in architecture and governance. He authored the award-winning book *Enterprise GENERATIVE AI Well-Architected Framework & Patterns*, acclaimed by industry veterans globally. Currently an enterprise solutions architect at Ameriprise Financial, a 130-year-old Fortune 500 organization, Biswas has over 19 years of IT experience across India, the United States, and Canada. He has held key roles at Thomson Reuters and IBM, focusing on enterprise architecture, architectural governance, DevOps, and DevSecOps.

Biswas excels in designing secure, scalable enterprise systems, facilitating multi-cloud adoption, and driving digital transformation. His expertise spans Software as a Service (SaaS) platform engineering, Amazon Web Services (AWS) cloud migration, generative AI implementation, machine learning, and smart IoT platforms. He has led large-scale, mission-critical enterprise projects, showcasing his ability to integrate advanced cutting-edge technologies into practical business strategies.

Introduction

The role of data engineering has become increasingly crucial. Data engineering is the foundation on which organizations build their reporting, analytics, and machine learning capabilities. Data engineering as a discipline transforms raw data into reliable and insightful informational resources that provide valuable insights to organizations. However, data engineering is much more than that. It is a complex and iterative process.

If an enterprise project can be seen as constructing a multi-storied building, then data engineering is where the execution happens—from architecting and designing the structure of the data solution to laying the data storage foundation; installing the data processing frameworks; developing pipelines that source, wrangle, transform, and integrate data; and developing reports and dashboards while ensuring quality and maintenance of the systems, pipelines, reports, and dashboards, among several other things.

This book is designed to serve you as a desk reference. Whether you are a beginner or a professional, this book will help you with the fundamentals and provide you with job-ready skills to deliver high-value proposition to the business. This book is written with an aim to bring back the knowledge gained from traditional textbook learning, providing an organized exploration of key data engineering concepts and practices. I encourage you to own a paper copy. This book will serve you well for the next decade in terms of concepts, tools, and practices.

Though the content is exhaustive, the book covers around 60% of data engineering topics. Data engineering is a complex and evolving discipline; there are so many more topics that are not touched on. As a reader, I encourage you to contact me and share your remarks and things you witnessed in the data space that you wish to share. I am thankful to those who have already written to me, and I wish to see more of these efforts from the readers and enthusiasts.

INTRODUCTION

Having said that, this book covers a wide range of topics that are practiced in modern data engineering, with a focus on building data pipelines for machine learning development. This book is organized into the following topics:

- Data wrangling: Techniques for cleaning, transforming, and preparing data for analysis using Pandas, Polars, and GPU-based CuDF library

- Data validation: Ensuring data quality and integrity throughout the pipeline using Pydantic, Pandera, and Great Expectations

- Concurrency computing: Leveraging parallel processing for improved performance using Dask and DaskML

- APIs: Designing and working with APIs for data exchange using FastAPI

- Streaming pipelines: Designing and working with real-time data flows efficiently using Kafka

- Workflow orchestration: Managing complex data workflows and dependencies using cron, Airflow, and Prefect

- Cloud computing: Utilizing cloud services for scalable data engineering solutions using AWS, Azure, and Google Cloud Platform (GCP)

The book covers major cloud computing service providers. The reason is that organizations are increasingly moving toward a multi-cloud approach. And so, being equipped with knowledge on diverse cloud environments is crucial for your career success. This book delivers that.

If you are a data engineer looking to expand your skillset and move up the ladder, a data scientist who seeks to understand how things work under the hood, a software engineer who is already doing the work, or a student who is looking to gain a position, this book is essential for you. Even if you are in a functional domain or a founder of a start-up, interested in setting up a team and good data engineering practice, this book will help you from start to end. This book is a guide that will help with learning for people at every stage (from novice to expert).

INTRODUCTION

I highly encourage you not to seek solutions from chat bots during learning. At the time of writing this (June 2024), many bot programs either provide legacy code or steer you into their own direction and approach. It may seem better because of the way information is presented, but it slows down your learning and professional development. Even to this date, I find search engines to be more effective, which take me to Q&A sites where one can witness how a solution has been arrived at and why other approaches didn't succeed. Furthermore, I find Google search to be effective in terms of locating a certain method or a function in an API documentation that may be better than the website itself.

At the end of this book, you will be equipped with skills, knowledge, and practices that can be derived from a decade of experience in data engineering. Many vendors and cloud computing service providers provide you with free credits if you register for the first time. I encourage you to make use of these free credits, explore various tools and services, and gain knowledge. And when you are done, please remember to clean up your workspace, shut down any virtual machines and databases, take a backup of your work, and delete your workspace in the cloud to prevent additional charges.

CHAPTER 1

Core Technologies in Data Engineering

Introduction

This chapter is designed to get you started. As we begin our journey into advanced data engineering practices, it is important to build a solid foundation in key concepts and tools. This chapter is designed to provide a refresher on Python programming and software version control using GitHub and a high-level review of structured query language (SQL). While these may seem not connected and different from each other, they play an important role in developing effective data engineering pipelines. Python's data ecosystem, software version controlling systems using Git, and working with relational database systems using SQL will certainly solidify your foundations and help you tackle data engineering challenges. These are challenging and complex topics, especially reviewing them together; I hope you persist, learn, and practice on your own.

By the end of the chapter, you will

- Understand some of the key Python programming concepts relevant to advanced software engineering practices.

- Gain an appreciation toward software version control systems and git; create git repositories; and perform essential version control operations like branching, forking, and submitting pull requests.

- Have enough knowledge to set up your own code repository in GitHub.

- Review fundamental SQL concepts for querying and manipulating data in database systems.

- Identify advanced SQL concepts and best practices for writing efficient and optimized SQL code.

Engineering data pipelines is not easy. There is a lot of complexity and it involves leveraging analytical thinking. The complexity includes but is not limited to varying data models, intricate business logic, tapping into various data sources, and so many other tasks and projects that make up the profession. Often, data engineers do build machine learning models to identify anomalies and build advanced imputations on the data. I have taken some of the topics in my experience that would serve value to the data engineer or software developer who is engineering pipelines.

Python Programming

There are two major versions of the Python programming language, Python 2 and Python 3. Python 2 has reached end of life for support. Python 3 is the de facto language version for Python programming. The Python programming language (version 3.*) is still the most preferred language for development, comes with extensive community-backed data processing libraries, and is widely adopted. In this section, let us take some time to look at some of the Python programming concepts that may help, in addition to your basic programming knowledge. These are discussed at a high level. I encourage you to perform further investigation into these classes, methods, and arguments that can be passed into these methods.

F Strings

Python f strings allow you to easily insert variables and expressions within a print string. When compared with traditional string formatting, f strings are a lot more readable and easy to understand. You can create an f string by placing the character F in front of the string, before quotes. The character is case insensitive, meaning you can either insert an F or an f. Here is an example:

```
Greetings = f"Hello, it is a beautiful day here, at the park today"
```

Now you can insert variables and expressions within the f string. We have shown the preceding f string without inserts for demonstration purposes. Here is how you can perform that:

```
Fruit = "berries"
Cost = 1.5
Number = 3

print(f"the {fruit} costs {cost} per piece, totals to $ {cost*number} dollars")
```

will return

```
the oranges costs 1.5 per piece, totals to $ 4.5 dollars
```

Python Functions

In programming, functions are a set of instructions that can be executed together to accomplish a specific task or operation. It may or may not accept inputs and may or may not return output. Let us look at an example of a function:

```
def add(a,b):
    c = a+b
    return c
```

When you call this function

```
print(add(2,3)
```

it will return

5

Let us look at a bit more substantial function. Here, we have a function that takes Celsius as input and calculates Fahrenheit as output:

```
def f_calc(celsius):
    return ((9/5 * celsius) + 32)
```

This is the same as the following function. Both are syntactically equal. We will discuss this way of defining functions in another section:

```
def f_calc(celsius: int) -> float:
    return ((9/5 * celsius) + 32)
```

In the preceding code, we have passed a variable named celsius of type integer. Inside the function we are calculating a value and returning that value in floating point type. This is how functions can accept input parameters, process them, and provide output.

Advanced Function Arguments

*args

Let us look at using the "*args" parameter in a function. The arguments parameter in functions can be used to pass a variable number of arguments to a function. If you have the syntax "*args" once, you can pass several numbers of variables, without having to define each one of them. Let us look at two examples in contrast to understand this concept better:

```
def add(a,b):
    c = a+b
    return c
```

When you run this

```
add(2,3)
```

it will return

5

What if we do not know how many input parameters we need? We can use arguments here:

```
def add_args(*args):
    total_sum = 0
    for i in args:
        total_sum += i
    return total_sum
```

So when you run this

```
add_args(3,4,5,6,7)
```

it will return

```
25
```

Note You can also pass a list to the function and still obtain the functionality. But with arguments, you can call an empty function and return none.

**kwargs

This is quite similar to arguments, and the difference is that you can have a key associated with each item in the argument:

```
def add_kwargs(**kwargs):
    total_sum = 0
    for i in kwargs.values():
    total_sum += i
    return total_sum
```

Use of args and kwargs can be highly beneficial in enhancing the flexibility of a function, complementing the decorator implementation and passing a varying number of parameters as inputs.

Lambda Functions

In Python, lambdas are short for lambda expressions. They are inline functions that evaluate one single expression. That single expression is evaluated when the expression is called. Let us look at this with an illustration:

```
calculate_fahrenheit = lambda celsius: 9/5*celsius + 32
```

So when you call this function

```
calculate_fahrenheit(27)
```

it will return

80.6

Decorators in Python

Decorators in Python are a way to add additional functionality to your existing function. The decorator itself is a function. In the parameter column, it would take another function, as an input parameter. In calculus terms, it is equal to the function of a function. If math is not your cup of tea, then let's look at a simple example.

Here is a test decorator:

```python
def testdec(func):
    def wrapper():
        print("+++Beginning of Decorator Function+++")
        func()
        print("+++End of Decorator Function+++")
    return wrapper
```

Let us try to use this decorator on a sample function:

```python
@testdec
def hello():
    print("Hello!")
```

When we run this function

```python
hello()
```

it will provide us

```
+++Beginning of Decorator Function+++
Hello!
+++End of Decorator Function+++
```

Type Hinting

Type hinting is a Python feature that lets you specify the data types of the variable declarations, arguments in functions, and the variables that these functions return. Type hints make it easier to understand what data types to expect and which function returns what type of data. The idea behind type hinting is better readability, code maintainability, and being able to debug and troubleshoot errors faster. Type hints are not mandatory; you can still pass an argument without specifying the data type (Python will still check them during runtime).

Here is a simple example demonstrating the type hinting functionality in Python:

```
def some_function(a: float, b: float) -> float:
    c = a+b
    return c
```

Typing Module

The typing module supports type hints for advanced data structures with various features. Here is an example for a dictionary:

```
From typing import Dict

var2: Dict[int, str] = {234: "JohnDoe"}
```

Notice the usage of "Dict." It serves the same purpose as "dict," a dictionary type offered by core Python. "Dict" does not provide additional functionality except you have to specify the data type for both key and value aspects of a dictionary. If you are using Python 3.9 or later, you can use either "Dict" or "dict."

Let us look at another example called the Union, where a variable could be one of few types:

```
from typing import Union

var4: list[Union[int, float]] = [3, 5, 5.9, 6.2]
```

In Python 3.10+, you can also use the | operator. Here are some examples:

```
var5: list[int | float] = [3, 5, 5.9, 6.2]
var6: list[int | str | float] = [5, 6, "foo", "bar", 7.8]
```

Generators in Python

A generator is like a regular function in Python and uses a yield keyword instead of a return keyword to return the results. The yield keyword helps return an iterator that would provide a sequence of values when iterated over. The return keyword on the other hand returns a one-time result of the function.

Let us look at a simple example:

```
def generator_square(n):
    x = 0
    while x <= n:
        yield x*x
        x += 1

for i in generator_square(10):
    print(i)
```

In this example, we have a function that generates a square of a given number. Instead of returning a list one time, this function, when called, will return only one instance of a function. This function returns a generator object. And so, when you iterate over the function, you will receive values one at a time in each iteration.

Enumerate Functions

The Oxford dictionary of English defines enumerate as "to list or mention of a number of things, one by one." This is all you need to remember. The enumerate function, when applied to a data type, returns an enumerate object. These enumerate objects return the data type with an index that starts from 0 and the value corresponding to the index.

Here is a basic illustration:

```
a = ['apples','oranges','melons','bananas']
print(a)
```

would yield

```
['apples', 'oranges', 'melons', 'bananas']
```

However, if you enumerate the same list, you would obtain something like this:

```
for i in enumerate(a):
    print(i)
```

would yield

```
(0, 'apples')
(1, 'oranges')
(2, 'melons')
(3, 'bananas')
```

Here is an illustration for using a tuple with type hints for better readability:

```
var10: tuple[int, str] = ([1,"apple"],[2,"orange"],[3,"melon"],
[4,"banana"])
```

```
print(var10)
([1, 'apple'], [2, 'orange'], [3, 'melon'], [4, 'banana'])
```

```
for i,j in enumerate(var10):
    print(i,"-->",j)
```

would yield

```
0 --> [1, 'apple']
1 --> [2, 'orange']
2 --> [3, 'melon']
3 --> [4, 'banana']
```

When it comes to a dictionary, the dictionary already has a key and it is not necessary to add an index. However, it is not a limiting factor. We can still iterate over a dictionary using enumerate but it needs to be over each (key, value) tuple. We need to use the "items()" method to obtain each tuple from the dictionary.

Here is an illustration iterating over a dictionary:

```
sample_dict = {1:"apple", 2:"orange",3:"melon",4:"banana"}
print(sample_dict)
```

would yield

```
{1: 'apple', 2: 'orange', 3: 'melon', 4: 'banana'}
```

Here is another illustration:

```
for i,j in sample_dict.items():
    print(f"Key is {i} and Value is {j}")
```

would yield

```
Key is 1 and Value is apple
Key is 2 and Value is orange
Key is 3 and Value is melon
Key is 4 and Value is banana
```

Let us use enumerate, where the variable "i" stands for index and variable "j" stands for each key-value pair in a dictionary:

```
for i, j in enumerate(sample_dict.items()):
    print(i,j)
```

would yield

```
0 (1, 'apple')
1 (2, 'orange')
2 (3, 'melon')
3 (4, 'banana')
```

Let us try another variant here:

```
for index, (key, value) in enumerate(sample_dict.items()):
    print(index, key, value)
```

would yield

```
0 1 apple
1 2 orange
2 3 melon
3 4 banana
```

Here is an illustration of iterating over a string object:

```
name = "JohnDoe"
for i,j in enumerate(name):
    print(i,"-->",j)
```

would yield

```
0 --> J
1 --> o
2 --> h
3 --> n
4 --> D
5 --> o
6 --> e
```

List Comprehension

List comprehension is a method of defining and creating a list through pythonic expressions. List comprehension returns a list. It can be seen as a substitute for lambda expressions.

Let us look at a simple example:

```
readings_in_km = [10,12,13,9,15,21,24,27]

readings_in_miles = [(i/1.6) for i in readings_in_km]

print(readings_in_miles)

[6.25, 7.5, 8.125, 5.625, 9.375, 13.125, 15.0, 16.875]
```

We were able to dynamically (and elegantly) create a list of readings that were in miles from readings that were measured in kilometers.

Let us look at another illustration, where we square or cube a number based on whether the number is odd or even. This would mean there needs to be an iterator loop and a decision loop to check each item in the list. Here is how the code may look without list comprehensions:

```
a: list[int] = [1,2,3,4,5,6,7,8,9,10]
b: list[int] =[]

for x in a:
    if x % 2 == 0:
        b.append(x**2)
    else:
        b.append(x**3)
print(b)
```

Let us try to apply list comprehension to this operation:

```
a: list[int] = [1,2,3,4,5,6,7,8,9,10]
b = [x**2 if x % 2 == 0 else x**3 for x in a]
```

Here is another example creating tuples within lists. Here is the regular way of performing this:

```
var13: list[int, int] = []

a_list: list[int] = [1,2,3]
b_list: list[int] = [3,1,3]

for x in a_list:
    for y in b_list:
        if x+y <= 4:
            var13.append((x,y))
```

Here is the same operation using list comprehensions:

```
var12 = [(x, y) for x in a_list for y in b_list if x+y <= 4]
print(var12)
```

would yield

[(1, 1), (1, 2), (1, 3), (2, 1), (2, 2), (3, 1)]

Random Module

The random module is Python's inbuilt module that is used to generate random numbers or pseudo-random numbers including integers, floats, and sequences. Let's look at some methods that you may utilize more often as part of building your data pipelines.

random( )

This method returns a random float in the range between 0.0 and 1.0:

```
import random

var14 = random.random()
print(var14)
```

randint( )

This method returns an integer in the range between a minimum and a maximum threshold supplied by the user:

```
import random
print(random.randint(1,33))
```

getrandbits( )

The getrandbits() method from the random library would generate a random integer in the range between 0 and $2^k - 1$. This is beneficial when you need random integers bigger than randint or even in cases where you need to generate random Boolean returns.

Let us look at both:

```
import random

print(random.getrandbits(20))
print(random.getrandbits(20))
print(random.getrandbits(20))
```

would yield

619869
239874
945215

Here is the illustration for generating random Boolean returns:

```
print(bool(random.getrandbits(1)))
print(bool(random.getrandbits(1)))
print(bool(random.getrandbits(1)))
print(bool(random.getrandbits(1)))
```

would yield

True
False
True
True

As you can see, you can either generate the value in k random bits, or you can cast it to a Boolean with a parameter of 1 and you would get a random Boolean generator as well.

choice( )

The choice function selects a random element from a user-supplied sequence (such as a list, a tuple, etc.):

```
import random

fruit = ['apple','banana','melon','orange']
print(random.choice(fruit))
```

would return

'apple'

If you run the choice method again

```
print(random.choice(fruit))
```

it would return something different. Here is another output:

`'orange'`

shuffle( )

The shuffle() method shuffles the sequence, as it changes the way the characters are currently placed. For instance, we already have

```
print(fruit)
fruit = ['apple','banana','melon','orange']
```

Now

```
random.shuffle(fruit)
```

And so

```
print(fruit)
```

`['melon', 'banana', 'apple', 'orange']`

sample( )

The sample method from the random library returns k unique elements chosen from a population list or a set. This method generates samples without replacement, meaning once an item is chosen, it cannot be chosen again:

```
Import random

alpha = list("abcdefghijklmnopqrstuvwxyz")

print(alpha)
['a', 'b', 'c', 'd', 'e', 'f', 'g', 'h', 'i', 'j', 'k', 'l', 'm', 'n', 'o',
 'p', 'q', 'r', 's', 't', 'u', 'v', 'w', 'x', 'y', 'z']

print(random.sample(alpha, 7))
['n', 'v', 'x', 'u', 'p', 'i', 'k']

print(random.sample(alpha, 7))
['s', 'y', 'i', 't', 'k', 'w', 'u']
```

```
print(random.sample(alpha, 7))
['n', 'u', 'm', 'z', 'o', 'r', 'd']
print(random.sample(alpha, 7))
['e', 'y', 'a', 'i', 'o', 'c', 'z']
```

seed()

The seed() method from the random library would set the seed for generating the random numbers. And so, if you specify a seed with a certain integer, the idea is that you would obtain similar random numbers and, in that, you can reproduce the same output every time. It helps generate random samples in machine learning, simulation models, etc.

Git Source Code Management

Git is a software version control system that tracks updates and changes to software programs. A version control system is adopted between colleagues for collaborating in software development. Prior to current practices, developers would collaborate by sharing code over email or by instant messaging platforms. Even if you are an independent/indie developer, you may benefit from using a version control system so you can pick up where you left and refer to an earlier version of code, if required.

Foundations of Git

Git is a distributed version control system, wherein there exists a remote copy and a server copy of the same code. The code database is called a repository or simply repo. You obtain a copy of a repo from the server and start working with it. The process is called branching. Then, you can submit a proposal to merge your branch to the centralized copy (master branch), which is commonly referred to as pull request. Upon submission it goes to a person who is administering the repo, and they will do what is called approving a pull request.

GitHub

Let us look at GitHub, one of the leading git platforms. GitHub is considered the world's largest distributed version control system that hosts and tracks changes of software code while offering a plethora of features for development.

Setup and Installation

Visit https://github.com/ and click "Sign up" for a new account. You will be asked to sign up by providing a username and password and your email. You may choose to personalize your GitHub or wish to skip personalization. Now, open your terminal and check whether a git and Secure Shell (SSH) client is installed using the following command:

```
sudo apt update
sudo apt install git ssh
```

Alternatively you may wish to visit the website, git-scm, and obtain the git package under the Downloads section. Here is the website:

https://git-scm.com/downloads

Once the download and installation is complete, let us proceed to set up the git in our workstations. Here is how to get started.

Let us set up our username and our email as this information will be used by git commit every time you commit changes to files. Here are the commands:

```
git config --global user.name "<your-unique-username>"
git config --global user.email <your-email>
```

Remember to supply the unique username you had created during the registration process and please use quotes for registering the username. For the email, you need to provide the email you had utilized for the GitHub registration process. And make sure to use the "--global" keyword option as git will always use these credentials for any of your interactions with the git system. If you wish to override this information with a different set of credentials, you may run these same commands but without the "--global" option.

CHAPTER 1   CORE TECHNOLOGIES IN DATA ENGINEERING

You can access and write files to your GitHub repository right from your own terminal using Secure Shell protocol. You generate an SSH key and passphrase, sign in to your GitHub account, and add the SSH key and passphrase you generated from your terminal to the GitHub account. This way you have established a secured handshake between your workstation and your GitHub account.

Let us generate a new SSH key:

```
ssh-keygen -t ed25519 -C "<your-email-here>"
```

```
$ ssh-keygen -t ed25519 -C "cheddarfondue@icloud.com"
Generating public/private ed25519 key pair.
Enter file in which to save the key (/home/parallels/.ssh/id_ed25519):
Enter passphrase (empty for no passphrase):
Enter same passphrase again:
Your identification has been saved in /home/parallels/.ssh/id_ed25519
Your public key has been saved in /home/parallels/.ssh/id_ed25519.pub
The key fingerprint is:
SHA256:ZOjK47Hfl/hieWgsJnVA2vjWo04xnq6QX2bGkcGzBgQ cheddarfondue@icloud.com
The key's randomart image is:
+--[ED25519 256]--+
|     E..         |
|    . ...        |
|     .== o       |
|     oooB        |
|      .OoS       |
|     o =+=+      |
|    o =oX+ = .   |
|    +.@+.O +     |
|     =*++.=.     |
+----[SHA256]-----+
$
```

Figure 1-1. Generating a new SSH key in the terminal

For now, please skip (by pressing Enter) when asked for a passphrase among other options. Let the default location be as is. Here is how that may look for you.

Let us start the SSH agent in your terminal:

```
eval "$(ssh-agent -s)"
```

This command will start the SSH daemon process in the background. Now, let us add the SSH private key to the SSH agent using the following command:

```
ssh-add ~/.ssh/id_ed25519
```

Now open a new tab on your web browser and log in to your GitHub (in case you were signed off); click the icon on the top-right corner of your screen and choose the option called "Settings" from the list of options. Here is how that may look:

CHAPTER 1   CORE TECHNOLOGIES IN DATA ENGINEERING

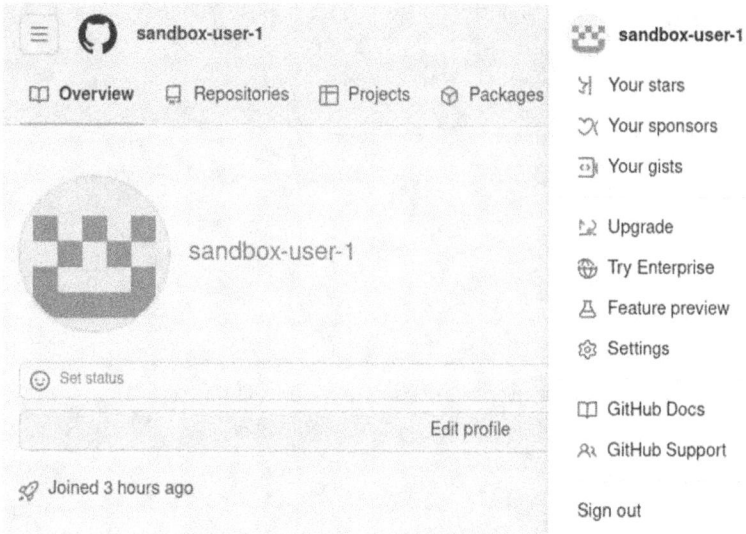

Figure 1-2. *Accessing the GitHub account's settings*

Once you are inside the GitHub settings, navigate and click the option called "SSH and GPG Keys" and choose the option "New SSH Key" on the top-right side of the browser.

You will get an option to add the SSH public key that was created in the previous steps. To perform that you will have to copy the SSH public key first.

Open your terminal and obtain the SSH public key using the following command:

```
cat ~/.ssh/id_ed25519.pub
```

You may obtain an output like the following:

```
ssh-ed25519 DH37589D364RIUHFIUSDHsr4SFFsdfx4WFRoc9Ks/7AuOsNpYy4xkjH6X youremail@emailprovider.com
```

Copy this public key and navigate to the web browser where you have GitHub open—specifically, Settings ➤ SSH and GPG Keys—to add a new SSH key. Paste the key that you just copied. Here is how that looks:

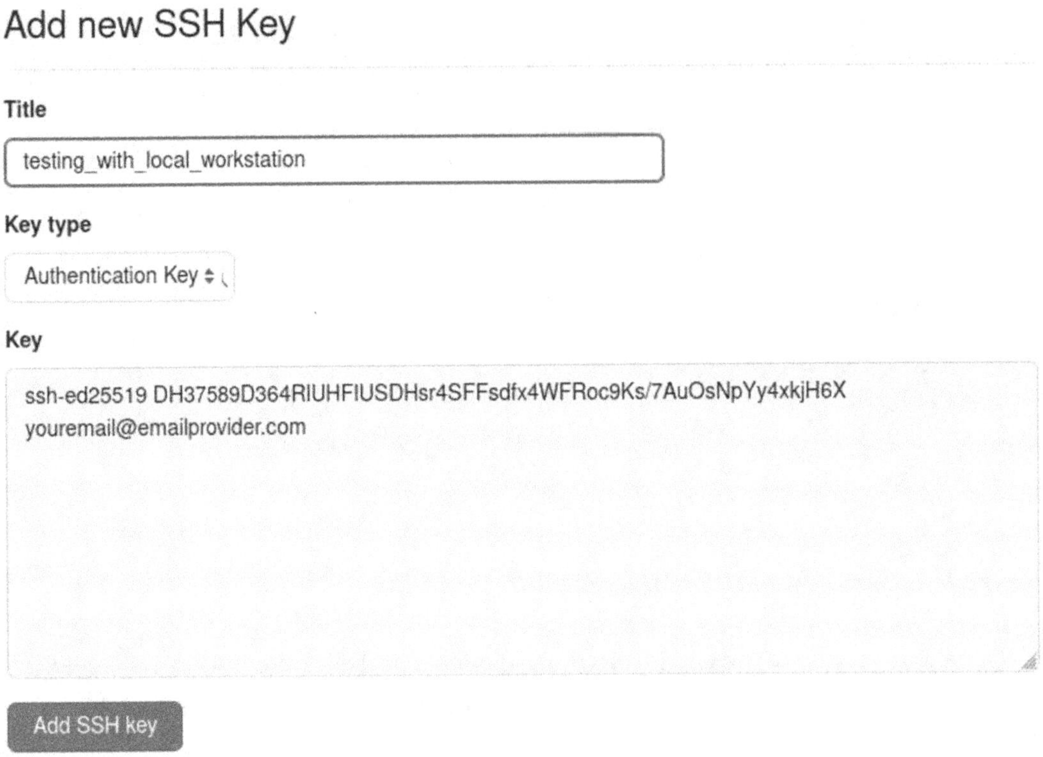

Add new SSH Key

Title

testing_with_local_workstation

Key type

Authentication Key ⇵

Key

ssh-ed25519 DH37589D364RIUHFIUSDHsr4SFFsdfx4WFRoc9Ks/7AuOsNpYy4xkjH6X youremail@emailprovider.com

[Add SSH key]

Figure 1-3. Adding a new SSH key

You can provide a title as you wish, something of your convenience, and leave the key type as "Authentication Key." Click Add SSH key.

Now that we have added the SSH public key, let us test our connection from our shell to a remote GitHub server by using the following command:

```
ssh -T git@github.com
```

You may receive a warning. Choose yes. Here is how that may look for you:

```
$ ssh -T git@github.com
The authenticity of host 'github.com (20.207.73.82)' can't be established.
ED25519 key fingerprint is SHA256:+DiY3wvvV6TuJJhbpZisF/zLDA0zPMSvHdkr4UvCOqU.
This key is not known by any other names
Are you sure you want to continue connecting (yes/no/[fingerprint])? yes
Warning: Permanently added 'github.com' (ED25519) to the list of known hosts.
Hi sandbox-user-1! You've successfully authenticated, but GitHub does not provid
e shell access.
$
```

Figure 1-4. Testing the SSH connection

CHAPTER 1   CORE TECHNOLOGIES IN DATA ENGINEERING

You may ignore the last line "GitHub does not provide shell access." Overall, if you see the preceding message, it is certain that you have set up GitHub correctly.

Let us create a new repository and use GitHub to version control the same. To create a new repository, open a terminal and change directory into a folder where you can test your git functionalities. Or create a new folder and perform the git initialization:

```
mkdir firstrepo
cd firstrepo
```

Now, within the folder, initialize a git repository by using the following command:

```
git init
```

The preceding command will create a hidden folder and initialize a database. You can change the directory into the hidden folder and notice the files. Here is how that may look for you:

```
$ git init
hint: Using 'master' as the name for the initial branch. This default branch name
hint: is subject to change. To configure the initial branch name to use in all
hint: of your new repositories, which will suppress this warning, call:
hint:
hint:   git config --global init.defaultBranch <name>
hint:
hint: Names commonly chosen instead of 'master' are 'main', 'trunk' and
hint: 'development'. The just-created branch can be renamed via this command:
hint:
hint:   git branch -m <name>
Initialized empty Git repository in /home/parallels/Documents/git/firstrepo/.git/
$ cd .git
$ ls
HEAD  branches  config  description  hooks  info  objects  refs
$
```

Figure 1-5. *Initializing a new git repository*

It may be useful to note that a HEAD file is a text file that points to the current branch, whereas the config file, another text file, stores local configuration about your repository. Without changing anything exit off of the hidden folder.

If you want to know the history of the repository, then you may use this command:

```
git log
```

You may not obtain any logs just yet as you recently created this repo. Another useful command is the git status that helps obtain the status of your repo:

```
git status
```

This will provide the status of the current git repo. At the moment, we do not have any files added or changed, so let us try to add a new file.

Here is a Python script that you can use:

```python
import random

def somefunct(n: int) -> None:
    for i in range(n):
        print(random.randint(1,100))

somefunct(5)
```

When you run the code, it will print five random integers between 0 and 100. Now let us check to see if the status of our git repo has changed:

```
$ git status
On branch master

No commits yet

Untracked files:
  (use "git add <file>..." to include in what will be committed)
        mycode.py

nothing added to commit but untracked files present (use "git add" to track)
```

Figure 1-6. *Checking the git repo status*

Now let us add the Python code to git so the git system can track the code's progress:

```
git add mycode.py
```

The git add begins to enable git tracking features on the file "mycode.py"; from the workflow perspective, the git system has picked this file from your local workstation and moved it to a staging area, where the git tracking features are enabled.

Up until now, we have just moved the file from the local file system to a staging area where git can identify this file and start tracking. This file is still not in the git repository yet. If you type in "git log," you would get a response something like this:

```
fatal: your current branch 'master' does not have any commits yet
```

It is true. The file has not been committed yet. So let us try to commit this file using the following command:

```
git commit -m "created a function that returns k random numbers"
```

This commit has moved the code into an actual git repository situated in the local directory. The message in the quote is actually a commit message describing what changes has been made in this commit.

If you would like to push your local git repository to a remote git repository, then you would want the git push command. It is also called syncing, where you would upload the local git repo with files to the remote git repo.

Note It is important to know that so far we have been working only on the local workstation. This repository does not exist in the remote GitHub. Hence, we have to push the current branch and set it as master. We will look at other examples where we begin working with a remote branch.

Core Concepts of Git

What we have seen so far is considered as "hello world" in git. Let's take this a bit further. Imagine we have a code repo in our git repository. How can we use git to collaborate with our team? Let us explore some core concepts in git.

Imagine you have a folder that consists of two Python codes. Let's say the hello.py has the following:

```
def hello():
    print("hello world")
```

And the other code main1.py has the following:

```
from hello import hello
hello()
print("this is from the main1")
```

You want to work on this project to add another function. Hence, you need to get a copy of this existing code from the GitHub server to your local workstation. Well, there are several ways to accomplish this. Let us look at them in detail.

Cloning

Cloning in git refers to obtaining a copy of this repository from the remote GitHub server to your local workstation. The idea is to obtain a local copy and work locally. This is usually done at the beginning of a new project or when joining an existing project as a newest team member. Cloning is mandatory. Here are some examples:

```
git clone git@github.com:sandbox-user-1/testrepo.git
```

```
git clone https://github.com/sandbox-user-1/testrepo.git
```

Branching

A branch is a parallel version of the codebase from the main codebase called the main branch (or master branch). Branching, as the name indicates, enables you to work on new features and troubleshooting code and propose the changes back to the main branch through pull request. Branching is for temporary changes and collaboration.

Here's how to create a new branch:

```
git clone git@github.com:sandbox-user-1/testrepo.git
git branch <name-of-your-branch>
```

By default, the pointer will be in the main branch. If you want to access the development branch, you need to change to it manually. Here's how:

```
git checkout <name-of-your-branch>
```

or

```
git switch <name-of-your-branch>
```

Forking

A fork is also a copy of the repository except that you create your own main or master branch within your git environment. Imagine the team has a master branch. When you branch, you get a development branch (or whatever you want to name it). When you fork, you get a master or main branch in your git account. Forking is done in special cases, external projects, and independent work done in parallel.

CHAPTER 1 CORE TECHNOLOGIES IN DATA ENGINEERING

In the GitHub web graphical user interface (GUI), you may open the repository that needs to be forked and click the "Fork" button that is on the top-right corner. Here is how that may look for you:

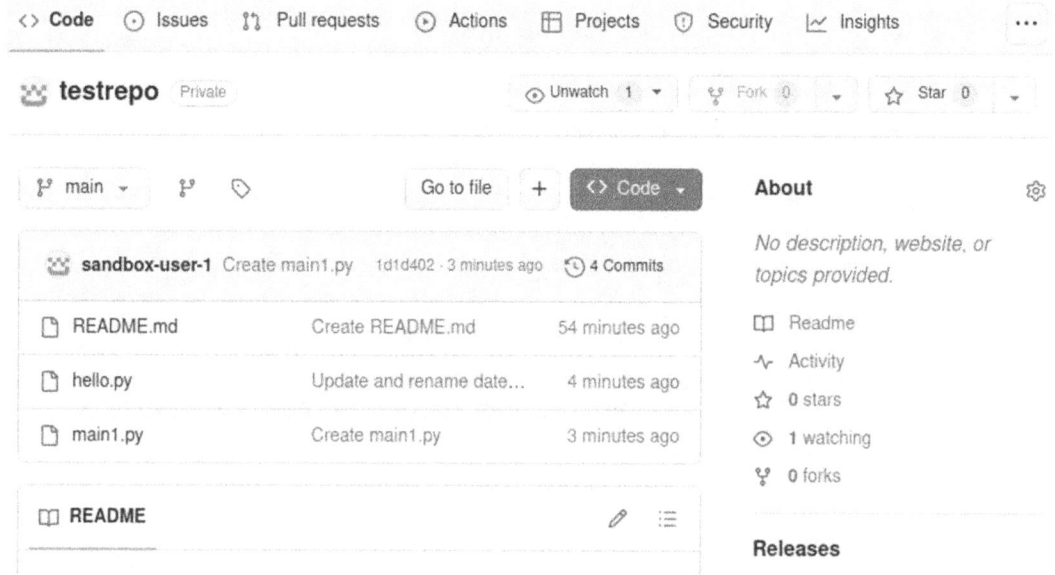

Figure 1-7. Forking a repository on GitHub

Note The Fork button is grayed, because I cannot "fork" my own repository into my own account.

Pull request

Pull requests are a way of proposing changes to the current codebase. By proposing a pull request, you have developed a feature, or you have completed some work in the code and are now proposing that your code be merged into the main repo.

To create a pull request

1. You already have a branch you are working with or create your own branch.

2. You make the necessary changes to the code.

25

3. You are now going to commit the changes:

   - git add <list of files> or "."

   - git commit -m "<your message in quotes>"

4. You would push these changes into the origin repository (your branch or your fork):

   - git push origin <new-feature>

```
$ git push origin newfeature
Enumerating objects: 7, done.
Counting objects: 100% (7/7), done.
Delta compression using up to 2 threads
Compressing objects: 100% (4/4), done.
Writing objects: 100% (5/5), 674 bytes | 674.00 KiB/s, done.
Total 5 (delta 0), reused 0 (delta 0), pack-reused 0
remote:
remote: Create a pull request for 'newfeature' on GitHub by visiting:
remote:      https://github.com/sandbox-user-1/testrepo/pull/new/newfeature
remote:
To github.com:sandbox-user-1/testrepo.git
 * [new branch]      newfeature -> newfeature
```

Figure 1-8. Pushing changes to a git branch

Here is how git push may look on the terminal:

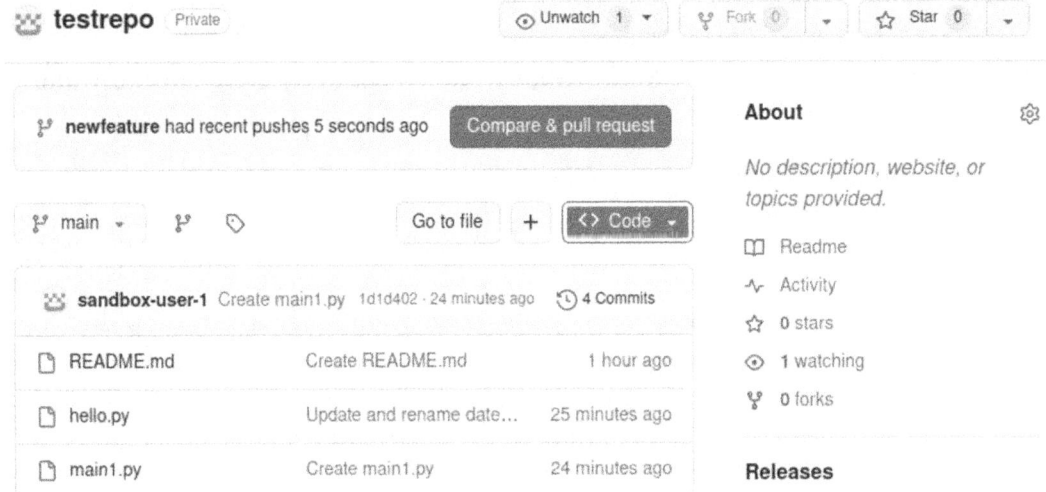

Figure 1-9. Comparing changes and creating a pull request on GitHub

CHAPTER 1   CORE TECHNOLOGIES IN DATA ENGINEERING

Now you are ready to perform a pull request. In the GitHub GUI, you would see something pop up saying recently received a pull request. Simply click "Compare & pull request."

Now you will have to enter the title and message for this pull request. GitHub will show you the changes made to the file by highlighting the new code in green. You can click Create pull request.

Here is how that looks:

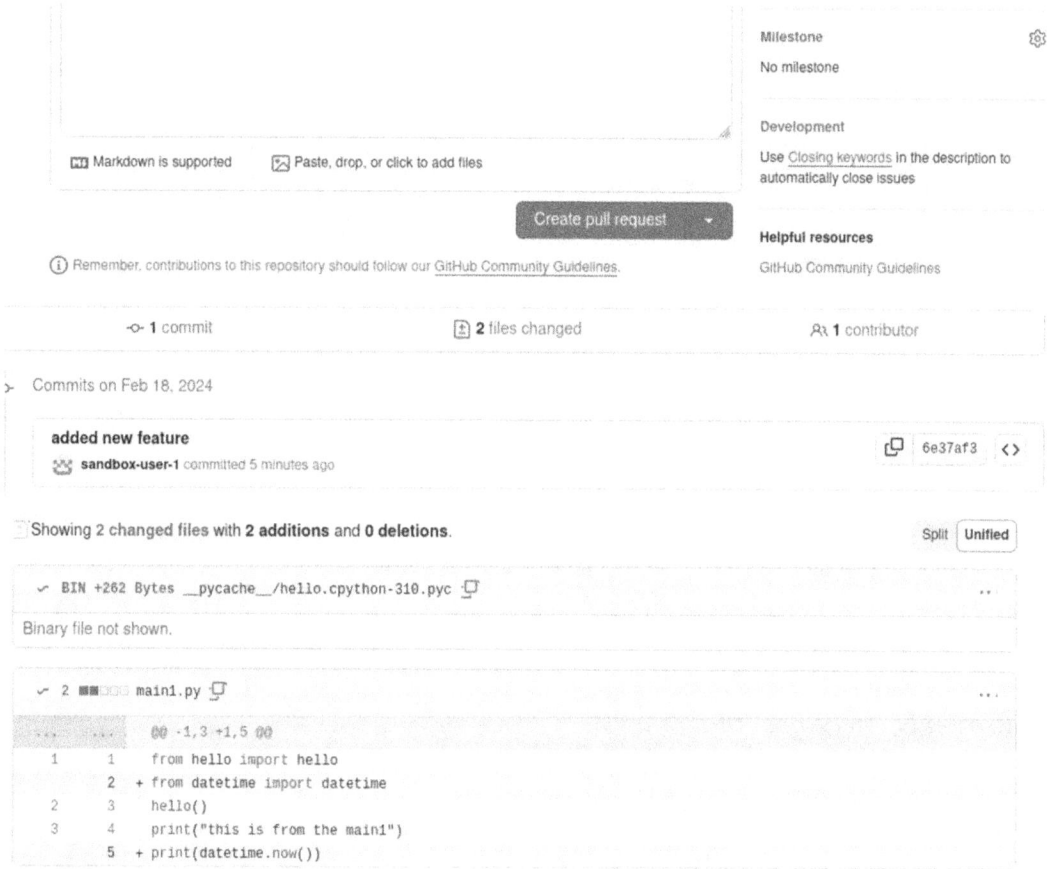

Figure 1-10.   Process of comparing changes and creating a pull request

27

Once you create a pull request, you now have to merge the pull request to remote main.

Here is how that may look:

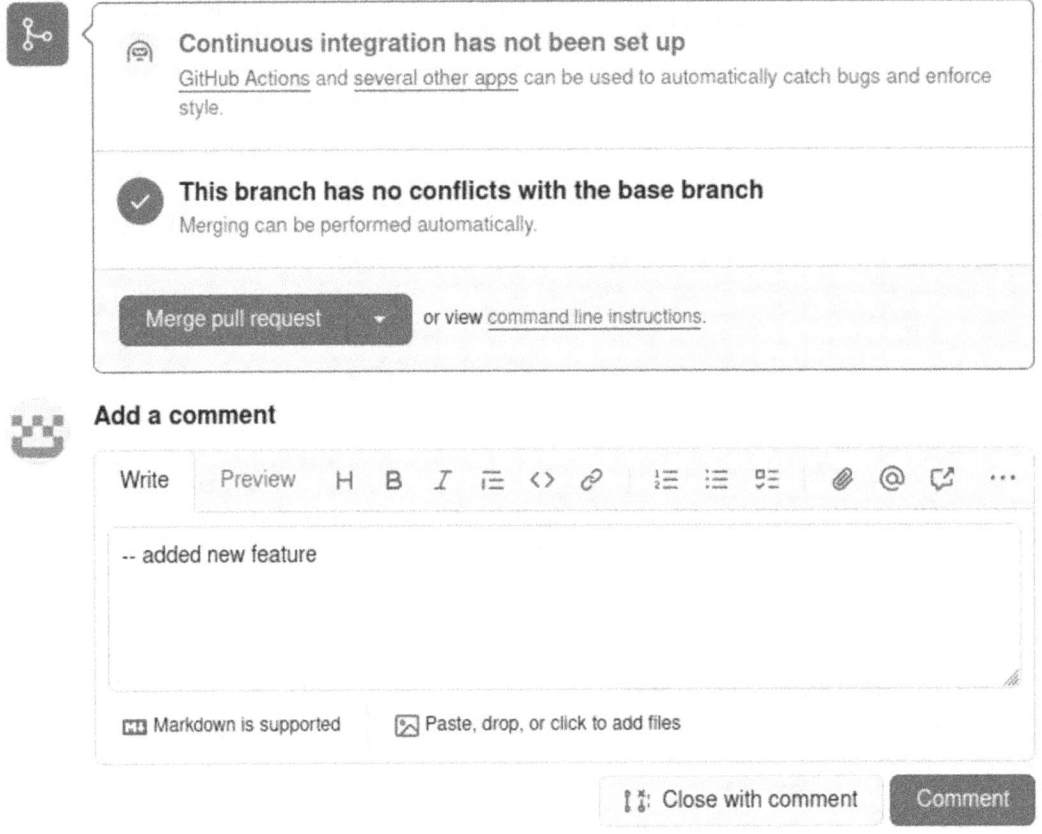

Figure 1-11. Merging a pull request

Click "Merge pull request," followed by "Confirm pull request."

CHAPTER 1 CORE TECHNOLOGIES IN DATA ENGINEERING

Once done, you may delete the branch. Here is how that may look:

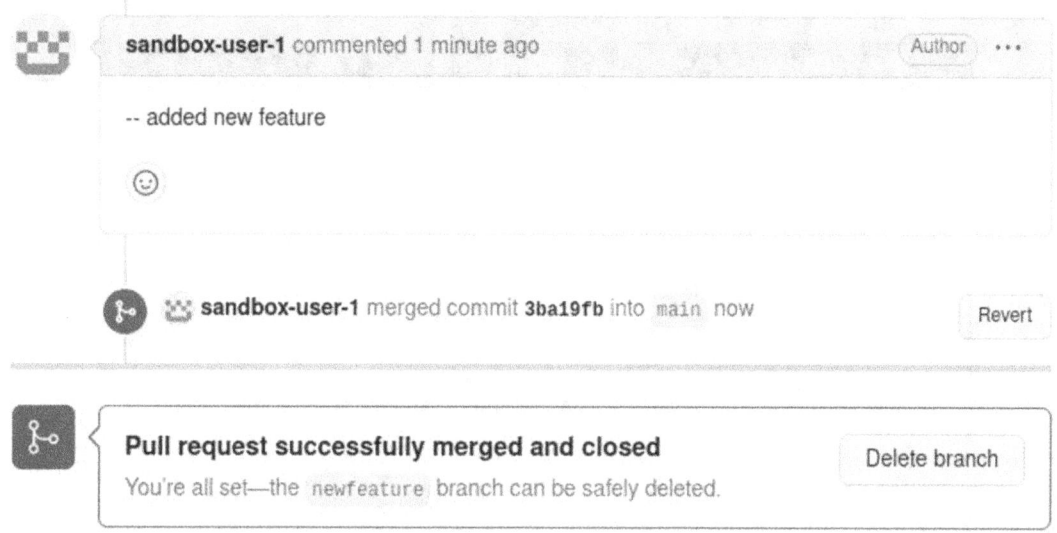

Figure 1-12. Deleting a branch after merging

Gitignore

Data pipelines sometimes contain sensitive information like passwords, temporary files, and other information that should not be committed in the repository. We use the ".gitignore" feature to intentionally ignore certain files.

Create a ".gitignore" file in the root directory of your folder, and specify various files that GitHub should ignore.

Here is an illustration of a simple .gitignore that you can incorporate in your data pipelines as well:

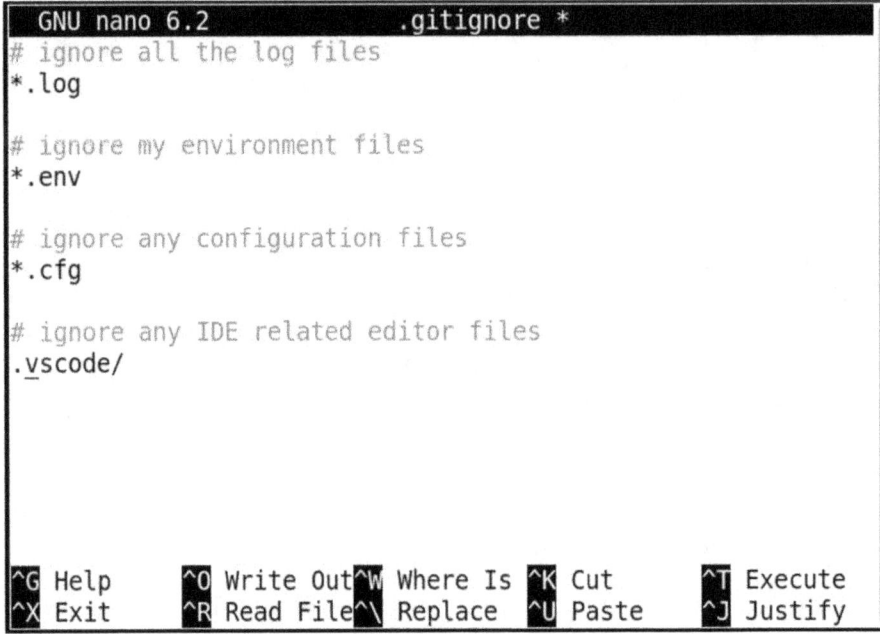

Figure 1-13. *Example of a .gitignore file*

Once created, let us stage this file and add to the local repository:

```
git add .gitignore
git commit -m "Added .gitignore file"
```

Note It is advisable to exercise caution not to push secrets in your repository. If you already have some files that need to be not tracked by git, then you may remove these files by `git rm –cached secret_passwords.txt`

We have only scratched the surface on git here. I encourage you to visit http://git-scm.org to learn more about git features.

SQL Programming

SQL is at the heart of every data engineering profession. All our historical data and enterprise systems are equipped with relational database systems. SQL is the way to access, query, and work with data from these relational database systems. SQL stands for structured query language. Let us look at some basic syntax. We will be using Postgres database syntax to demonstrate some concepts and functionalities that I feel would be relevant in everyday tasks.

Essential SQL Queries

The starting point for any SQL query is

```
SELECT * FROM TableName
```

which would return the table with all the rows and columns.

```
SELECT
    fieldname1,
    fieldname2
FROM
    TableName
```

would return all rows for those two fields.

```
SELECT
    fieldname1,
    fieldname2
FROM
    TableName
LIMIT
    10
```

would return all columns but only ten rows.

Usually, we may not write SQL code that performs full table scan unless it is a "relatively" small table or necessary due to other means. Typical SQL queries would carry field names and some sort of filters with conditional expressions.

Conditional Data Filtering

Here is a query that returns a product category with two conditions, the starting alphabet letter and character count of the product category:

```
SELECT
     product_id,
     LENGTH(product_id) as charlength
FROM
     productTable
WHERE
     product_id LIKE 'H%' AND
     LENGTH(product_id) BETWEEN 2 and 4
ORDER BY
     charlength
```

Joining SQL Tables

In certain cases, we would have to work with multiple tables to obtain certain data. This would involve joining these tables. There are few types of joins, namely, left, right, inner, outer, cartesian, and self-joins. Here is an example:

```
SELECT
     productTable.fieldname1,
     productTable.fieldname2,
     inventory.fieldname1,
     inventory.fieldname4
FROM
     productTable
LEFT JOIN
     inventory
ON
     productTable.id = inventory.product_id
```

Self-Join in SQL

Another important concept that helps you in your development work is the concept of self-joins. It is basically joining a given table with the exact same table. Let's say you have an employee table with employee id, name, and manager id, where every employee is assigned an id, their name, and the id of their supervisor. We need to obtain a table where we have the employee name and manager name.

Here is how that may look:

```
SELECT
     Emp.id,
     Emp.name,
     Mang.name
FROM
     Employee Emp
INNER JOIN
     Employee Mang
ON
     Emp.id = Mang.manager_id
```

In the preceding code, we are referencing the same table with a different alias and join based on their id value. There is also the concept of common table expression (CTE), where the result of a query is temporarily persisted to perform a specific operation with another query.

Common Table Expressions

A common table expression houses subqueries that store a result temporarily, using which a main query can be referenced with an appropriate SQL command and executed. CTEs are a way of simplifying complex queries and enhance readability.

Here is a generic CTE:

```
WITH productRaw AS (
    SELECT
         p_name AS name,
         p_qty AS qty,
         p_price AS price
```

```
        FROM
            productsTable
),
calculateGT AS (
    SELECT
        name,
        qty,
        price,
        (qty*price) AS grand_total
    FROM
        productRaw
)
SELECT
    name,
    qty,
    price,
    grand_total
FROM
    calculateGT
```

Of course, there may be optimized ways of calculating the grand total given price and quantity of each product. The idea is to show how you can leverage persisting results of multiple queries temporarily and utilize the result in generating the final result table.

Views in SQL

In SQL, we can persist the results of these queries as objects within the database. These are done in the form of views. There are two types of views, standard views and materialized views.

Standard View

Standard views are views that you can set up within the database that display the results of a specific query with multiple joins, filters, functions, etc. When you query a standard view (instead of running a large blurb of complex SQL code), it will always return the most recent results.

Here is how a standard view may look:

```
CREATE VIEW prod_inventory_report AS
SELECT
    productTable.fieldname1,
    productTable.fieldname2,
    inventory.fieldname1,
    inventory.fieldname4
FROM
    productTable
LEFT JOIN
    inventory
ON
    productTable.id = inventory.product_id;
```

Let us query this view:

```
SELECT
    *
FROM
    prod_inventory_report
```

Materialized View

Materialized views store the results of the view in a physical table (standard views do not physically store the results, but rather call the underlying query every time). And so, you created a physical table in the name of materialized view, so the performance is significantly better. However, you have to set up triggers to refresh materialized views every so often.

Here is how a materialized view may look (similar syntax to a standard view):

```
CREATE MATERIALIZED VIEW IF NOT EXISTS prod_inventory_report AS
SELECT
    productTable.fieldname1,
    productTable.fieldname2,
    inventory.fieldname1,
    inventory.fieldname4
```

```
FROM
    productTable
LEFT JOIN
    inventory
ON
    productTable.id = inventory.product_id
WITH DATA
WITH REFRESH ON DEMAND
WITH TABLESPACE pg_default
```

Temporary Tables in SQL

Temporary tables are another useful tool in SQL. These tables are created during a specific database session and persist only until the session ends. The idea of their limited lifespan and the automatic drop of these tables upon session conclusion is why they are named as temporary tables.

Here is how that may look:

```
CREATE TEMPORARY TABLE prod_report
(
    product_name VARCHAR(80),
    product_quantity INTEGER,
    inventory_city VARCHAR(50),
    inventory_state VARCHAR(50)
)
ON COMMIT PRESERVE ROWS;
```

In the case of temporary tables, both the schema and the data (if you insert data after creation of this table) persist only during that given database session. When you mention preserve rows, they are deleted at the end of the session. You can also mention delete rows that would delete immediately.

Window Functions in SQL

Window functions calculate aggregations for a given column in a table; however, they retrain their identities. Aggregations in SQL usually involve obtaining just the metric and it will ignore the fields. Window functions retain the identities of the table.

SQL Aggregate Functions

Let us look at an example:

```
SELECT
    Department,
    EmpID,
    Salary,
    avg(Salary) OVER (PARTITION BY Department)
FROM
    Employees;
```

This example would obtain the fields and calculate the mean salary for each department. The results may look like

```
Department | EmpID | Salary | avg
-----------+-------+--------+-------
    IT     |  11   |  5200  | 4633.0
    IT     |   7   |  4200  | 4633.0
    IT     |   9   |  4500  | 4633.0
    DS     |   5   |  3500  | 3700.0
    DS     |   2   |  3900  | 3700.0
    HR     |   3   |  4800  | 4900.0
    HR     |   1   |  5000  | 4900.0
```

As you can see, the window function computed the statistical average salary and grouped by each department while retaining the table structure and adding a new column for the computed average.

SQL Rank Functions

Let us look at another example. Let us assign a rank for each row within each department based on their earnings.

Here is how that looks:

```
SELECT
    Department,
    EmpID,
    Salary,
    RANK() OVER (PARTITION BY Department ORDER BY Salary DESC)
FROM
    Employees;
```

This may return a table like

```
Department | EmpID | Salary | rank
-----------+-------+--------+-------
    IT     |  11   |  5200  |  1
    IT     |   9   |  4500  |  2
    IT     |   7   |  4200  |  3
    DS     |   2   |  3900  |  1
    DS     |   5   |  3500  |  2
    HR     |   1   |  5000  |  1
    HR     |   3   |  4800  |  2
```

Query Tuning and Execution Plan Optimizations

SQL is an extremely powerful tool in the data domain. However, it is important to establish an understanding of how SQL queries work under the hood and so not to overwork the SQL engine and the server that hosts the SQL database. In this section, let me share with you some of the ways you can minimize the execution time of a query, improve query performance, and reduce the consumption of hardware resources.

It is important to understand, for a given project, for each database, and for each schema, the most commonly queried tables and the fields that you impose conditions on. Now, create indexes on those fields, for these most commonly utilized tables.

You can also create an index that involves more than one field, typically two or three fields. It is called composite index and helps improve the query performance.

When you have a bigger query that involves joins and other operations, use the idea of common table expressions to simplify the complex logic.

Do minimize any queries that involve full table scan, by using "Select *"; define the fields as much as possible.

If you do have a complex query, consider using materialized views to precompute and store the results of these queries. When the result of a complex query is heavy on the read operation, then the materialized view is the best fit here.

Consider using temp tables, in cases where you have to write results to a table temporarily but you do not want that table to persist in the database.

It may be good to minimize using a subquery in your SQL query; if you do end up writing a subquery, always keep your sort operations within the master query, and try to avoid specifying ORDER BY in a subquery.

When specifying a condition that includes an OR statement, here is an approach for you to consider: write two separate queries and union the results of the queries.

Conclusion

So far, we have covered a lot of ground from refreshing Python programming to getting hands-on with git and GitHub and reviewing the SQL essentials. The concepts and techniques you have learned here will serve as a foundation to tackle complex data engineering projects. Python's data ecosystem will enable you to write effective data pipelines for data transformation and analysis tasks. Git will allow you to collaborate with your team, participate in collaborative software development, and maintain a clean and documented history of your data pipeline as it evolves. SQL skills will prove indispensable for working with the relational database systems that form the foundation of almost all data architectures. I encourage you to practice, explore, and apply what you have learned. Having said that, let us venture into the exciting topic of data wrangling, in the next few chapters.

CHAPTER 2

Data Wrangling using Pandas

Introduction

Data wrangling is the process of transforming, manipulating, and preparing a dataset from disparate data sources. Data munging and data manipulation are other terms that are used to describe the same process. In this chapter, we will look at Pandas, the oldest Python data analytics library. My sincere opinion is that Pandas library played a significant role in the adoption of Python language for data analysis and manipulation tasks. In this chapter, we will look at Pandas 2.0, a major release of Pandas, exploring its data structures, handling missing values, performing data transformations, combining multiple data objects, and other relevant topics.

By the end of the chapter, you will learn

- The core data structures of Pandas 2.0, Series and DataFrame, and how they can be used in data wrangling
- Various indexing techniques to access and slice Pandas data structures
- How to handle missing values effectively and perform data transformations through reshaping, pivoting, and aggregating data
- How to combine several Pandas objects using merge, join, and concatenation operators

CHAPTER 2   DATA WRANGLING USING PANDAS

Pandas is a Python library that was built for data manipulation and transformation purposes. Pandas is built on top of NumPy, another Python library for scientific and numerical computations.

Pandas is largely driven by user community, open source, and free to modify, distribute, and use for commercial purposes. Born in 2011, Pandas has evolved significantly over the past decade, with a major version having been released recently, with rewritten codebase, additional functionalities, and its ability to interact well with other libraries. With over 100 million downloads per month, Pandas library is the de facto standard for data manipulation and transformation (source: `http://library.io`).

Even if one is not using Pandas and NumPy in their production, the knowledge of the data structures and methods will certainly help with getting frequented with other data manipulation libraries. Pandas sets standards for various other Python libraries.

Data Structures

Pandas essentially provides two data structures, series and data frames, respectively. A series is an n-dimensional array, whereas a data frame is a two-dimensional labeled data structure that supports tabular structure like spreadsheets or an output of a SQL table.

Series

The series data structure enables one to handle both labeled and unlabeled arrays of data. The data can be integers, floating point numbers, strings of text, or any other Python object.

The methods provided by the Pandas library are called objects. To call the Series object, one has to initialize the constructor and supply appropriate parameters. An empty initialization would look like

`pd.Series()`

Here is a simple initialization of a series:

```
a = [1,2,3,4,5]
b = pd.Series(data=np.random.rand(5), index=a)
```

One can also generate Series using a Python dictionary. Here is an example:

```
simple_dict1 = {
"library" : "pandas",
"version" : 2.0,
"year" : 2023
}

simple_series = pd.Series(simple_dict1)
```

One can also get and set values in a given series. Here's how to get a value:

```
simple_series["year"]
```

And we would obtain

```
2023
```

To modify a value in a series, one can also do the following:

```
simple_series["year"] = 2024
```

And we would obtain

```
simple_series
library     pandas
version        2.0
year          2024
dtype: object
```

In this output, the dtype is NumPy's data type object, which contains the type of the data, in this case, an object.

Data Frame

A data frame is the most commonly used data structure in Pandas. A Pandas data frame is a two-dimensional structured array of objects that can contain integers, floats, strings of text, or other Python objects (Series, another data frame, etc.).

While writing production code, one may generate a data frame from an external dataset stored in a different format. However, it may be essential to understand the key attributes of a data frame. Some of them are as follows:

> Data: The data points or the actual data. It could be one or more Series or dictionary data types. The data may contain integers, floats, strings, or any other Python object.
>
> Index: The row labels of the data points and the index of the DataFrame itself.
>
> Columns: The column labels of the data points.
>
> Dtype: The data type of the data frame.

A simple initialization would look like

```
data = {
  "i": [121, 144, 169],
  "j": [11, 12, 13]
}
df_data = pd.DataFrame(data=data, index=["a","b","c"])
print(df_data)
     i    j
a   121  11
b   144  12
c   169  13
```

Indexing

Indexing is simply a row reference in the data frame. Indexing helps identify the Pandas data structure. One can also say that indexing is a way of referring to a pointer that identifies a specific row or a subset of rows in a Pandas data frame. It is a way of improving the performance of data retrieval.

Essential Indexing Methods

Let us look at an example of indexing in a Pandas data frame. First, we will load the dataset and rename certain verbose columns:

```
df = pd.read_csv(
    "/your-path-to-the-file/winequality/winequality-red.csv"
    ,sep=";"
    ,usecols=['residual sugar', 'chlorides', 'total sulfur dioxide',
    'density', 'pH', 'alcohol', 'quality']
)

df = df.rename({
    'residual sugar': 'rSugar',
    'total sulfur dioxide': 'total_so2'
    },
    axis=1
)
```

Here's how to return a single row from the data frame:

```
print(df.iloc[5])
```

will yield

```
rSugar         1.8000
chlorides      0.0750
total_so2     40.0000
density        0.9978
pH             3.5100
alcohol        9.4000
quality        5.0000
Name: 5, dtype: float64
```

By passing a list as a parameter, one can return multiple rows in the following manner:

```
print(df.loc[[5,9,22]])
```

CHAPTER 2 DATA WRANGLING USING PANDAS

will yield

	rSugar	chlorides	total_so2	density	pH	alcohol	quality
5	1.8	0.075	40.0	0.9978	3.51	9.4	5
9	6.1	0.071	102.0	0.9978	3.35	10.5	5
22	1.6	0.106	37.0	0.9966	3.17	9.5	5

A list can also be passed as a parameter in the following manner:

```
pinot_noir = [1,4,7,22,39]
```

```
df.iloc[pinot_noir]
```

will yield

	rSugar	chlorides	total_so2	density	pH	alcohol	quality
1	2.6	0.098	67.0	0.9968	3.20	9.8	5
4	1.9	0.076	34.0	0.9978	3.51	9.4	5
7	1.2	0.065	21.0	0.9946	3.39	10.0	7
22	1.6	0.106	37.0	0.9966	3.17	9.5	5
39	5.9	0.074	87.0	0.9978	3.33	10.5	5

Furthermore, one can also pass a range as a parameter, in the following manner:

```
df.loc[5:25]
```

will yield

	rSugar	chlorides	total_so2	density	pH	alcohol	quality
5	1.8	0.075	40.0	0.9978	3.51	9.4	5
6	1.6	0.069	59.0	0.9964	3.30	9.4	5
7	1.2	0.065	21.0	0.9946	3.39	10.0	7
8	2.0	0.073	18.0	0.9968	3.36	9.5	7
9	6.1	0.071	102.0	0.9978	3.35	10.5	5
10	1.8	0.097	65.0	0.9959	3.28	9.2	5
11	6.1	0.071	102.0	0.9978	3.35	10.5	5
12	1.6	0.089	59.0	0.9943	3.58	9.9	5
13	1.6	0.114	29.0	0.9974	3.26	9.1	5
14	3.8	0.176	145.0	0.9986	3.16	9.2	5
15	3.9	0.170	148.0	0.9986	3.17	9.2	5

16	1.8	0.092	103.0	0.9969	3.30	10.5	7
17	1.7	0.368	56.0	0.9968	3.11	9.3	5
18	4.4	0.086	29.0	0.9974	3.38	9.0	4
19	1.8	0.341	56.0	0.9969	3.04	9.2	6
20	1.8	0.077	60.0	0.9968	3.39	9.4	6
21	2.3	0.082	71.0	0.9982	3.52	9.7	5
22	1.6	0.106	37.0	0.9966	3.17	9.5	5
23	2.3	0.084	67.0	0.9968	3.17	9.4	5
24	2.4	0.085	40.0	0.9968	3.43	9.7	6
25	1.4	0.080	23.0	0.9955	3.34	9.3	5

Conditional filtering would look something like the following:

```
print(df.loc[df["quality"] > 7])
```

would yield

	rSugar	chlorides	total_so2	density	pH	alcohol	quality
267	3.6	0.078	37.0	0.99730	3.35	12.8	8
278	6.4	0.073	13.0	0.99760	3.23	12.6	8
390	1.4	0.045	88.0	0.99240	3.56	12.9	8
440	2.2	0.072	29.0	0.99870	2.88	9.8	8
455	5.2	0.086	19.0	0.99880	3.22	13.4	8
481	2.8	0.080	17.0	0.99640	3.15	11.7	8
495	2.6	0.070	16.0	0.99720	3.15	11.0	8
498	2.6	0.070	16.0	0.99720	3.15	11.0	8
588	2.0	0.060	50.0	0.99170	3.72	14.0	8
828	2.3	0.065	45.0	0.99417	3.46	12.7	8
1061	1.8	0.071	16.0	0.99462	3.21	12.5	8
1090	1.9	0.083	74.0	0.99451	2.98	11.8	8
1120	2.5	0.076	17.0	0.99235	3.20	13.1	8
1202	1.8	0.068	12.0	0.99516	3.35	11.7	8
1269	1.8	0.044	87.0	0.99080	3.50	14.0	8
1403	1.7	0.061	13.0	0.99600	3.23	10.0	8
1449	2.0	0.056	29.0	0.99472	3.23	11.3	8
1549	1.8	0.074	24.0	0.99419	3.24	11.4	8

CHAPTER 2   DATA WRANGLING USING PANDAS

Conditional indexing is also possible within data frames using the "query" method. Here is an example:

```
hq_wine = df.loc[df["quality"] > 7]

print(hq_wine.query('(rSugar < pH)'))
```

would yield

	rSugar	chlorides	total_so2	density	pH	alcohol	quality
390	1.4	0.045	88.0	0.99240	3.56	12.9	8
440	2.2	0.072	29.0	0.99870	2.88	9.8	8
481	2.8	0.080	17.0	0.99640	3.15	11.7	8
495	2.6	0.070	16.0	0.99720	3.15	11.0	8
498	2.6	0.070	16.0	0.99720	3.15	11.0	8
588	2.0	0.060	50.0	0.99170	3.72	14.0	8
828	2.3	0.065	45.0	0.99417	3.46	12.7	8
1061	1.8	0.071	16.0	0.99462	3.21	12.5	8
1090	1.9	0.083	74.0	0.99451	2.98	11.8	8
1120	2.5	0.076	17.0	0.99235	3.20	13.1	8
1202	1.8	0.068	12.0	0.99516	3.35	11.7	8
1269	1.8	0.044	87.0	0.99080	3.50	14.0	8
1403	1.7	0.061	13.0	0.99600	3.23	10.0	8
1449	2.0	0.056	29.0	0.99472	3.23	11.3	8
1549	1.8	0.074	24.0	0.99419	3.24	11.4	8

One can use both .loc and .iloc methods to retrieve specific rows of the data frame. The .iloc method is a positional index and attempts to access rows and columns by referencing integer-based location to select a position, whereas the .loc is a label-based index and accesses the rows and columns by labels.

Let's say we have a data frame with indexes that are not in a specific order, for instance:

```
idx_series = pd.Series(["peach", "caramel", "apples", "melon", "orange", "grapes"], index=[74, 75, 76, 1, 2, 3])
```

This would generate a Pandas series as follows:

```
74        peach
75       caramel
76        apples
1         melon
2         orange
3         grapes
dtype: object
```

When using the loc method, Pandas would attempt to locate the label that is passed as a parameter. Here, the label "1" has the value melon:

`idx_series.loc[1]`

`Out[107]: 'melon'`

In contrast, the iloc method would attempt to locate the integer position that is passed as the parameter. The integer referencing starts with 0 and ends in length of the list – 1. And so, the first row would yield caramel:

`idx_series.iloc[1]`

`Out[108]: 'caramel'`

Multi-indexing

Multi-indexing simply enables to select values from both rows and columns, by referencing more than one index. One can create a multi-index from arrays, tuples, and data frames using appropriate methods. The simplest method is to define a multi-index using arrays and pass them directly into a data frame or a series. Pandas will automatically integrate them into its data structure. Here's an example:

First, let's create the multi-index:

```
import numpy as np

multi_index_arr = [
    np.array(
            ['chunk1','chunk1','chunk1','chunk1','chunk1',
             'chunk2','chunk2','chunk2','chunk2','chunk2']
            ),
```

CHAPTER 2   DATA WRANGLING USING PANDAS

```
    np.array(
            ['a','b','c','d','e',
             'f','g','h','i','j']
            )
]
```

Now pass this array into the data frame. For convenience, a data frame with random numbers has been initialized:

```
df = pd.DataFrame(
        np.random.randn(10,5),
        index=multi_index_arr
    )

print(df)
```

Here is the result of this data frame:

```
                   0         1         2         3         4
chunk1 a    0.681923 -0.298217  1.269076  0.202422  0.419290
       b   -1.397923 -1.308753  1.624954 -0.968342 -2.025347
       c    0.003117 -0.677547 -0.262260  0.684654  1.064416
       d   -0.811899  1.571434  0.169619  0.424022  0.108066
       e   -0.366923  0.365224 -0.173412 -1.852635  0.019401
chunk2 f    1.102141  1.069164 -0.446039 -1.744530  1.069642
       g    1.335156  1.166403 -0.880927  0.031362  0.268140
       h    0.381478 -2.620881 -0.669324 -0.892250  2.069332
       i   -0.063832 -1.076992 -0.944367  0.616954  1.183344
       j   -0.110984  0.858179 -0.899983 -0.829470 -1.078892
```

This can also be extended to columns:

Let's define the column index:

```
multi_index_col = [
   np.array(
           ['chunk1a','chunk1a','chunk1a',
            'chunk2a','chunk2a','chunk2a']
           ),
```

```
np.array(
    ['k','l','m',
     'n','o','p']
    )
]
```

Now reference both the row and column multi-indexes to the data frame. For simplicity, we are initializing a data frame with random numbers:

```
df = pd.DataFrame(
    np.random.randn(10,6),
    index=multi_index_arr,
    columns=multi_index_col
)

print(df)
```

Here's how this looks like:

```
              chunk1a                            chunk2a
                    k         l         m         n         o         p
chunk1 a     0.933805  1.117525 -1.159911 -1.503417  1.863563  0.179210
       b    -1.690026  0.995953 -0.763078  1.580060  1.132970 -0.930042
       c    -0.325951 -0.148638 -1.481013  1.211841 -1.341596  0.582081
       d    -0.661698  0.838978 -0.734689  0.341315 -1.776477 -1.236906
       e     0.460951 -2.022154 -0.236631  1.863633  0.466077 -1.603421
chunk2 f    -0.122058  0.141682  1.123913  0.300724 -2.104929 -1.036110
       g    -0.332314  0.000606  0.239429  0.869499 -0.482440 -1.219407
       h    -1.035517 -0.698888  2.486068 -1.379440 -0.503729  1.170276
       i    -0.213047  2.714762  0.463097 -1.864837 -0.603789  0.351523
       j    -0.346246  0.720362  1.328618 -0.271097  1.034155  0.665097
```

And so, if we wish to retrieve a subset, here's how to do it:

```
print(df.loc['chunk1']['chunk1a'])
```

would yield

```
          k         l         m
a  0.933805  1.117525 -1.159911
b -1.690026  0.995953 -0.763078
c -0.325951 -0.148638 -1.481013
d -0.661698  0.838978 -0.734689
e  0.460951 -2.022154 -0.236631
```

Here's how to access a single element:

```
print(df.loc['chunk1','a']['chunk2a','p'])
```

would yield

```
0.1792095
```

Time Delta Index

Time delta can be defined as the difference in time lapsed between given inputs. They are expressed in units of time, namely, seconds, minutes, days, etc.

Here is an example of creating time delta indexes:

```
timedelta_index = pd.timedelta_range(start='0 minutes', periods=50, freq='36S')
```

will yield

```
TimedeltaIndex(['0 days 00:00:00', '0 days 00:00:36', '0 days 00:01:12',
                '0 days 00:01:48', '0 days 00:02:24', '0 days 00:03:00',
                '0 days 00:03:36', '0 days 00:04:12', '0 days 00:04:48',
                '0 days 00:05:24', '0 days 00:06:00', '0 days 00:06:36',
                '0 days 00:07:12', '0 days 00:07:48', '0 days 00:08:24',
                '0 days 00:09:00', '0 days 00:09:36', '0 days 00:10:12',
                '0 days 00:10:48', '0 days 00:11:24', '0 days 00:12:00',
                '0 days 00:12:36', '0 days 00:13:12', '0 days 00:13:48',
                '0 days 00:14:24', '0 days 00:15:00', '0 days 00:15:36',
                '0 days 00:16:12', '0 days 00:16:48', '0 days 00:17:24',
                '0 days 00:18:00', '0 days 00:18:36', '0 days 00:19:12',
```

```
            '0 days 00:19:48', '0 days 00:20:24', '0 days 00:21:00',
            '0 days 00:21:36', '0 days 00:22:12', '0 days 00:22:48',
            '0 days 00:23:24', '0 days 00:24:00', '0 days 00:24:36',
            '0 days 00:25:12', '0 days 00:25:48', '0 days 00:26:24',
            '0 days 00:27:00', '0 days 00:27:36', '0 days 00:28:12',
            '0 days 00:28:48', '0 days 00:29:24'],
           dtype='timedelta64[ns]', freq='36S')
```

Let us initialize a series with our newly created time delta index.

Here is how the code may look:

```
series_a = pd.Series(np.arange(50), index=timedelta_index)
```

would yield

```
0 days 00:00:00     0
0 days 00:00:36     1
0 days 00:01:12     2
0 days 00:01:48     3
0 days 00:02:24     4
0 days 00:03:00     5
0 days 00:03:36     6
0 days 00:04:12     7
0 days 00:04:48     8
0 days 00:05:24     9
0 days 00:06:00     10
0 days 00:06:36     11
0 days 00:07:12     12
0 days 00:07:48     13
0 days 00:08:24     14
0 days 00:09:00     15
0 days 00:09:36     16
0 days 00:10:12     17
0 days 00:10:48     18
0 days 00:11:24     19
0 days 00:12:00     20
0 days 00:12:36     21
```

```
0 days 00:13:12    22
0 days 00:13:48    23
0 days 00:14:24    24
0 days 00:15:00    25
0 days 00:15:36    26
0 days 00:16:12    27
0 days 00:16:48    28
0 days 00:17:24    29
0 days 00:18:00    30
0 days 00:18:36    31
0 days 00:19:12    32
0 days 00:19:48    33
0 days 00:20:24    34
0 days 00:21:00    35
0 days 00:21:36    36
0 days 00:22:12    37
0 days 00:22:48    38
0 days 00:23:24    39
0 days 00:24:00    40
0 days 00:24:36    41
0 days 00:25:12    42
0 days 00:25:48    43
0 days 00:26:24    44
0 days 00:27:00    45
0 days 00:27:36    46
0 days 00:28:12    47
0 days 00:28:48    48
0 days 00:29:24    49
Freq: 36S, dtype: int64
```

Here's how to retrieve a single value:

```
series_a['0 days 00:27:36']
```

would yield

46

CHAPTER 2 DATA WRANGLING USING PANDAS

Here's how to look at a subset of the data with respect to time range:

```
s['0 days 00:12:30':'0 days 00:27:30']
```

would yield

```
0 days 00:12:36    21
0 days 00:13:12    22
0 days 00:13:48    23
0 days 00:14:24    24
0 days 00:15:00    25
0 days 00:15:36    26
0 days 00:16:12    27
0 days 00:16:48    28
0 days 00:17:24    29
0 days 00:18:00    30
0 days 00:18:36    31
0 days 00:19:12    32
0 days 00:19:48    33
0 days 00:20:24    34
0 days 00:21:00    35
0 days 00:21:36    36
0 days 00:22:12    37
0 days 00:22:48    38
0 days 00:23:24    39
0 days 00:24:00    40
0 days 00:24:36    41
0 days 00:25:12    42
0 days 00:25:48    43
0 days 00:26:24    44
0 days 00:27:00    45
Freq: 36S, dtype: int64
```

While the simplest form of index is referencing a specific row, Pandas library offers much more sophisticated indexing such as multi-index, time delta index, period index, and categorical index.

CHAPTER 2   DATA WRANGLING USING PANDAS

Data Extraction and Loading

One of the powerful features in Pandas is its ability to both import and export data in a myriad of formats. It is also a good time to introduce various data formats. Data formats and file formats have evolved to meet the growing demands of big data. Current estimates suggest that globally we are creating, consuming, and copying around 10 zettabytes per month in various formats.

CSV

CSV stands for comma-separated values; the CSV format is one of the oldest and the most common types of storing data. It is also known as tabular store, where data is stored in rows and columns.

The data is stored in a text file. Each line of the file corresponds to a row. As the title suggests, the data points may be separated using commas. There will be the same number of data points per row. It is expected that the first line would contain the column definitions also known as header or metadata. CSV data files are known for their simplicity and ability to parse quickly.

Here is how one can import data in CSV format:

```
import pandas as pd

new_df = pd.read_csv("path to the csv file", usecols = ["col1", "col4", "col7",...]
```

where the usecols parameter would help select only the columns required for the project.

Here's how to export the data:

```
final_df = pd.to_csv(index=False)
```

Here's how to create a compressed CSV file:

```
final_df = pd.to_csv('final_output.zip', index=False, compression=compression_opts)
```

JSON

JSON stands for JavaScript Object Notation. It is more frequently used to transmit data between the app (web app, mobile app, etc.) and the web server. Furthermore, it is the de facto data format for APIs to communicate to requests and responses.

The data is stored in key-value pairs, and data is organized in nested structure. JSON supports data types like integers, strings, arrays, and Booleans, to name a few.

Here is how one can import a JSON document using Pandas:

```
json_df = pd.read_json('path to file.json')
```

Here's how to export:

```
final_df.to_json('path to file.json')
```

HDF5

HDF5 stands for Hierarchical Data Format, version 5. Unlike CSV and JSON, HDF5 is a compressed format, where data is optimized for efficient storage, and provides faster read and write operations.

While working with huge amounts of data in HDF5, one can attempt to access just the subset of the huge dataset without having to load the entire dataset in memory. And so, HDF5 is physical memory friendly.

HDF5 supports large and complex heterogeneous data. As the name suggests, the HDF5 data format has a hierarchical structure. The structure contains one or more groups, wherein each group may contain various types of data along with their metadata:

```
new_df = pd.read_hdf('path to file.h5', key = 'the key to specific dataset')
```

```
print(new_df)
```

If the entire file contains a single data frame, then the key parameter can be ignored.

Here is how to write a data frame to an existing HDF5 store:

```
exportstore = pd.HDFStore("an_existing_store.h5", "w")
```

```
exportstore.put('new data', new_df)
```

```
print(exportstore.keys())
```

Feather

Originally developed by Apache Arrow, for faster data exchange, Feather is a fast and efficient data format that is developed for data science and analysis. To access Feather, one needs to install PyArrow and Feather-format Python packages as prerequisites to work with Feather data format.

Here is an example of reading a Feather file:

```
import pandas as pd
import numpy as np

feather_df = pd.read_feather('location to your file.feather',
use_threads=True)
```

Here, the usage of "use_threads" is to enable parallelization of file reading operations using multiple threads.

Here's how to export a data frame to Feather:

```
final_df.to_feather('location to destination file.feather')
```

Parquet

Parquet is developed by the Apache Parquet project and is an open source, columnar-style file format. Parquet is designed for big data processing, data warehousing, and analytics applications. Parquet has cross-language support and supports various compression algorithms for data exchange and transport. Parquet provides faster performance, where only a subset of a large dataset is required to be read for a given request or a query. Parquet is considered to be a good format for reading heavy data pipelines.

Here is an example:

```
import pandas as pd
import pyarrow

new_df = pd.read_parquet("location to file.parquet", engine='pyarrow')

print(df)
```

Here is another example that uses fastparquet as the loading engine:

```
import pandas as pd
import fastparquet

new_df = pd.read_parquet("location to file.parquet", engine='fastparquet')

print(df)
```

Here's how to export a data frame to Parquet:

```
final_df.to_parquet('location to the destination file.parquet',
'compression=snappy')
```

There are options to choose the type of compression, like snappy, gzip, lz4, zstd, etc. These are just various algorithms that attempt to reduce the file size.

ORC

ORC stands for Optimized Row Columnar. It is considered more efficient than flat file formats. ORC is a columnar file format, where the column values are stored next to each other. ORC file formats support various compression algorithms. Each ORC file contains one or more stripes of data. Each stripe contains a header, data stored in rows, and a footer. The footer contains metadata and summary statistics of the data points stored in the given stripe. ORC is considered to be a good format for writing heavy data pipelines:

```
import pandas as pd

new_df = pd.read_orc("location to your file.orc", columns=["col1",
"col5",...])

final_df.to_orc("location to destination file.orc", engine='pyarrow')
```

Avro

Avro is a file format that is row based. It is safe to say that Avro evolved as part of Apache Hadoop for big data processing. Avro schema is defined in JSON and supports binary encoding for faster access and efficient storage. Binary encoding is a process of serializing and deserializing data for processing and storing large amounts of data efficiently. Avro also provides a container file, a special kind of a file that contains multiple data inside it and enables storing various types of data together (image, audio, text, etc.).

Pandas does not have native support for handling Avro files; however, third party libraries can be used to load the Avro data. Here is an example:

```
import fastavro
import pandas as pd

with open('location to Avro file', 'rb') as readAvro:
    avro_read = fastavro.reader(readAvro)
    avro_data = list(avro_read)

new_df = pd.Dataframe(avro_data)

print(new_df)
```

Pickle

Just like how one can serialize and deserialize data for efficient storage, Python pickle enables serializing and deserializing a Python object. When a Python object is said to be picked, the pickle module serializes the given Python object before writing to a file. When it is said to be unpicked, the file is deserialized and so gets reconstructed back to its original form. Unlike other file formats, pickle is limited to Python language alone.

Here's how to export a data frame to pickle:

```
pd.to_pickel(given_data_frame, "location to file.pkl")
```

Here's how to read pickle into a data frame:

```
unpkl_df = pd.read_pickle("location to file.pkl")
```

Chunk Loading

Chunk loading refers to importing an entire large dataset in relatively smaller chunk sizes, as opposed to attempting to load the entire dataset in one go. By leveraging the concept of chunk loading, you can process several large datasets (larger than your memory datasets) without running into memory problems. Here is one way of leveraging chunk loading:

```
import pandas as pd

df = pd.DataFrame()

partition: int = 5000
# specifying partition chunks as 5000 rows
# will load every 5000 rows at every iteration
file = "/location-to-folder/filepath.csv"

for chunk in pd.read_csv(file, chunksize=partition):
    df = pd.concat(chunk,ignore_index=True)
```

Missing Values

Background

In data analysis and data processing tasks, missing data is a common phenomenon. Missing data or missing values happen when there is no value available for a data point in an event or an observation. You might witness an empty space with no value populated in certain places in a dataset. Missing values can occur for a variety of reasons.

The most common idea is the error that may occur during the data entry stages. During the process of manual data entry, it is not uncommon to forget to enter a value; another case is when data values have been entered incorrectly and, during quality checks, such data values are removed, therefore carrying a missing value label for a given observation and a variable.

Such data entry errors can also occur in automated machines or sensor mechanisms. There may be certain cases when the equipment performing reading may be under maintenance or any other issues that could have led to not reading the value at a given point in time, leading to missing value occurrences.

One of the common sources of data are surveys, whether it is done in person or online. Surveys are not easy; one may choose not to answer questions or to provide a different value from one's usual response (e.g., responding strongly to a survey scale of 1–5). Besides, the relationships between various data points in a given survey also matter. There may be cases where a missing value for one question may lead to missing values for several downstream questions. For instance, imagine we have a survey about motor vehicle insurance and the respondent does not own a motor vehicle; this may lead to skipping most questions.

CHAPTER 2   DATA WRANGLING USING PANDAS

Missing Values in Data Pipelines

Missing values can also occur during the process of data cleaning and transportation. When a filter or a conditional transformation is employed on a dataset, the output of such operation may lead to missing values, indicating those specific variables did not match the filtering criteria. Similarly, when data is migrated or transported from a source system to a staging layer in an analytical system, there may be cases of missing values due to inconsistencies in data formats or lack of support for a specific encoding type.

For instance, if a source system is using an ISO-8859 encoding and a target system is using a UTF-8 encoding, then characters may be misinterpreted during the migration process; in some fields, data may be truncated to certain characters or cut off, resulting in either missing values or data that does not make sense.

In Python programming, there are few ways of representing a missing value in a dataset. Let us look at them in detail.

None

None is a common way of representing missing values in data analysis. None has its own data type called "NoneType." None specifically denotes no value at all, which is different from an integer taking a 0; None is not the same as an empty string.

NaN

NaN stands for not a number. Not a number or NaN means the value is missing at a given place. NaN has a float type. Here is an example.

Let us create a data frame:

```
df = pd.DataFrame(np.random.randn(2,3), index=[1,2,3,4,5], columns=["a","b","c"])
```

The data frame would look like

```
       a         b         c
1   1.720131 -0.623429  0.196984
2  -1.093708 -0.190980 -1.063972
```

Now let us reindex the data frame:

```
df.reindex([1,2,3,4,5])
```

would yield us

```
          a         b         c
1  1.720131 -0.623429  0.196984
2 -1.093708 -0.190980 -1.063972
3       NaN       NaN       NaN
4       NaN       NaN       NaN
5       NaN       NaN       NaN
```

NaT

NaT means not a time. NaT represents missing values for the DateTime datatype.

NA

NA means not available. In the Pandas library, NA has been used historically to denote missing values for all data types. Instead of using None, NaN, or NaT for various data types, NA may be employed to represent missing values in integer, Boolean, and string data types.

Handling Missing Values

Pandas library offers few methods and treatments to work with missing values. Let's look at some of these.

isna() Method

The Pandas isna() method works to return a Boolean value indicating whether there is missing value in the data. The method returns true if missing values exist and false if there are no missing values for each element in the data frame. It works with NA and NaN types.

When we use this method to the preceding example

```
df2.isna()
```

would yield us

	a	b	c
1	False	False	False
2	False	False	False
3	True	True	True
4	True	True	True
5	True	True	True

Note You can also use "`isnull()`" in place of "`isna()`" for better readability.

---

notna() Method

The notna() can be seen as an inverse of the isna() method. Instead of identifying missing values, this method detects non-missing values; and so, the method would return true if the values are not missing and return false if the values are indeed missing. Here is how that looks for our same example:

df2.notna()

would return

	a	b	c
1	True	True	True
2	True	True	True
3	False	False	False
4	False	False	False
5	False	False	False

When to Use Which Method?

The isna() method identifies missing values, while the notna() method identifies non-missing values.

My take is that if you are looking to handle missing data, analyze missing data, and perform statistical imputations, then my suggestion is the isna() method. You can sum and count the number of missing values by using the sum() and count() methods, respectively.

Here is how that looks:

```
df2.isna().sum()

df2.isna().count()

df2["b"].isna().sum()
```

If you are looking to disregard the missing value rows completely and just need only the data points that do not have missing values, then my suggestion is the notna() method:

```
fillna() with values

fillna(0)

interpolate()
```

Note Use Multivariate Imputer with Iterative Imputer (and also Simple Imputer) imputation when one is absolutely sure there appears to be a clerical error that can be easily guessed: `https://scikit-learn.org/stable/modules/impute.html`.

For instance, a study that includes test subjects in a certain age range, along with checking a few other boxes that can be backtracked with reasonable estimation.

Data Transformation

Data transformation is one of the most crucial operations in building a quality dataset for machine learning pipelines. In broad terms, it can be seen as the process of converting data from one shape to another. The goal of the data transformation is to prepare data or create a dataset for downstream consumption.

CHAPTER 2 DATA WRANGLING USING PANDAS

Data Exploration

When in a new project or a new task, it is important to start somewhere. The first step is often the crucial step. And so, the goal of the first step should try to be as close as learning more about the dataset. Fortunately, there are some statistical tools available for data scientists and practitioners in Pandas. Let's look at some of these in detail.

Here is a dataset for illustration:

```
data = {'product': ['Apple', 'Orange', 'Melon', 'Peach', 'Pineapple'],
        'price': [22, 27, 25, 29, 35],
        'quantity': [5, 7, 4, 9, 3]
        }
```

Let's convert this into a data frame:

```
df = pd.DataFrame(data)
```

Now

```
df['price'].describe()
```

would yield

```
count     5.000000
mean     27.600000
std       4.878524
min      22.000000
25%      25.000000
50%      27.000000
75%      29.000000
max      35.000000
Name: price, dtype: float64
```

In case of a product variable

```
df['product'].describe()
```

CHAPTER 2   DATA WRANGLING USING PANDAS

would yield

```
count         5
unique        5
top       Apple
freq          1
Name: product, dtype: object
```

One can also obtain individual values for transformation purposes. Here are some examples.

Calculating the average price of a fruit

```
df['price'].mean()
```

would yield

27.6

Here is another example of obtaining a unique set of products:

```
df['product'].unique()
```

would yield

```
Out[17]: array(['Apple', 'Orange', 'Melon', 'Peach', 'Pineapple'],
dtype=object)
```

Creating a new variable on the go

```
df['total_cost'] = df['price']*df['quantity']
```

would yield

```
     product  price  quantity  total_cost
0      Apple     22         5         110
1     Orange     27         7         189
2      Melon     25         4         100
3      Peach     29         9         261
4  Pineapple     35         3         105
```

67

CHAPTER 2 DATA WRANGLING USING PANDAS

Let's add another column including the sale date. These dates are randomly generated. And our dataset looks like this:

	product	price	quantity	total_cost	sale_date
0	Apple	22	5	110	2023-12-09 22:59:06.997287
1	Orange	27	7	189	2024-01-03 11:03:12.484349
2	Melon	25	4	100	2023-12-10 01:59:22.598255
3	Peach	29	9	261	2023-12-21 05:28:01.462895
4	Pineapple	35	3	105	2023-12-17 23:38:42.783745

By running the describe function for sale_date, we get

```
df['sale_date'].describe()
```

```
count                              5
mean     2023-12-18 17:49:41.265306112
min         2023-12-09 22:59:06.997287
25%      2023-12-10 01:59:22.598255104
50%      2023-12-17 23:38:42.783745024
75%      2023-12-21 05:28:01.462895104
max         2024-01-03 11:03:12.484349
Name: sale_date, dtype: object
```

Furthermore, there is the option of sorting the dataset using the sort_values method():

```
df.sort_values('price')
```

would yield

	product	price	quantity	total_cost	sale_date
0	Apple	22	5	110	2023-12-09 22:59:06.997287
2	Melon	25	4	100	2023-12-10 01:59:22.598255
1	Orange	27	7	189	2024-01-03 11:03:12.484349
3	Peach	29	9	261	2023-12-21 05:28:01.462895
4	Pineapple	35	3	105	2023-12-17 23:38:42.783745

Combining Multiple Pandas Objects

Pandas library offers various methods and strategies to combine one or more of its data structures, whether they are series or data frames. Not only that, Pandas also provides the ability to compare two datasets of the same data structure to understand the differences between them.

One of the most important functionalities to know is the merge method provided by Pandas library. The merge() method offers relational databases like joins. These joins offered by the merge() method are performed in-memory.

The merge() method offers left join, right join, inner join, outer join, and cross join options.

Let us look at them in detail. For illustrative purposes, let us assume we are strictly focused on data frames with rows and columns.

Left Join

Given two data frames, the left join of these data frames would mean to keep everything on the data frame that is on the left and add any matching information from the data frame that is on the right. Let us imagine we have two data frames, a and b, respectively. The process starts off with keeping everything on the data frame a and looking for potential matches from the data frame b. The matching is usually searched based on the common column or a key in the data frame. When there is a match, then bring that matched column from b and merge it with a. If there is a matching row but only partially filled-in information, then the remaining would be populated with NA.

Here is how that looks. Let us initialize two data frames:

```
left_df1 = pd.DataFrame({
    'ID': [1, 2, 3, 4, 5],
    'Name': ['Alice', 'Bob', 'Charlie', 'David', 'John'],
    'DeptID': [101, 102, 103, 104, 103]
})
```

would yield us

```
   ID    Name  DeptID
0   1   Alice     101
1   2     Bob     102
```

CHAPTER 2   DATA WRANGLING USING PANDAS

```
2   3   Charlie     103
3   4   David       104
4   5   John        103
```

```python
right_df1 = pd.DataFrame({
    'DeptID': [101, 102, 103],
    'DeptName': ['HR', 'IT', 'Marketing']
})
```

would yield us

```
    DeptID   DeptName
0      101         HR
1      102         IT
2      103  Marketing
```

Left Join Using merge()

Now let us perform the left join using the merge() method:

```python
leftjoin = pd.merge(
    left_df1,
    right_df1,
    on="DeptID",
    how="left"
)

print(leftjoin)
```

would yield us

```
   ID     Name  DeptID   DeptName
0   1    Alice     101         HR
1   2      Bob     102         IT
2   3  Charlie     103  Marketing
3   4    David     104        NaN
4   5     John     103  Marketing
```

In the preceding example, you can see that everything from the left_df has been kept, while only the matched columns from right_df have been merged. There is only one row on the left_df for which there appears to be no matching rows from the right_df and which is filled with NaN, as a missing value.

Right Join

Right join can be seen as the inverse of left join.

Given two data frames, the right join of these data frames would mean to keep everything on the data frame that is on the right and add any matching information from the data frame that is on the left.

In our earlier example we had two data frames, left_df and right_df, respectively.

The right join would start off with keeping everything on the data frame right_df and look for potential matches from the data frame left_df. The matching is usually searched based on the common column or a key in the data frame. When there is a match, then bring that matched column from left_df and merge it with right_df. If there is a matching row but only partially filled-in information, then the remaining would be populated with NA.

The data frame right_df has the following data:

```
   DeptID   DeptName
0    101         HR
1    102         IT
2    103  Marketing
```

The data frame left_df has the following data:

```
   ID     Name  DeptID
0   1    Alice     101
1   2      Bob     102
2   3  Charlie     103
3   4    David     104
4   5     John     103
```

Let us try to perform the right join on these data frames:

```
rightjoin = pd.merge(
      left_df1,
      right_df1,
      on="DeptID",
      how="right"
)

print(rightjoin)
```

would yield

```
   ID    Name  DeptID   DeptName
0   1   Alice     101         HR
1   2     Bob     102         IT
2   3 Charlie     103  Marketing
3   5    John     103  Marketing
```

In the case of the right join, the table that is on the right remains unchanged. The table that is on the left would be scanned to identify matching information based on the department id column. The matched columns would be then brought over and be annexed with the table on the right.

Outer Join

In the relational database world, these are called full outer joins. Outer join is where, given two data frames, both of them are kept without losing any data. Outer joins have the most number of missing values, but they preserve any and all data that is in both data frames:

```
outerjoin = pd.merge(
      left_df1,
      right_df1,
      on="DeptID",
      how="outer"
)

print(outerjoin)
```

would yield us

```
   ID    Name  DeptID    DeptName
0   1   Alice     101          HR
1   2     Bob     102          IT
2   3 Charlie     103   Marketing
3   5    John     103   Marketing
4   4   David     104         NaN
```

The outer join attempted to capture all data from both tables and assigned a missing value (not a number) where a match is not available.

Inner Join

Inner join is the inverse of full outer join. These joins are where, given two data frames, only the matched columns would be returned as the resultant data frame. The partial matches from both the data frames would be excluded, and only complete matches will be returned.

Here is how that looks for the same example:

```python
innerjoin = pd.merge(
    left_df1,
    right_df1,
    on="DeptID",
    how="inner"
)

print(innerjoin)
```

would yield

```
Out[174]:
   ID    Name  DeptID    DeptName
0   1   Alice     101          HR
1   2     Bob     102          IT
2   3 Charlie     103   Marketing
3   5    John     103   Marketing
```

CHAPTER 2   DATA WRANGLING USING PANDAS

As you can see the row where the partial match was available has been excluded from the resultant dataset.

Cross Join

Cross join is also referred to as cartesian join. It is where every row in the given data frame is assigned or merged with rows from another data frame. The common key is not required.

Here is how this looks.

Let us initialize two data frames.

```
df_one = pd.DataFrame({
    'ID': [1, 2],
    'Name': ['John', 'Doe']
})
```

would yield us

```
   ID  Name
0   1  John
1   2  Doe
```

```
df_two = pd.DataFrame({
    'ItemID': [101, 102],
    'Item': ['Laptop', 'Cable']
})
```

would yield us

```
   ItemID    Item
0     101  Laptop
1     102   Cable
```

Now let us perform the cross join:

```
cross_join = pd.merge(
      df_one,
      df_two,
      how="cross"
   )
print(cross_join)
```

would yield us

```
   ID  Name  ItemID    Item
0   1  Alice    101  Laptop
1   1  Alice    102   Cable
2   2    Bob    101  Laptop
3   2    Bob    102   Cable
```

Data Reshaping

Data reshaping is the process of transforming the structure of data while preserving the values and observations contained in the dataset. The idea of performing data reshaping is to identify new information and insights while changing the dimensions of it, without having to change the values.

pivot()

The most common idea of data reshaping is the concept of pivoting a table. Made famous by Microsoft Excel, pivoting a table enables rearranging of rows and column data. Common applications include preparing data for dashboards, further analysis, etc.

Here is an example of a pivot table. Let's say we have the following dataset:

```
data = {
    'Date':
        ['2023-06-01', '2023-06-01', '2023-06-01', '2023-06-02',
         '2023-06-02', '2023-06-02'],
    'City':
        ['Cupertino', 'Portland', 'Fort Worth', 'Cupertino', 'Portland',
         'Fort Worth'],
    'Temperature':
        [56, 65, 69, 59, 61, 66]
}

df = pd.DataFrame(data)
```

would yield us

```
        Date        City  Temperature
0  2023-06-01   Cupertino           56
1  2023-06-01    Portland           65
2  2023-06-01  Fort Worth           69
3  2023-06-02   Cupertino           59
4  2023-06-02    Portland           61
5  2023-06-02  Fort Worth           66
```

Now let us reshape the data so that the column values in the city become column categories:

```
pivoted_df = df.pivot(
      index='Date',
      columns='City',
      values='Temperature'
)
print(pivoted_df)
```

would yield us

```
City         Cupertino  Fort Worth  Portland
Date
2023-06-01          56          69        65
2023-06-02          59          66        61
```

And so, the dates no longer repeat themselves, each of the cities in the column has become a category in itself, and the temperature data is displayed in better presentable format. The pivot() function is very beneficial in cases where you need to reshape the data without having to perform any type of aggregations.

pivot_table()

The pivot_table() method is similar yet provides more functionality when compared with the pivot() method. pivot_table() helps with aggregating numeric data while reshaping the entire dataset. The pivot_table() method provides spreadsheet pivoting-like functionality.

Let us look at an example:

```
data = {
    'Month':
        ['January', 'January', 'January', 'January',
         'February', 'February', 'February', 'February'],
    'Region':
        ['North', 'East', 'West', 'South',
         'North', 'East', 'West', 'South' ],
    'Product':
        ['Product A', 'Product A', 'Product A', 'Product A',
         'Product B', 'Product B', 'Product B', 'Product B'],
    'Sales':
        [200, 150, 180, 210, 190, 175, 225, 250]
}

df = pd.DataFrame(data)

print(df)
```

would yield us

```
     Month Region    Product  Sales
0  January  North  Product A    200
1  January   East  Product A    150
2  January   West  Product A    180
3  January  South  Product A    210
4 February  North  Product B    190
5 February   East  Product B    175
6 February   West  Product B    225
7 February  South  Product B    250
```

```
pivot_table_df = df.pivot_table(
    index='Product',
    columns='Region',
    values='Sales',
    aggfunc='mean'
)
```

CHAPTER 2 DATA WRANGLING USING PANDAS

```
print(pivot_table_df)
```

```
Region      East  North  South  West
Product
Product A   150   200    210    180
Product B   175   190    250    225
```

In the preceding output, the pivot_table() method converted the region categories into separate columns and converted the product categories into indexes (separate rows), and for each product and region combination, it aggregated the total sales in terms of taking the statistical mean. This way, you can obtain the performances of product sales for a given region.

stack()

The stack() method is somewhat similar and closely related to the idea of pivoting the table. The stack() method also reshapes the data by converting the column categories into indexes. Let us look at an example to observe the functionality of the stack() method.

Here is a dataset that contains quarterly sales of two products in a given two years:

```
data1 = {
    'Product': ['Product A', 'Product B'],
    '2023-Q1': [100, 80],
    '2023-Q2': [150, 130],
    '2023-Q3': [200, 180],
    '2023-Q4': [250, 230],
    '2024-Q1': [300, 280],
    '2024-Q2': [350, 330],
    '2024-Q3': [400, 380],
    '2024-Q4': [450, 430]
}
df1 = pd.DataFrame(data1).set_index('Product')
print(df1)
```

would yield

	2023-Q1	2023-Q2	2023-Q3	...	2024-Q2	2024-Q3	2024-Q4
Product				...			
Product A	100	150	200	...	350	400	450
Product B	80	130	180	...	330	380	430

Here is how we can stack quarterly sales for each product:

quartSales = df1.stack()

print(quartSales)

would yield

Product		
Product A	2023-Q1	100
	2023-Q2	150
	2023-Q3	200
	2023-Q4	250
	2024-Q1	300
	2024-Q2	350
	2024-Q3	400
	2024-Q4	450
Product B	2023-Q1	80
	2023-Q2	130
	2023-Q3	180
	2023-Q4	230
	2024-Q1	280
	2024-Q2	330
	2024-Q3	380
	2024-Q4	430

The resultant data frame contains an index that was previously column fields. Now, the resultant data frame has two indexes: one is the product type and the other is the sales quarter column.

CHAPTER 2   DATA WRANGLING USING PANDAS

unstack()

unstack() is the inverse operation of the stack() method. However, both these methods are mutually exclusive. You can call the unstack() method natively without having to use the stack() method. Let us look at an example:

```
d
data2 = {
    'Product':
        ['Product A', 'Product A', 'Product B', 'Product B'],
    'Month':
        ['January', 'February', 'January', 'February'],
    'Sales':
        [120, 110, 150, 140]
}

df3 = pd.DataFrame(data2).set_index(['Product','Month'])

print(df3)
```

would yield

```
                   Sales
Product   Month
Product A January    120
          February   110
Product B January    150
          February   140
```

Here is how we can widen the narrow table using the unstack() method:

```
wideSales = df3.unstack()

print(wideSales)
```

```
              Sales
Month      February January
Product
Product A       110     120
Product B       140     150
```

80

melt()

The melt() method would perform the inverse function of the pivot() function. The melt() method would convert the wide data format into a narrow data format.

Let us look at the functionality of the melt() method with the following example:

```
melt_data = {
    'Product':
        ['Product A', 'Product B', 'Product C'],
    'January':
        [300, 200, 400],
    'February':
        [350, 250, 450],
    'March':
        [400, 300, 500]
}

melt_df = pd.DataFrame(melt_data)

print(melt_df)
```

would yield

```
     Product  January  February  March
0  Product A      300       350    400
1  Product B      200       250    300
2  Product C      400       450    500
```

Now, let us apply the melt() method, where we would like to identify the product and measure the product against the month and sales variables:

```
melted_df = melt_df.melt(
    id_vars=['Product'],
    var_name='Month',
    value_name='Sales'
)
```

```
print(melted_df)
```

```
    Product     Month  Sales
0  Product A   January    300
1  Product B   January    200
2  Product C   January    400
3  Product A  February    350
4  Product B  February    250
5  Product C  February    450
6  Product A     March    400
7  Product B     March    300
8  Product C     March    500
```

The preceding resultant data frame has successfully converted the wide data frame into a narrow data frame.

crosstab()

crosstab() is a powerful statistical tool that is used to compute cross-tabulation of two or more factors. The crosstab() function builds a frequency table for given columns.

Here is an example:

```
ctdata = {
    'Person':
        [1, 2, 3, 4,
         5, 6, 7, 8, 9],
    'State':
        ['NY', 'WA', 'CO', 'NY',
         'WA', 'CO', 'NY', 'WA', 'CO'],
    'Likes':
        ['Mountains', 'Mountains', 'Mountains', 'Mountains',
         'Mountains', 'Oceans', 'StateParks', 'StateParks', 'StateParks']
}

ctdf = pd.DataFrame(ctdata)

print(ctdf)
```

would yield us

```
   Person State      Likes
0       1    NY  Mountains
1       2    WA  Mountains
2       3    CO  Mountains
3       4    NY  Mountains
4       5    WA  Mountains
5       6    CO     Oceans
6       7    NY  StateParks
7       8    WA  StateParks
8       9    CO  StateParks
```

Now, let us build a frequency table with the columns "State" and "Likes," respectively.

```
crossstab_df = pd.crosstab(
     ctdf['State'],
     ctdf['Likes'],
     rownames=["Region"],
     colnames=["FavoriteActivity"]
)

print(crossstab_df)
```

would yield us the following:

```
FavoriteActivity  Mountains  Oceans  StateParks
Region
CO                        1       1           1
NY                        2       0           1
WA                        2       0           1
```

The preceding data frame has provided the count of persons. You can also obtain the percentage of persons who liked a certain activity by using the "normalize" argument:

```
normalActMatrix = pd.crosstab(
     ctdf['State'],
     ctdf['Likes'],
```

CHAPTER 2 DATA WRANGLING USING PANDAS

```
        rownames=["Region"],
        colnames=["FavoriteActivity"],
        normalize=True
    )
print(normalActMatrix)
```

would yield us

```
FavoriteActivity  Mountains    Oceans   StateParks
Region
CO                 0.111111  0.111111     0.111111
NY                 0.222222  0.000000     0.111111
WA                 0.222222  0.000000     0.111111
```

You can also add the "margins" parameter to get the row and column sum of the frequency table:

```
freqTable = pd.crosstab(
        ctdf['State'],
        ctdf['Likes'],
        rownames=["Region"],
        colnames=["FavoriteActivity"],
        margins=True
    )
print(freqTable)
```

would yield

```
FavoriteActivity  Mountains  Oceans  StateParks  All
Region
CO                        1       1           1    3
NY                        2       0           1    3
WA                        2       0           1    3
All                       5       1           3    9
```

which can also be used along with the "normalize" parameter:

```
normalFreqTable = pd.crosstab(
       ctdf['State'],
       ctdf['Likes'],
       rownames=["Region"],
       colnames=["FavoriteActivity"],
       margins=True,
       normalize=True
   )

print(normalFreqTable)
```

would yield us

```
FavoriteActivity  Mountains    Oceans  StateParks       All
Region
CO                 0.111111  0.111111    0.111111  0.333333
NY                 0.222222  0.000000    0.111111  0.333333
WA                 0.222222  0.000000    0.111111  0.333333
All                0.555556  0.111111    0.333333  1.000000
```

factorize()

The factorize() method is another important method to know in the data reshaping topics. The factorize() method helps numerically categorize a given categorical variable (that may be a string). Some downstream analytic models may require a numerical encoding of a categorical variable. Let us look at an example:

```
factdata = {
   'Person':
       [1, 2, 3, 4, 5],
   'Likes':
       ['Mountains', 'Oceans', 'Oceans', 'Parks', 'Parks']
}

fact_df = pd.DataFrame(factdata)

print(fact_df)
```

CHAPTER 2   DATA WRANGLING USING PANDAS

would yield us

```
   Person       Likes
0       1   Mountains
1       2      Oceans
2       3      Oceans
3       4       Parks
4       5       Parks
```

The "factorize()" method returns two objects, the labels and the unique values. Here is how that looks in our example:

```
labels, uniques = pd.factorize(fact_df['Likes'])

print("The labels are:", labels)

print("The unique values are:", uniques.values)
```

would yield us

```
The labels are: [0 1 1 2 2]
The unique values are: ['Mountains' 'Oceans' 'Parks']
```

Let us incorporate the labels in our existing data frame:

```
fact_df['Likes_Coded'] = labels

print(fact_df)
```

would yield us

```
   Person       Likes  Likes_Coded
0       1   Mountains            0
1       2      Oceans            1
2       3      Oceans            1
3       4       Parks            2
4       5       Parks            2
```

compare()

The compare() method enables you to compare two data frames or series and identify their differences. Let us take a simple example where we observe differences between two data frames.

Here is a simple example:

```
ctdata = {
    'Person':
        [1, 2, 3, 4, 5, 6, 7, 8, 9],
    'State':
        ['NY', 'WA', 'CO', 'NY',
         'WA', 'CO', 'NY', 'WA', 'CO'],
    'Likes':
        ['Mountains', 'Mountains', 'Mountains', 'Mountains',
         'Mountains', 'Oceans', 'StateParks', 'StateParks', 'StateParks']
}

ct_df = pd.DataFrame(ctdata)

print(ct_df)
```

would yield

```
   Person State       Likes
0       1    NY   Mountains
1       2    WA   Mountains
2       3    CO   Mountains
3       4    NY   Mountains
4       5    WA   Mountains
5       6    CO      Oceans
6       7    NY  StateParks
7       8    WA  StateParks
8       9    CO  StateParks
```

Let us make a copy:

```
ct_df1 = ct_df.copy()
```

CHAPTER 2 DATA WRANGLING USING PANDAS

Now, alter a few values:

```
ct_df1.loc[3,'State'] = 'NC'

ct_df1.loc[5,'State'] = 'CA'
```

Now we have a different dataset named "ct_df1".
Here is how that looks:

```
   Person State      Likes
0       1    NY  Mountains
1       2    WA  Mountains
2       3    CO  Mountains
3       4    NC  Mountains
4       5    WA  Mountains
5       6    CA     Oceans
6       7    NY StateParks
7       8    WA StateParks
8       9    CO StateParks
```

Now, let us perform the comparison:

```
ct_df.compare(ct_df1)
```

would yield us

```
   State
   self other
3    NY    NC
5    CO    CA
```

Let us try to keep the entire dataset while performing the comparison. This can be achieved by using the "keep_shape" parameter:

```
ct_df.compare(ct_df1, keep_shape=True)
```

would result in

	Person		State		Likes	
	self	other	self	other	self	other
0	NaN	NaN	NaN	NaN	NaN	NaN
1	NaN	NaN	NaN	NaN	NaN	NaN
2	NaN	NaN	NaN	NaN	NaN	NaN
3	NaN	NaN	NY	NC	NaN	NaN
4	NaN	NaN	NaN	NaN	NaN	NaN
5	NaN	NaN	CO	CA	NaN	NaN
6	NaN	NaN	NaN	NaN	NaN	NaN
7	NaN	NaN	NaN	NaN	NaN	NaN
8	NaN	NaN	NaN	NaN	NaN	NaN

Notice the usage of missing values in the dataset. During comparison, if the values are similar on the respective rows and columns, then missing values are assigned. The values that are different between the data frames are only displayed.

groupby()

The groupby() method is an important method and function to know in data analysis using the Pandas library. The groupby() method performs aggregations and summarizations by grouping data. The groupby() method usually involves one or more of the following three steps:

- Split, a process of partitioning a dataset based on a certain criteria
- Apply, a process of applying functions on all the datasets that were split
- Combine, a process of merging the split dataset back to the original dimensions

Let us look at a detailed example:

```
groupbydata = {
    'Person':
        ['Alice', 'Bob', 'Charlie', 'David'],
```

```
    'Dept':
        ['Marketing', 'Marketing', 'DigitalM', 'DigitalM'],
    'Sales':
        [20, 12, 35, 42]
}
grp_df = pd.DataFrame(groupbydata)

print(grp_df)
```

would yield us

```
    Person       Dept  Sales
0    Alice  Marketing     20
1      Bob  Marketing     12
2  Charlie   DigitalM     35
3    David   DigitalM     42
```

Now let us apply the groupby() to this dataset. Here we are looking to obtain total sales performed by each department. This would mean grouping the dataset and computing the sum for each of the categories.

Here is how that looks:

```
grp_df1 = grp_df.groupby('Dept')['Sales'].sum()

print(grp_df1)
```

would yield us

```
Dept
DigitalM     77
Marketing    32
Name: Sales, dtype: int64
```

Conclusion

So far, you have gained a solid understanding of the core concepts and techniques that form the pillars of effective data manipulation using Pandas. Through this chapter you have discovered the core data structures, indexing methods, data reshaping strategies, advanced data wrangling methods, combining multiple Pandas objects, and working with missing values. The concepts and ideas that were discussed here will serve you good even if you do not actively use Pandas library; these concepts will lay the foundation for you to learn various other data wrangling libraries. You may witness in the upcoming chapters that some of these methods and methodologies you saw here may repeat themselves in some fashion.

CHAPTER 3

Data Wrangling using Rust's Polars

Introduction

So far, Pandas library has been the most famous library for data manipulation and analysis. However, Pandas may face performance challenges when dealing with massive datasets. When dealing with large datasets, we need a solution that can perform well without having to add additional memory and compute. This is where Polars excels. Polars is a fast data manipulation library programmed in Rust programming and integrates well with Python's data analysis ecosystem. Polars supports lazy evaluation and automatic query optimization, enabling data engineers to handle even the most demanding data wrangling tasks easily.

By the end of this chapter, you will

- Excel in the key concepts of Polars, its architecture, lazy evaluation, and eager evaluation.

- Understand Polars' data structures such as DataFrames, Series, and LazyFrames.

- Extract and load data in various formats.

- Perform data transformations, filtering, and aggregation using Polars' methods.

- Combine multiple data objects using Polars and investigate advanced topics like data reshaping, missing values and unique value identifications, etc.
- Write SQL-like statements in Polars and query your flat files on your file system using Polars command line interface (CLI).

Introduction to Polars

Polars is a promising, up-and-coming Python library within the data processing ecosystem of Python language. The data structures of Polars library are based on the Apache Arrow project, meaning Polars stores its data internally using Apache Arrow arrays, comparable to how Pandas library uses NumPy internally to store its data in the form of NumPy arrays.

Polars library is written in Rust programming language. Rust is a recent programming language that demonstrates C/C++-like efficiency. Rust provides strong support for concurrency and parallel operations. Polars library on Python enables multi-threading and parallelization on almost every command and every line of code. No special instruction needs to be provided to enable parallelization—no need for decorators or tweaking different settings. Parallel processing and multi-threading are enabled by default.

It may be possible to use Polars library along with other libraries within the Python data analysis ecosystem. However, it may be the case that Python libraries are usually single threaded. If the data processing task can be parallelized, one may leverage other libraries such as multiprocessing to speed up the entire data pipeline.

The syntax for Polars library in Python can appear to be very similar when compared to Pandas library. However, there are few differences that will be discussed in the coming topics. To start off, Polars does not support index or multi-index for its data structures, unlike Pandas. And so, when writing Polars code, one may not be able to use methods like the loc or iloc method that are available in Python.

Polars library supports both lazy evaluation and eager evaluation, whereas Python supports only eager evaluation. Polars automatically optimizes the query plan and finds opportunities to optimally leverage memory while accelerating the execution of a query.

Lazy vs. Eager Evaluation

To understand and use Polars library effectively, it may be beneficial to understand lazy evaluation and eager evaluation in programming.

Eager evaluation is the process of evaluating the expression immediately as soon as the expression is seen in the code. Eager evaluation is the default behavior in most programming languages. Python is no exception.

Lazy evaluation is simply a process of running a code if and only if that code requires it to be executed during runtime.

Here is an illustration. Let's look at the following Python code:

```
def ap(a: int, b: int) -> None:
    c = a + b
    print("the numbers are",a,"and", b)
    print("the sum is", c)
```

So when this function is called

```
ap(4,3)
```

it would then yield

```
the numbers are 4 and 3
the sum is 7
```

Here, this program is a simple case of eager evaluation. Every line in that function is processed. When the function is called with parameters, all the three lines are evaluated and the statements are printed.

In contrast, in the case of lazy evaluation, the line that contains c=a+b will be deferred from execution, until the time it is actually needed. Can you guess when c=a+b is actually needed? It would be the last line where we are printing the sum. That is when the addition operation is actually executed. The function requires the value of c to be computed.

Note To illustrate this even further, let's take an example of visiting a restaurant to get a meal. Eager evaluation is like visiting a buffet restaurant with standard offerings and no customizations. Every dish is cooked and ready to be served, regardless of how much the customer would consume or whether anyone would even consume that dish at all.

Lazy evaluation is like visiting a restaurant, where the server person would ask questions about which sides and items you like, how you like them, whether you are allergic to something, and several other things. The chef would then prepare the dish and include items based on the specific responses of the customer.

Eager evaluation computes all the results up front, consumes system resources to compute everything regardless of they are ultimately needed, and may experience a slower initial performance but subsequent access may be faster. **Lazy evaluation** computes only the results that are required, conserves system resources by executing only what is necessary, and may seem faster initially but accessing subsequent results may be slower.

It is recommended to use lazy evaluation wherever that may be available in Polars. In addition to default parallelizations, multi-threading, "close-to-the-metal" programming architecture, Rust-like efficiencies, and automatic query optimizations, one may also benefit from executing only the required code that may go easy on the computing resources.

Data Structures in Polars

To set up Polars in a system, it may be easy to merely install from a Python package installer:

```
pip install polars
```

or

```
pip install polars[all]
```

or

```
conda install -c conda-forge polars
```

Here is how to use the library in a project:

```
import polars as pl
```

CHAPTER 3 DATA WRANGLING USING RUST'S POLARS

Polars offers data structures similar to those of Pandas library. They are Series, DataFrames, and LazyFrames (the one with lazy evaluation). It may be safe to say that data frames are the most commonly used data structure in the Python ecosystem.

Polars Series

Here is an example of Series:

```
a = pl.Series([1,4,7,9,16,25])
```

```
print(a)
```

One can also name the series, in the following manner:

```
b = pl.Series(
      "firstSeries",
      [1,4,7,9,16,25]
    )
```

```
print(b)
```

would yield

```
shape: (6,)
Series: 'firstSeries' [i64]
[
 1
 4
 7
 9
 16
 25
]
```

Here is how one can append data elements:

```
b1 = pl.Series(
      "secondSeries",
      [36,47]
    )
```

```
b = b.append(b1)
```

```
print(b)
```

would yield

```
shape: (8,)
Series: 'firstSeries' [i64]
[
1
4
7
9
16
25
36
47
]
```

Series support several other methods as well; here is an example of obtaining the total number of unique values:

```
a1 = pl.Series(
        [1,4,7,9,16,25,5,25,9,5]
        )
```

```
print(a1.n_unique())
```

would yield

7

Polars Data Frame

A data frame in Polars is similar to the concept of a data frame in Pandas. There are options to reshape the data and perform aggregations and transformations.

Here is a simple example of initializing a data frame:

```
df = pl.DataFrame(
    {
        "fruits":
            ["Apple","Peach","Melon","Pineapple","Orange"],
        "price":
            [1.5, 0.75, 3.0, 3.25, 1.33],
        "quantity":
            [4, 3, 1, 2, 6]
    }
)
print(df)
```

would yield

```
shape: (5, 3)
┌───────────┬───────┬──────────┐
│ fruits    ┆ price ┆ quantity │
│ ---       ┆ ---   ┆ ---      │
│ str       ┆ f64   ┆ i64      │
╞═══════════╪═══════╪══════════╡
│ Apple     ┆ 1.5   ┆ 4        │
│ Peach     ┆ 0.75  ┆ 3        │
│ Melon     ┆ 3.0   ┆ 1        │
│ Pineapple ┆ 3.25  ┆ 2        │
│ Orange    ┆ 1.33  ┆ 6        │
└───────────┴───────┴──────────┘
```

Polars, by default, has prints with aligned rows and columns with borders. This may resemble a pretty print library in Python. Like Pandas, one can view the head and tail of the data frame simply by calling head(n) and tail(n), respectively.

The option describe() would return the summary statistics for the data frame. Here is how the syntax looks like:

```
print(df.describe())
```

would yield

shape: (9, 4)

describe	fruits	price	quantity
str	str	f64	f64
count	5	5.0	5.0
null_count	0	0.0	0.0
mean	null	1.966	3.2
std	null	1.097511	1.923538
min	Apple	0.75	1.0
25%	null	1.33	2.0
50%	null	1.5	3.0
75%	null	3.0	4.0
max	Pineapple	3.25	6.0

Polars Lazy Frame

A lazy frame is a data structure that enables lazy evaluation in Polars. A lazy frame enables the query optimization feature, in addition to parallelism. It is recommended to leverage on lazy evaluation where possible to gain maximum performance from Polars. A lazy frame cannot be necessarily referred to as one of the native data structures provided by Polars; however, given a data frame, there are ways to convert the same into a lazy frame.

Let's look at the example discussed in the previous section. Though it is a Polars data frame, it is not set up for lazy evaluation.

Here is how it can be set up for lazy evaluation:

```
plf1 = pl.DataFrame(
    {
        "fruits":
            ["Apple","Peach","Melon","Pineapple","Orange"],
```

```
            "price":
                [1.5, 0.75, 3.0, 3.25, 1.33],
            "quantity":
                [4, 3, 1, 2, 6]
    }
).lazy()
```

By adding the lazy(), the data frame is now set up to be a lazy frame.

Here is how to enable an existing Pandas data frame to a lazy frame. Let's initialize a Pandas data frame to begin with:

```
import pandas as pd

pandas_df1 = pd.DataFrame(
    {
        "a":
            ["alpha","beta","gamma","delta","eta"],
        "b":
            [1, 2, 3, 4, 5]
    }
)

print(pandas_df1)
```

would yield

```
       a  b
0  alpha  1
1   beta  2
2  gamma  3
3  delta  4
4    eta  5
```

Here's how to create a lazy frame:

```
pandas_lazyframe = pl.LazyFrame(pandas_df1)

print(pandas_lazyframe)
```

would yield

```
naive plan: (run LazyFrame.explain(optimized=True) to see the
optimized plan)
DF ["a", "b"]; PROJECT */2 COLUMNS; SELECTION: None
```

One can also overwrite the Pandas data frame to a lazy frame.

Data Extraction and Loading
CSV

Polars supports importing and exporting many of the common data formats. Here is how to work with the CSV file:

Here's how to import a CSV:

```
polars_df = pl.read_csv("path to file.csv")
```

The preceding code is meant to perform eager evaluation. To enable lazy computation and leverage the lazy frame, one needs to use the scan_csv method. Scanning delays the parsing of the file unless it is absolutely required:

```
polars_df_lazy = pl.scan_csv("path to file.csv")
```

Here's how to export a Polars data frame to CSV:

```
polars_df.write_csv("destination to file.csv")
```

JSON

Polars supports read and write operations for both JSON and ndJSON.

ndJSON stands for newline-delimited JSON and enables faster inserts and reads from files.

Here's how to read JSON:

```
json_df = pl.read_json("path to file.json")
```

Here's how to read a ndJSON into a data frame:

```
ndjson_df = pl.read_ndjson("path to the ndjson.file")
```

To enable lazy evaluation, Polars offers the option that is limited to ndJSON files:

```
lazyjson_df = pl.scan_ndjson("path to the file.json")
```

Here's how to write a Polars data frame into a JSON file:

```
jsonwrite_df = pl.write_json("path to location.file")
```

And here's for ndJSON:

```
jsonwrite_df = pl.write_ndjson("path to location.file")
```

Parquet

As noted earlier, Parquet is a columnar data store and it is considered faster than many of the data sources. Here is how one can read Parquet files from Polars:

```
parquet_df = pl.read_parquet("path to file.parquet")
```

Here's how to use lazy evaluation on a Parquet file:

```
parquet_df = pl.scan_parquet("path to file.parquet")
```

Here is how to write a Polars data frame into a Parquet file:

```
parquet_df = pl.write_parquet("path to destination file.parquet")
```

Polars also can provide lazy evaluation and read and write data in various other formats like Avro, ORC, Feather, etc.

Data Transformation in Polars

To perform data analysis and transformation using Polars, it is important to understand the core components of Polars. These are contexts and expressions. Let us look at expressions first. Expressions refer to the actual transformations and data manipulations that are performed on various columns, to obtain new values. Examples could be applying a formula to an entire column like converting a Fahrenheit to Celsius or extracting a subset of data from the column, to name a few.

CHAPTER 3   DATA WRANGLING USING RUST'S POLARS

Polars Context

A context is simply a context where the expressions get evaluated and executed. There are three main contexts in Polars: selection contexts, filtering contexts, and group-by contexts. Let's look at them in detail.

Selection Context

A selection context is where the idea of selection applies to a given Polars data frame. Some of the operations that can be performed in a selection context are

```
dataframe.select([" ... "])
dataframe.with_columns([" ... "])
```

Let us look at an example:

```python
import polars as pl

df = pl.DataFrame(
    {
    "A": [1, 2, 3],
    "B": [4, 5, 6]
    }
)

print(df)
```

When we look at the data, we see

```
shape: (3, 2)
┌─────┬─────┐
│ A   │ B   │
│ --- │ --- │
│ i64 │ i64 │
╞═════╪═════╡
│ 1   │ 4   │
│ 2   │ 5   │
│ 3   │ 6   │
└─────┴─────┘
```

If we need to select a specific column, we can use

```
print(df.select(['A']))
```

which would yield

shape: (3, 1)

```
┌─────┐
│ A   │
│ --- │
│ i64 │
╞═════╡
│ 1   │
│ 2   │
│ 3   │
└─────┘
```

However, this statement has very limited scope. If you want to perform further transformations on the selected column, you need to include an additional expression.

Here is how that looks:

```
timesTwo = df.select(
    ( pl.col(['A']) * 2 )
    .alias("NewCol_A")
    )
```

```
print(timesTwo)
```

would yield

shape: (3, 1)

```
┌──────────┐
│ NewCol_A │
│ ---      │
│ i64      │
╞══════════╡
│ 2        │
│ 4        │
│ 6        │
└──────────┘
```

Filter Context

A filter context is where a given expression filters a dataset and provides a data frame that matches the user-supplied conditions. The operations that can be performed in this context are

```
dataframe.filter([" ... "])
dataframe.where([" ... "])
```

Here is an example of using filter in the preceding dataset:

```
conFilter = df.filter(pl.col('A')>1)
```

```
print(conFilter)
```

would yield us

shape: (2, 2)

```
┌─────┬─────┐
│ A   │ B   │
│ --- │ --- │
│ i64 │ i64 │
╞═════╪═════╡
│ 2   │ 5   │
│ 3   │ 6   │
└─────┴─────┘
```

Group-By Context

A group-by context enables performing aggregation and summaries on the existing Polars data frame while also performing computations on top of the aggregated data frame. The operation that can be performed in this context is

```
dataframe.group_by([...]).agg([...])
```

CHAPTER 3 DATA WRANGLING USING RUST'S POLARS

Let us look at an example for the group-by function:

```
df1 = pl.DataFrame(
    {
        "Region":
            ["CityBranch", "CampusBranch", "CityBranch",
             "SuburbBranch", "CampusBranch", "SuburbBranch",
             "CityBranch", "CampusBranch"],
        "Product":
            ["A", "B", "B",
             "A", "B", "A",
             "A", "B"],
        "Sales":
            [100, 200, 150, 300, 250, 400, 350, 180]
    }
)

print(df1)
```

would yield us

shape: (8, 3)

Region	Product	Sales
str	str	i64
CityBranch	A	100
CampusBranch	B	200
CityBranch	B	150
SuburbBranch	A	300
CampusBranch	B	250
SuburbBranch	A	400
CityBranch	A	350
CampusBranch	B	180

CHAPTER 3 DATA WRANGLING USING RUST'S POLARS

Now let us perform the group_by operation on this dataset:

```
groupbydf = df1\
    .group_by(["Region", "Product"])\
    .agg(
        [
            pl.col("Sales")
                .sum()
                .alias("Total"),
            pl.col("Sales")
                .max()
                .alias("Max Sales"),
            pl.count("Sales")
                .alias("TransactionsCount")\
        ]
    )
print(groupbydf)
```

would yield us

shape: (4, 5)

Region	Product	Total	Max Sales	TransactionsCount
str	str	i64	i64	u32
CityBranch	B	150	150	1
CityBranch	A	450	350	2
SuburbBranch	A	700	400	2
CampusBranch	B	630	250	3

In the preceding example, the group_by method performs aggregation based on both the product and region columns. Within the aggregation method, the total sales is calculated by using the sum() method, the biggest sale is calculated by using the max() method, and the total number of transactions is computed by using the count() method.

You can also leverage the idea of lazy evaluation when working with these contexts and expressions. Here is how that looks:

```
df1_lazy = pl.LazyFrame(
    {
        "Region":
            ["CityBranch", "CampusBranch", "CityBranch",
             "SuburbBranch", "CampusBranch", "SuburbBranch",
             "CityBranch", "CampusBranch"],
        "Product":
            ["A", "B", "B",
             "A", "B", "A",
             "A", "B"],
        "Sales":
            [100, 200, 150,
             300, 250, 400,
             350, 180]
    }
)

print(df1_lazy)
```

would yield

```
naive plan: (run LazyFrame.explain(optimized=True) to see the optimized plan)
DF ["Region", "Product", "Sales"]; PROJECT */3 COLUMNS; SELECTION: None
```

Basic Operations

Let us look at performing basic operations while working with expressions in Polars.

CHAPTER 3   DATA WRANGLING USING RUST'S POLARS

Here, we will revisit the earlier example.

```
df3 = pl.DataFrame(
    {
        "A": [1, 2, 3],
        "B": [4, 5, 6]
    }
)

print(df3)
```

would yield us

shape: (3, 2)

```
┌─────┬─────┐
│ A   │ B   │
│ --- │ --- │
│ i64 │ i64 │
╞═════╪═════╡
│ 1   │ 4   │
│ 2   │ 5   │
│ 3   │ 6   │
└─────┴─────┘
```

Here are some arithmetic operations performed on these columns:

```
modified_df3 = df3.select(
    (pl.col("A") + 12)
        .alias("A+12"),
    (pl.col("B") - 3)
        .alias("B-3")
)

print(modified_df3)
```

would yield us

shape: (3, 2)

```
┌──────┬──────┐
│ A+12 │ B-3  │
│ ---  │ ---  │
│ i64  │ i64  │
╞══════╪══════╡
│ 13   │ 1    │
│ 14   │ 2    │
│ 15   │ 3    │
└──────┴──────┘
```

Let us look at the with_columns() method within the selection context:

```
modified_df3_wc = df3.with_columns(
        (pl.col("A") + 12)
        .alias("A+12"),
        (pl.col("B") - 3)
        .alias("B-3")
)
print(modified_df3_wc)
```

would return

shape: (3, 4)

```
┌─────┬─────┬──────┬─────┐
│ A   │ B   │ A+12 │ B-3 │
│ --- │ --- │ ---  │ --- │
│ i64 │ i64 │ i64  │ i64 │
╞═════╪═════╪══════╪═════╡
│ 1   │ 4   │ 13   │ 1   │
│ 2   │ 5   │ 14   │ 2   │
│ 3   │ 6   │ 15   │ 3   │
└─────┴─────┴──────┴─────┘
```

The difference between the df.select() method and df.with_columns() method is that the latter would create new columns, whereas the former would only display results unless that is stored in a separate data frame.

CHAPTER 3 DATA WRANGLING USING RUST'S POLARS

Let us look at a logical operator:

```
logical_df3 = df3.select(
        (pl.col('A')>1)
        .alias("Is A > 1?")
)
print(logical_df3)
```

would yield us

shape: (3, 1)

```
┌──────────┐
│ Is A > 1?│
│ ---      │
│ bool     │
╞══════════╡
│ false    │
│ true     │
│ true     │
└──────────┘
```

You can also create a new column by using the with_columns() method:

```
logical_df3_wc = df3.with_columns(
        (pl.col('A')>1)
        .alias("Is A > 1?")
    )
print(logical_df3_wc)
```

would return

shape: (3, 3)

```
┌─────┬─────┬──────────┐
│ A   │ B   │ Is A > 1?│
│ --- │ --- │ ---      │
│ i64 │ i64 │ bool     │
╞═════╪═════╪══════════╡
│ 1   │ 4   │ false    │
│ 2   │ 5   │ true     │
│ 3   │ 6   │ true     │
└─────┴─────┴──────────┘
```

String Operations

Polars uses Apache Arrow as its backend, which enables a contiguous memory model for all its operations. Contiguous memory is where the elements are stored sequentially so that the CPU can retrieve these elements faster and have more efficient operations. String processing can get expensive as the data grows over time.

Aggregation and Group-By

Polars offers powerful aggregations and summarizations, in both the eager API and lazy API. Let us look at some of the aggregations by loading a dataset:

```
polars_dataset = pl.read_csv(
        "/path-to-the-folder/polars_dataset.csv",
        separator=",",
        infer_schema_length=0
        ) \
        .with_columns(
            pl.col("birthdate")
            .str.to_date(strict=False)
        )
```

print(polars_dataset)

would yield

shape: (1_000, 7)

```
| firstname | lastname  | gender | birthdate  | type       | state                | occupation      |
| ---       | ---       | ---    | ---        | ---        | ---                  | ---             |
| str       | str       | str    | date       | str        | str                  | str             |
| Rosa      | Houseago  | F      | null       | Stateparks | California           | Accountant      |
| Reese     | Medford   | M      | null       | Stateparks | District of Columbia | Electrician     |
| Gianna    | Rylands   | F      | 1973-07-02 | Mountains  | Kentucky             | Doctor          |
| Abran     | McGeown   | M      | 1962-12-08 | Baseball   | Virginia             | Accountant      |
| Lanita    | Yantsev   | F      | 1976-06-07 | Stateparks | Illinois             | Electrician     |
| ...       | ...       | ...    | ...        | ...        | ...                  | ...             |
| Selina    | Heyworth  | F      | 1951-04-10 | Baseball   | Ohio                 | IT Professional |
| Rubi      | Licari    | F      | null       | Oceans     | Michigan             | Lawyer          |
| Rubin     | Stanworth | M      | 1956-09-08 | Baseball   | Indiana              | Electrician     |
| Benjie    | Amort     | M      | 1957-07-07 | Museums    | Kentucky             | Doctor          |
| Gallard   | Samuels   | M      | 1961-07-09 | Oceans     | California           | Doctor          |
```

Combining Multiple Polars Objects

Polars supports a wide range of data transformations and joins, similar to Pandas. Let us look at them in detail.

Left Join

A left join returns all rows from the left table, scans the rows from the right table, identifies matches, and annexes them to the left table. Let us look at an example.

Here we initialize two data frames:

```
import polars as pl

df1 = pl.DataFrame({
    'ID': [1, 2, 3, 4, 5],
    'Name': ['Alice', 'Bob', 'Charlie', 'David', 'John'],
    'DeptID': [101, 102, 103, 104, 103]
```

```
})

df2 = pl.DataFrame({
    'DeptID': [101, 102, 103],
    'DeptName': ['HR', 'IT', 'Marketing']
})
```

Now, we perform the left join by joining using the department id:

```
leftjoin = df1.join(
        df2,
        on = "DeptID",
        how = "left",
        coalesce=True
    )
print(leftjoin)
```

If you specify coalesce = True, the column on which the join operation is performed will not occur on the final result. This would yield us

shape: (5, 4)

ID	Name	DeptID	DeptName
i64	str	i64	str
1	Alice	101	HR
2	Bob	102	IT
3	Charlie	103	Marketing
4	David	104	null
5	John	103	Marketing

Chapter 3 Data Wrangling Using Rust's Polars

If coalesce is set to false, here is how that may change our dataset:

```
leftjoin = df1.join(
        df2,
        on = "DeptID",
        how = "left",
        coalesce=False
    )

print(leftjoin)
```

would yield us

```
shape: (5, 5)
┌─────┬─────────┬────────┬──────────────┬───────────┐
│ ID  │ Name    │ DeptID │ DeptID_right │ DeptName  │
│ --- │ ---     │ ---    │ ---          │ ---       │
│ i64 │ str     │ i64    │ i64          │ str       │
╞═════╪═════════╪════════╪══════════════╪═══════════╡
│ 1   │ Alice   │ 101    │ 101          │ HR        │
│ 2   │ Bob     │ 102    │ 102          │ IT        │
│ 3   │ Charlie │ 103    │ 103          │ Marketing │
│ 4   │ David   │ 104    │ null         │ null      │
│ 5   │ John    │ 103    │ 103          │ Marketing │
└─────┴─────────┴────────┴──────────────┴───────────┘
```

Outer Join

Outer join is such that when a match is found between two given tables, it would retrieve any and all rows that are matched:

```
outerjoin = df1.join(
        df2,
        on = "DeptID",
        how = "full",
        coalesce=True
    )
```

print(outerjoin)

would yield us

shape: (5, 4)

ID	Name	DeptID	DeptName
i64	str	i64	str
1	Alice	101	HR
2	Bob	102	IT
3	Charlie	103	Marketing
4	David	104	null
5	John	103	Marketing

Inner Join

Inner join is the inverse of outer join and only returns rows that have matching values from both tables:

```
innerjoin = df1.join(
        df2,
        on = "DeptID",
        how = "inner",
        coalesce=True)
```

print(innerjoin)

would yield us

shape: (4, 4)

CHAPTER 3   DATA WRANGLING USING RUST'S POLARS

```
┌─────┬─────────┬────────┬───────────┐
│ ID  │ Name    │ DeptID │ DeptName  │
│ --- │ ---     │ ---    │ ---       │
│ i64 │ str     │ i64    │ str       │
╞═════╪═════════╪════════╪═══════════╡
│ 1   │ Alice   │ 101    │ HR        │
│ 2   │ Bob     │ 102    │ IT        │
│ 3   │ Charlie │ 103    │ Marketing │
│ 5   │ John    │ 103    │ Marketing │
└─────┴─────────┴────────┴───────────┘
```

Semi Join

Semi join filters rows that have a match on the right table:

```
semijoin = df1.join(
    df2,
    on = "DeptID",
    how = "semi",
    coalesce=True)
```

print(semijoin)

would yield us

shape: (4, 3)

```
┌─────┬─────────┬────────┐
│ ID  │ Name    │ DeptID │
│ --- │ ---     │ ---    │
│ i64 │ str     │ i64    │
╞═════╪═════════╪════════╡
│ 1   │ Alice   │ 101    │
│ 2   │ Bob     │ 102    │
│ 3   │ Charlie │ 103    │
│ 5   │ John    │ 103    │
└─────┴─────────┴────────┘
```

Anti Join

Anti join filters rows that do not have a match on the right table:

```
antijoin = df1.join(
        df2,
        on = "DeptID",
        how = "anti",
        coalesce=True)
```

print(antijoin)

would yield us

shape: (1, 3)

ID	Name	DeptID
i64	str	i64
4	David	104

Cross Join

Cross join would perform a cartesian product of given two tables.

Here is how that would look. Let us initialize two data frames:

```
df_one = pl.DataFrame(
    {
        'ID': [1, 2],
        'Name': ['John', 'Doe']
    }
)
df_two = pl.DataFrame(
    {
        'ItemID': [101, 102],
```

```
        'Item': ['Laptop', 'Cable']
    }
)
```

Now let us perform the cartesian product between these two data frames:

```
crossjoin = df_one.join(
        df_two,
        how = "cross",
        coalesce=True
    )

print(crossjoin)
```

would yield us

shape: (4, 4)

ID	Name	ItemID	Item
i64	str	i64	str
1	John	101	Laptop
1	John	102	Cable
2	Doe	101	Laptop
2	Doe	102	Cable

Advanced Operations

Let us look at some advanced methods in Polars data analysis library. For our analysis purposes, we will leverage our earlier dataset:

```
import polars as pl

polars_dataset = pl.read_csv(
        "/Users/pk/Documents/polars_dataset.csv",
        separator=",",
        infer_schema_length=0
```

```
    ) \
    .with_columns(
        pl.col("birthdate")
        .str.to_date(strict=False)
    )
print(polars_dataset)
```

Identifying Missing Values

Let us try to look at missing values on specified columns within the dataset. This would enable us to locate rows with missing data:

```
missingdata = polars_dataset.filter(
        pl.col("birthdate").is_null()
        )
```

print(missingdata)

would yield us

shape: (588, 7)

firstname	lastname	gender	birthdate	type	state	occupation
str	str	str	date	str	str	str
Rosa	Houseago	F	null	Stateparks	California	Accountant
Reese	Medford	M	null	Stateparks	District of Columbia	Electrician
Alicia	Farrant	F	null	Oceans	Florida	Accountant
Shawn	Fenlon	M	null	Stateparks	Michigan	IT Professional
Lenette	Blackly	F	null	Mountains	Texas	Accountant
...	...	...	...	...	...	...
Perla	Brixey	F	null	Stateparks	Washington	Doctor
Bobbi	Longea	F	null	Mountains	Massachusetts	Accountant
Skip	Urry	M	null	Oceans	Arizona	Doctor
Amory	Cromie	M	null	Baseball	Virginia	Lawyer
Rubi	Licari	F	null	Oceans	Michigan	Lawyer

Identifying Unique Values

In the preceding dataset, we wish to understand how many unique occupations are mentioned. Let us try to obtain that information in the following manner:

```
various_occupations = polars_dataset.group_by('occupation') \
    .agg(
        pl.count('occupation') \
        .alias('various_occupations')
    )

print(various_occupations)
```

would yield us

shape: (5, 2)

occupation	various_occupations
str	u32
Doctor	225
IT Professional	196
Accountant	178
Lawyer	200
Electrician	201

Pivot Melt Examples

The pivot function would transform the entire dataset into a wider format, with column categories becoming columns and rows, and the aggregation metric would be populated to the respective row and column.

Here is how that looks:

```
pivotdata = polars_dataset.pivot(
        index="state",
        columns="occupation",
```

```
        values="firstname",
        aggregate_function="first"
)
```

print(pivotdata)

would yield us

shape: (51, 6)

state	Accountant	Electrician	Doctor	IT Professional	Lawyer
str	str	str	str	str	str
California	Rosa	Margret	Devin	Joana	Yetta
District of Columbia	Ginger	Reese	Giles	Zorina	Nancey
Kentucky	Neila	Aurea	Gianna	Cesare	Orren
Virginia	Abran	Claudelle	Kalil	Gwendolen	Rog
Illinois	Gareth	Lanita	Axel	Darelle	Nancee
...	...	...	...	...	...
Hawaii	Barr	Lauretta	null	null	Kiel
Kansas	null	Sarena	null	Cherye	Helyn
Vermont	null	null	null	Kerrie	null
South Dakota	Thorvald	null	null	null	null
Maine	null	null	Rustin	null	null

Polars/SQL Interaction

While Polars library does not have an underlying SQL engine, Polars does support interacting with data objects using SQL-like statements. So, when you are issuing a SQL-like statement to a Polars data object, Polars would convert that to appropriate expressions and continue to provide you fast results (as it always does). You can execute "SQL queries" on flat files, JSON files, and even Pandas data objects. Let us look at an illustration involving a dataset that contains random health data.

CHAPTER 3 DATA WRANGLING USING RUST'S POLARS

Let us initialize this random dataset:

```
import polars as pl

df = pl.read_csv(
        "path-to-your-folder/random_health_data.csv"
)
print(df.head())
```

would yield

shape: (5, 7)

id	name	age	state	gender	exercise_daily?	copay_paid
i64	str	str	str	str	bool	str
1	Allayne Moffett	null	SC	Male	null	$206.76
2	Kerby Benjafield	null	NM	Male	true	$21.58
3	Raina Vallentin	null	MI	Female	true	$125.18
4	Kaela Trodden	null	OH	Female	false	$86.20
5	Faber Kloisner	null	MN	Male	false	$219.38

This fictitious dataset contains the name of the patient, demographic data points like gender and age, the state where they reside, and another column called copay paid. Notice the last column, called copay_paid, which is a string and contains a special character at the beginning. To convert this column into a numeric data type, we need to remove the special character, the dollar symbol, and convert the remaining to a decimal.

Here is how we can remove the special character from the dataset:

```
df = df.with_columns(
    pl.col("copay_paid") \
    .str \
    .slice(1) \
    .alias("copay_paid")
)
```

print(df.head())

would yield

shape: (5, 7)

id	name	age	state	gender	exercise_daily?	copay_paid
i64	str	str	str	str	bool	str
1	Allayne Moffett	null	SC	Male	null	206.76
2	Kerby Benjafield	null	NM	Male	true	21.58
3	Raina Vallentin	null	MI	Female	true	125.18
4	Kaela Trodden	null	OH	Female	false	86.20
5	Faber Kloisner	null	MN	Male	false	219.38

Now, we can convert the string data type to a decimal data type. Here is how we can do that:

```
df = df.with_columns(
    pl.col("copay_paid")\
    .str \
    .to_decimal() \
    .alias("copay"))
```

print(df.head())

would yield

shape: (5, 8)

CHAPTER 3   DATA WRANGLING USING RUST'S POLARS

```
┌─────┬──────────────────┬──────┬───────┬────────┬────────────────┬─────────────┬─────────────┐
│ id  │ name             │ age  │ state │ gender │ exercise_daily?│ copay_paid  │ copay       │
│ --- │ ---              │ ---  │ ---   │ ---    │ ---            │ ---         │ ---         │
│ i64 │ str              │ str  │ str   │ str    │ bool           │ str         │ decimal[*,2]│
╞═════╪══════════════════╪══════╪═══════╪════════╪════════════════╪═════════════╪═════════════╡
│ 1   │ Allayne Moffett  │ null │ SC    │ Male   │ null           │ 206.76      │ 206.76      │
│ 2   │ Kerby Benjafield │ null │ NM    │ Male   │ true           │ 21.58       │ 21.58       │
│ 3   │ Raina Vallentin  │ null │ MI    │ Female │ true           │ 125.18      │ 125.18      │
│ 4   │ Kaela Trodden    │ null │ OH    │ Female │ false          │ 86.20       │ 86.20       │
│ 5   │ Faber Kloisner   │ null │ MN    │ Male   │ false          │ 219.38      │ 219.38      │
└─────┴──────────────────┴──────┴───────┴────────┴────────────────┴─────────────┴─────────────┘
```

Let us now initialize the SQL context of Polars. The SQL context object is where the mapping between SQL-like statements and Polars expressions is stored. And so, this step is essential.

Here is how we can do that:

```
pl_sql = pl.SQLContext(
    random_health_data = df
)
```

You would notice the Polars data frame has been referenced to as "random_health_data." This will be our table name in writing SQL-like statements. Let us define our query:

```
query1 = """
    select
        state,
        count(name) as total_Patients,
        max(copay) as most_Expensive
    from
        random_health_data
    where
        state is not null
        and copay is not null
    group by
        state
```

```
    order by
        mostExpensive desc
    limit
        10
"""
```

Now, we can execute our SQL query using our SQL context. Here is how we can do that:

```
sqlresult = pl_sql.execute(
    query1,
    eager=True
)
print(sqlresult)
```

would yield

shape: (10, 3)

state	totalPersons	mostExpensive
str	u32	decimal[*,2]
CA	95	274.51
HI	8	274.03
VA	26	273.82
AZ	22	273.68
WI	11	273.41
NY	35	272.23
FL	62	271.62
IA	16	270.97
TN	16	270.54
MA	16	269.18

CHAPTER 3　DATA WRANGLING USING RUST'S POLARS

> **Note** You may have noticed "eager=True" during the execution of the SQL query within the SQL context. You can also set the option to False to enable lazy execution. This will return a Polars lazy frame. You can continue to use that lazy frame on downstream data manipulation tasks.

Polars CLI

Polars also has an option to access the library through a command line interface. One can execute Polars commands and SQL-like commands right from the command line. Here's how to get started with using Polars in a command line interface:

```
pip install polars-cli
```

And type in "polars" to access the command line interface. Here is how that may look:

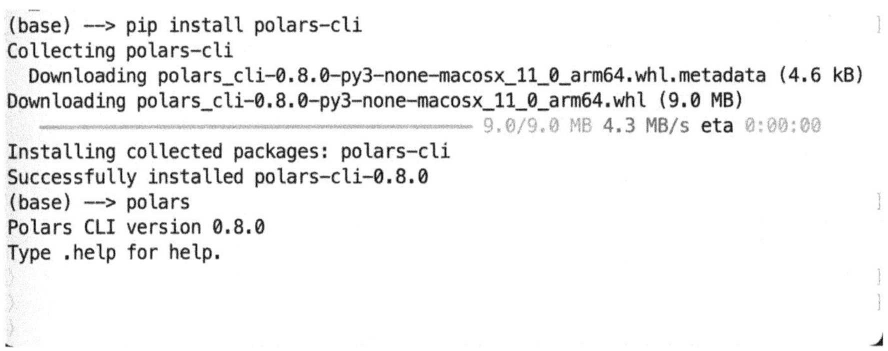

```
(base) --> pip install polars-cli
Collecting polars-cli
  Downloading polars_cli-0.8.0-py3-none-macosx_11_0_arm64.whl.metadata (4.6 kB)
Downloading polars_cli-0.8.0-py3-none-macosx_11_0_arm64.whl (9.0 MB)
   ──────────────────────────── 9.0/9.0 MB 4.3 MB/s eta 0:00:00
Installing collected packages: polars-cli
Successfully installed polars-cli-0.8.0
(base) --> polars
Polars CLI version 0.8.0
Type .help for help.

>
>
>
```

Figure 3-1. Polars command line interface

Here is how we can execute a SQL query on a flat file in your file system:

```
>select * from read_csv('./person.csv');
```

would yield

```
> select * from read_csv('./person.csv')
```

person_id	name	city	gender	birthdate
i64	str	str	str	str
109233	Fluffy	denver	f	1993-02-04
672345	Claws	boulder	m	1994-03-17
273864	Buffy	SLC	f	1989-05-13
498567	Fang	Los alamos	m	1990-08-27
489756	Bowser	albequerque	m	1979-08-31
132454	Chirpy	Tucson	f	1998-09-11
983674	Whistler	golden	null	1997-12-09
435768	Slim	aspen	m	1993-10-10

Figure 3-2. *Writing SQL queries on flat files using Polars CLI*

You can even specify the SQL query directly on your shell and use a pipe operator to call and execute the SQL command in the Polars CLI. Here is how that looks:

```
--> echo "select \
concat(firstname, ' ', lastname) as fullname, \
occupation \
from \
read_csv('./polars_dataset.csv') \
where type = 'Baseball' and state = 'New York' " | polars
```

would yield

fullname	occupation
str	str
Cynthia Paley	Doctor
Adam Seaborn	Doctor
Lilah Mercik	IT Professional
Alidia Wreiford	Lawyer
Rosita Guswell	Accountant

```
| Winston MacCahey   | IT Professional |
| Thorn Ayrs         | Accountant      |
| Cleon Stolle       | IT Professional |
| Gabriello Garshore | Doctor          |
```

Here is how that may look for you:

```
--> echo "select \
concat(firstname, ' ', lastname) as fullname, \
occupation \
from \
read_csv('./polars_dataset.csv') \
where type = 'Baseball' and state = 'New York' " | polars
```

fullname	occupation
str	str
Cynthia Paley	Doctor
Adam Seaborn	Doctor
Lilah Mercik	IT Professional
Alidia Wreiford	Lawyer
Rosita Guswell	Accountant
Winston MacCahey	IT Professional
Thorn Ayrs	Accountant
Cleon Stolle	IT Professional
Gabriello Garshore	Doctor

Figure 3-3. Execute SQL in Polars CLI

Conclusion

Congratulations on reaching this far! So far, you have covered a lot of ground with Python, SQL, git, Pandas, and Polars from these chapters. There is immense potential in Polars. So far, we looked at how to extract and load data in various formats, perform data transformation operations like filtering and aggregation, and combine multiple Polars objects using join operations. We also explored handling missing values, identifying unique values, and reshaping data with pivot operations. I hope you find it in you to adopt Polars in your data pipelines and demonstrate greater efficiency and value to the organization. While Polars may be the perfect solution for CPU-based data wrangling tasks, in the next chapter, we will look at GPU-based data wrangling libraries and how one can process massive amounts of data using GPU. I hope you stay excited toward the upcoming chapter.

CHAPTER 4

GPU Driven Data Wrangling Using CuDF

Introduction

As the volume of data continues to grow, there comes a time where CPU-based data processing may struggle to provide consistent results within reasonable time. This is where GPU computing comes into play, where the graphical processing units are leveraged to revolutionize data manipulation and analysis. In this chapter, we will explore CuDF, a GPU-accelerated data manipulation library that integrates well with the existing Python ecosystem. Before we jump into CuDF, let us also review the architecture of CPU- and GPU-based computations and concepts of GPU programming.

By the end of this chapter, you will

- Gain an appreciation of CPU vs. GPU computations and how they impact data processing.
- Familiarize yourself with GPU programming concepts.
- Set up CuDF on Google Colab and perform basic data loading operations.
- Apply advanced data transformations like grouping, aggregation, and window functions and explore feature engineering capabilities.
- Combine multiple CuDF objects using various joins and leverage CuDF's Pandas accelerator mode for GPU-accelerated Pandas code.

CHAPTER 4   GPU DRIVEN DATA WRANGLING USING CUDF

CPU vs. GPU

CPU stands for central processing unit. The CPU handles all the processing tasks in a computer. At a hardware level, the CPU has a microprocessor, which translates the program into machine language and executes the byte code. There is also a clock mechanism within the CPU, which sets the electrical pulse generating frequency for the CPU. The higher the clock frequency, the faster the processing.

Note   CPU also has cache, a memory unit that stores data required for a given iteration of a computation. The cache is supposed to be faster than the physical memory, and at present there appear to be three levels of cache. There is also a memory management unit that controls data movement between CPU and RAM during their communication and coordination for processing various tasks.

Among types of processing various tasks, the graphical processing is considered to be sophisticated and involves complex mathematics. An example could be to render accurate graphics, video with high detail, scientific simulation, or even a video game with a physics engine. These tasks require complex processing and computing complex mathematics functions simultaneously (in parallel). Graphical processing units evolved primarily to address the complex computation tasks that are massively parallel.

Like CPUs, graphical processing units also have a microprocessor, memory management unit, and other such components with them. The microprocessor consists of several compute cores. These cores, however, are smaller when compared with CPUs but they are powerful for parallel processing capabilities. The difference between CPU and GPU is that the CPU focuses on processing tasks and instructions that are more diverse and variant. For instance, the CPU may perform tasks like copying files from one place to another, executing a user-supplied program, loading a system program, and performing number crunching on a spreadsheet program. These tasks are not similar in nature by their context. The CPU would switch between each of them easily while completing the processing. The GPU, on the other hand, focuses on massively parallel processing of computations that are somewhat similar. The GPU would intake instructions from the CPU and execute them in parallel to speed up the processing and render the display.

Introduction to CUDA

CUDA stands for Compute Unified Device Architecture. CUDA is a software framework developed by NVIDIA for writing code specifically for NVIDIA GPUs. It enables the execution of kernels on NVIDIA GPUs, harnessing their parallel computing power for complex data processing. Although written in C language, CUDA also supports Python programming through libraries such as CuPy and PyCUDA. However, it's important to note that programs developed using the CUDA software framework are exclusive to NVIDIA GPUs and are not compatible with GPUs from other manufacturers.

> **Note** Currently in the hardware manufacturing market, there are several companies who manufacture GPUs. One of the famous manufacturers of GPUs is NVIDIA.
>
> There is also OpenCL, which is also a firmware like CUDA. OpenCL is more generic and can be used in non-NVIDIA GPUs as well.

Concepts of GPU Programming

GPU programming typically begins with receiving instructions from the CPU. The CPU is referred to as the host, which identifies the GPUs and their processing cores that are available and supplies the instructions for the GPUs. The GPUs are referred to as device units. For instance, if there is a large software program that is going to be executed, the CPU first identifies all the specific functions that are intended for the GPUs. Once these specific functions are identified, they are then executed on the GPUs.

Kernels

Kernels are packages of code (containing specific functions) written by developers. They are compiled into machine code using CUDA compiler. Kernels are functions written for execution on GPUs.

The process begins by the CPU calling the kernel function. In addition to calling the kernel, the host would also call the execution configuration, determining how many parallel threads are required for the kernel execution. The GPU executes the kernel using a grid of blocks, each block containing a set of threads. The scheduler portion of the

CHAPTER 4   GPU DRIVEN DATA WRANGLING USING CUDF

GPU block would send these threads to a streaming multiprocessor on the GPU. Once the execution is complete, the host will retrieve the results from the GPU's memory. This would have completed one full execution cycle.

Memory Management

During execution of kernel functions on GPUs, the threads who are executing (the same) kernel function may read and write data at various memory locations. These threads have access to local memory, shared memory, and global memory. Local memory is scoped primarily toward a given thread. Shared memory is memory that is scoped only to a given block but is accessed by all the threads underneath it. Global memory is something that is accessed by GPU and CPU. All threads in a given block may synchronize their execution for better memory access.

Introduction to CuDF

CuDF is a Python library that offers various data manipulation and transformation methods and functions exclusively to be executed on NVIDIA GPUs. CuDF leverages Apache Arrow for columnar format data structures. All CuDF's functions and methods are based on libcudf, a CUDA library meant for C++ programming.

CuDF is similar to Pandas, with the exception that it is exclusively designed to run on NVIDIA GPUs. This GPU-accelerated data manipulation and transformation library provides more efficient data processing (when compared with Pandas) on large datasets.

CuDF vs. Pandas

CuDF and Pandas provide most methods and functions that perform similar operations. However, there are some areas where these two Python libraries differ.

To start off, both CuDF and Pandas support similar data structures (Series, DataFrames, etc.). Both CuDF and Pandas support similar transformational operations on the data structures like various joins, filtering a data frame, indexing, group-by, and window operations. CuDF supports many of the existing Pandas data types and also provides support for special data types. It may be better to visit the documentation for the same.

In Pandas, nulls are converted to not a number, whereas in the case of CuDF, all data types can contain missing values.

Iterating over a data structure is not considered optimal in CuDF. You can still write code that would loop through a given column in a data frame and perform some kind of computation; however, it might provide poor performance. A better option is to either use an inbuilt CuDF method to perform the transformation or convert the CuDF data frame into either Pandas or Arrow data frame, perform the necessary operation, and convert the resultant data frame into a CuDF data frame.

When performing joins or group-by operations, Pandas would sort the data frame, whereas CuDF does not perform any kind of sorting or ordering of elements. If you require some sort of ordering elements, then you enable "sort=True" in the code.

Setup

Although the installation page provides more options for installing and working with CuDF, I recommend the following to be able to get up and running with CuDF.

You need a NVIDIA GPU with compute capability of 6.0 and above. First of all, the CuDF library is built on top of CUDA, a parallel processing framework exclusive for NVIDIA GPUs. NVIDIA offers various GPU products and you may be able to view them on their product catalog.

Second, compute capability is just a glorified term for saying newer models of their GPU. My understanding of compute capability is that the higher the version number, the better the GPU can perform and can process more complex operations and the more efficient it can do them.

Note Think of it like an iPhone, where the more recent versions have more capabilities when compared with the older iPhones. Compute capability is like saying a newer iPhone model.

You also need a Linux distribution. Though you may be able to work with Windows operating systems with WSL, it may be beneficial to have a Linux instance with gcc++ 9.0 and above available.

Note If you have a Windows operating system, then you may need to have a GPU that has compute capability of 7.0 or higher (an even more recent model).

CHAPTER 4   GPU DRIVEN DATA WRANGLING USING CUDF

You need appropriate NVIDIA drivers and CUDA drivers, where it may be required to have a CUDA driver version 11.2 or newer.

You also need the latest version of Python 3 installed on your system.

It may be better to uninstall any existing CUDA toolkits and CUDA packages that are loaded by default.

I recommend using a virtual environment specifically for executing CuDF pipelines. A better idea is to set up a conda or miniconda environment in your system and perform a conda install.

Here is how that looks:

```
conda create --solver=libmamba -n rapids-23.12 -c rapidsai -c conda-forge -c nvidia \
    rapids=23.12 python=3.10 cuda-version=12.0
```

For those who do not have access to one of the abovementioned, it may be beneficial to look at some cloud-based options. All the major cloud providers offer GPU-based computing options. While this will be covered in later chapters in detail, let us look at one such cloud provider example, so we can get our hands on with CuDF. I am choosing Google Cloud Platform-based Google Colab (arbitrarily).

To get started with Google Colab, simply visit the following portal and sign in with your Google account (or create an account if you do not possess one yet):

https://colab.google/

And click the New Notebook option:

Google Colaboratory

Colab is a hosted Jupyter Notebook service that requires no setup to use and provides free access to computing resources, including GPUs and TPUs. Colab is especially well suited to machine learning, data science, and education.

Open Colab   New Notebook

Figure 4-1. *Google Colab screen*

Google Colab offers a free option and a pro option. For our purposes, the free option may suffice.

CHAPTER 4   GPU DRIVEN DATA WRANGLING USING CUDF

You may get the following screen:

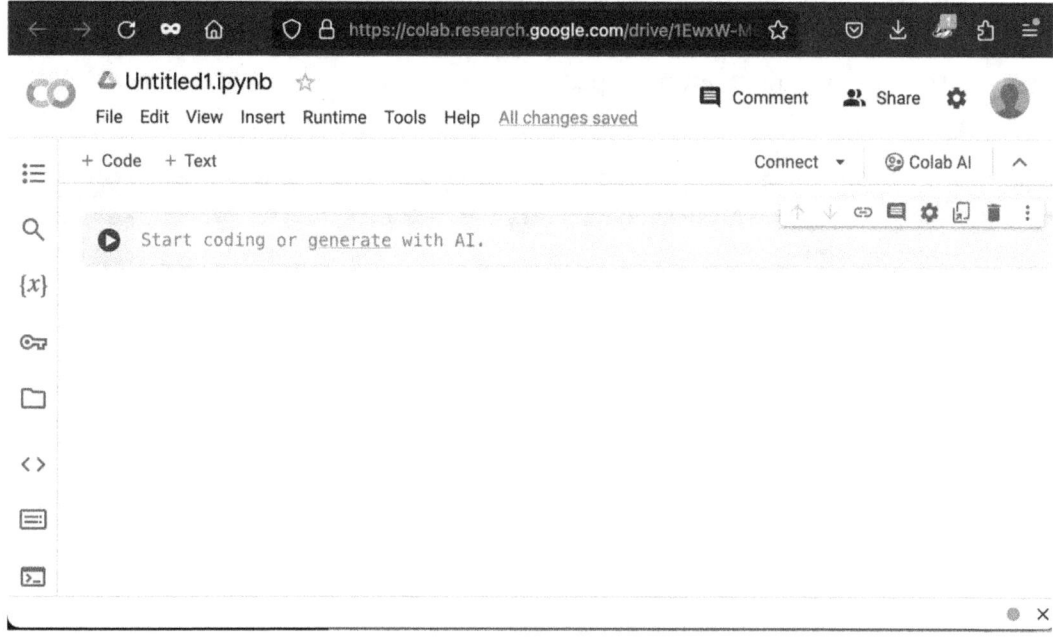

Figure 4-2. *Jupyter notebook environment within Google Colab*

At the time of learning, it may be wise to turn off the generate with AI option. If you do wish to turn it off, click "Tools" on the menu and choose the "Settings" option.

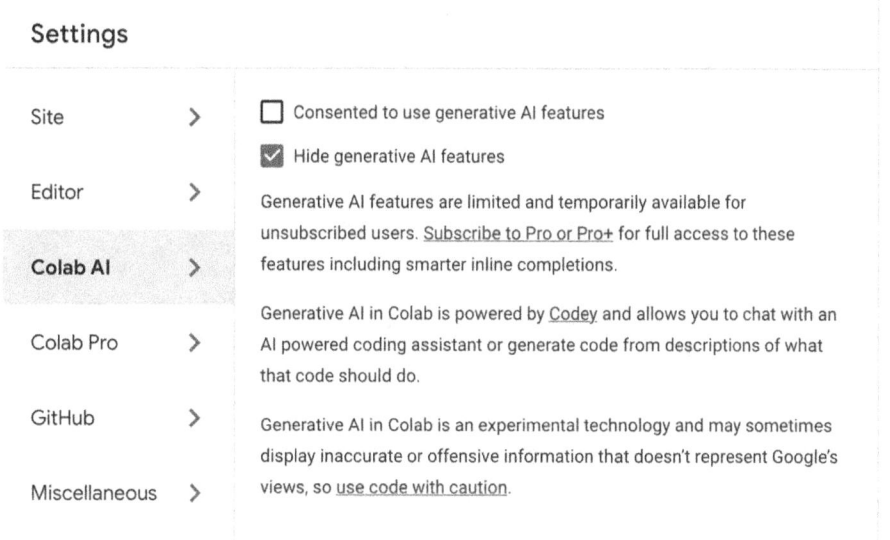

Figure 4-3. *Settings option within Google Colab*

139

CHAPTER 4 GPU DRIVEN DATA WRANGLING USING CUDF

Under the settings, locate "Colab AI" and check "Hide generative AI features."

Now, let us change the runtime of our Jupyter notebook from CPU to GPU. Let us navigate to the main page within Google Colab. From there, select "Runtime" on the menu and choose "Change runtime type."

Here is how that may look:

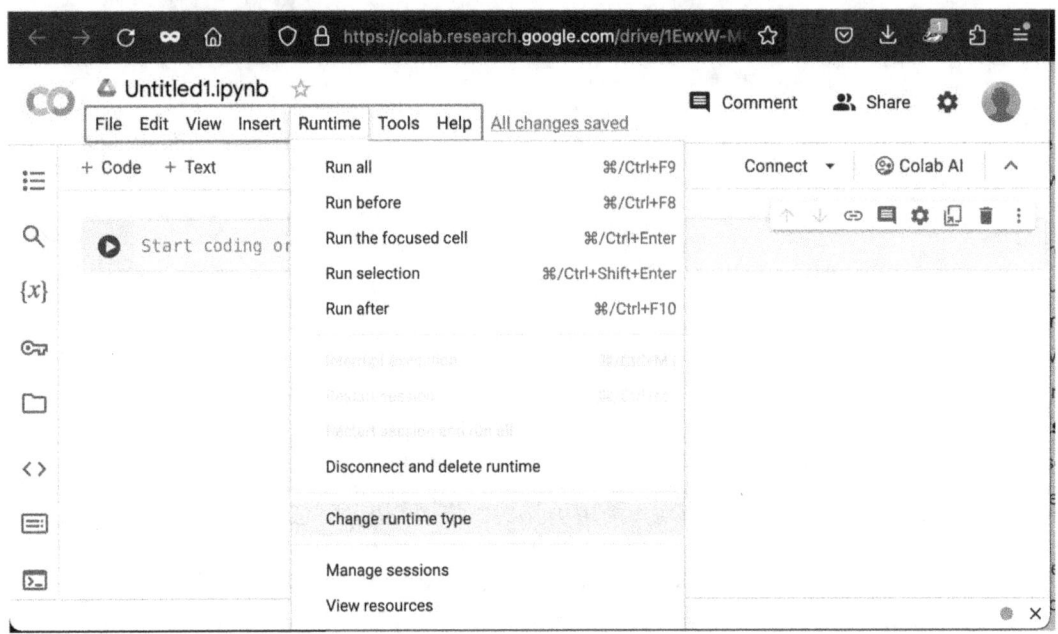

Figure 4-4. *Changing the runtime environment of the Google Colab notebook*

Choose "Python 3" for the runtime type and choose an option that contains the word "GPU" for the hardware accelerator. In this case, select the T4 GPU option. Note that the other GPU options are also NVIDIA offerings.

CHAPTER 4 GPU DRIVEN DATA WRANGLING USING CUDF

Here is how that may look for you:

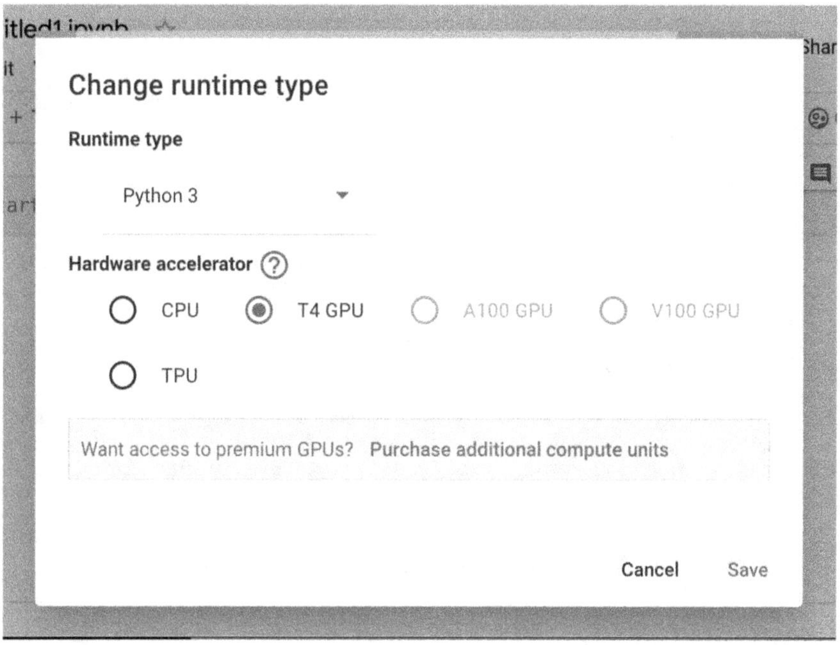

Figure 4-5. Changing the runtime type and hardware accelerator

Now, it is time to install CuDF. Copy and paste the following command in your code box. This might take some time so it may be better to wait for a little while:

```
!pip install cudf-cu11 --extra-index-url=https://pypi.ngc.nvidia.com
!rm -rf /usr/local/lib/python3.8/dist-packages/cupy*
!pip install cupy-cuda11x
```

In some cases, the installation may fail for reasons including conflict between package versions. If and when installation and setup appears to be challenging, do consider installing an older and stable version.

```
!pip install cudf-cu11 --extra-index-url=https://pypi.ngc.nvidia.com
!rm -rf /usr/local/lib/python3.8/dist-packages/cupy*
!pip install cupy-cuda11x

Looking in indexes: https://pypi.org/simple, https://pypi.ngc.nvidia.com
Collecting cudf-cu11
  Using cached cudf_cu11-24.6.0.tar.gz (2.6 kB)
  Installing build dependencies ... done
  Getting requirements to build wheel ... done
  Preparing metadata (pyproject.toml) ... done
Requirement already satisfied: cachetools in /usr/local/lib/python3.10/dist-pack
Collecting cubinlinker-cu11 (from cudf-cu11)
  Using cached cubinlinker_cu11-0.3.0.post2.tar.gz (513 bytes)
  Installing build dependencies ... done
  Getting requirements to build wheel ... done
  Preparing metadata (pyproject.toml) ... done
Collecting cuda-python<12.0a0,>=11.7.1 (from cudf-cu11)
  Using cached cuda_python-11.8.3-cp310-cp310-manylinux_2_17_x86_64.manylinux201
Requirement already satisfied: cupy-cuda11x>=12.0.0 in /usr/local/lib/python3.10
Requirement already satisfied: fsspec>=0.6.0 in /usr/local/lib/python3.10/dist-p
```

Figure 4-6. Installing CuDF within the Google Colab environment

Testing the Installation

Once we have the CuDF installed, it is time to test the installation. Please issue the following code and observe the output:

```
import cudf
a = cudf.Series([1, 2, 3, 4, 5])
print(a)
```

would yield

```
0    1
1    2
2    3
3    4
4    5
dtype: int64
```

Here is how the output may look:

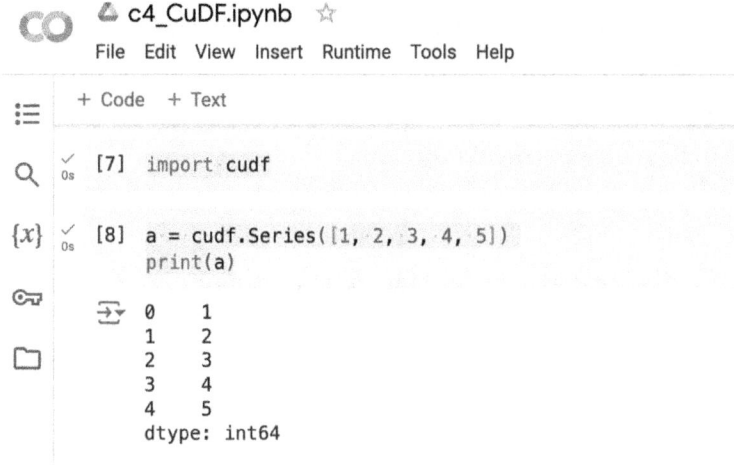

Figure 4-7. Initializing a CuDF series with sample data

File IO Operations

CuDF supports most of the commonly utilized data formats in big data and machine learning. Here is how the syntax may look:

```
csv_file = "/path-to-folder/path-to-file.csv"

cudf_df = cudf.read_csv(
    filepath
    ,sep=","
    ,header="infer")
```

CSV

If you plan to export a given CuDF data frame into a CSV, here's how to do it:

```
cudf.to_csv(
   "/path-to-file.csv"
   ,sep=","
   ,header= True
   ,index=True
)
```

Parquet

Parquet is one of the more efficient data stores, as we have discussed earlier. CuDF library provides methods to read a Parquet file as well as export a current data frame into a Parquet file, where GPU acceleration is enabled for these operations.

Here's how to read an external Parquet file:

```
import cudf

filename = "/some-file-path/some-file-name.parquet"

df = cudf.read_parquet(filename)
```

To export a CuDF data frame to Parquet, the syntax may look like

```
df.to_parquet(filename = "/some-file-path/some-file-name.parquet")
```

JSON

CuDF supports JSON file formats; however, many of the attributes for reading a JSON file do not appear to be GPU accelerated.

Here is an example:

```
import cudf

filename = "/filepath/file.json"

cudf_dataframe = cudf.read_json(filename)
```

Here's how to write out a JSON file:

```
cudf_dataframe = cudf.to_json("/filepath/filename.json")
```

Similarly, CuDF supports other data file formats like ORC, Avro, etc.

Basic Operations

Let us look at some basic operations with CuDF, by importing an example dataset. Here is how the code looks:

CHAPTER 4 GPU DRIVEN DATA WRANGLING USING CUDF

```
import cudf

cudf_df = cudf.read_csv("/content/sample_data/polars_dataset.csv")

print(cudf_df.head())
```

We can observe the following output:

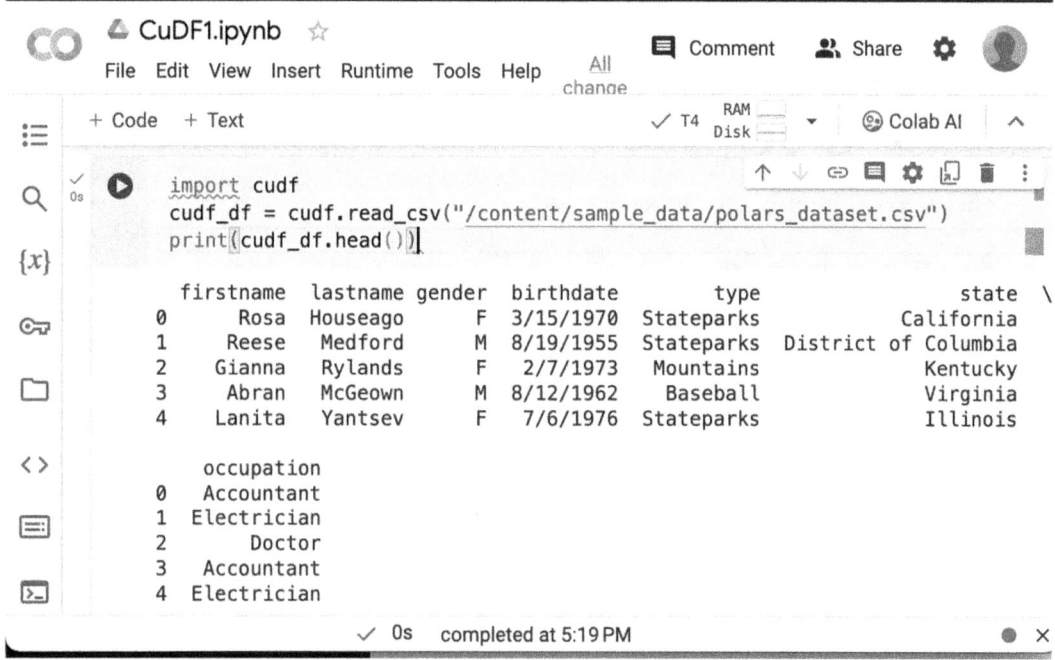

Figure 4-8. *Reading an external CSV file and examining first few rows*

Column Filtering

Let us try to select a subset of the data. Here's how to obtain only certain columns:

```
name_and_occupation = cudf_df[
    ["firstname","occupation"]
    ].head()

print(name_and_occupation)
```

CHAPTER 4   GPU DRIVEN DATA WRANGLING USING CUDF

would yield

```
   firstname    occupation
0       Rosa    Accountant
1      Reese   Electrician
2     Gianna        Doctor
3      Abran    Accountant
4     Lanita   Electrician
```

Row Filtering

Here's how to obtain only certain rows:

```
data_subset = cudf_df.loc[
   775:898,
    ["firstname","birthdate", "state", "occupation"]
   ].head()

print(data_subset)
```

Here is how that may look:

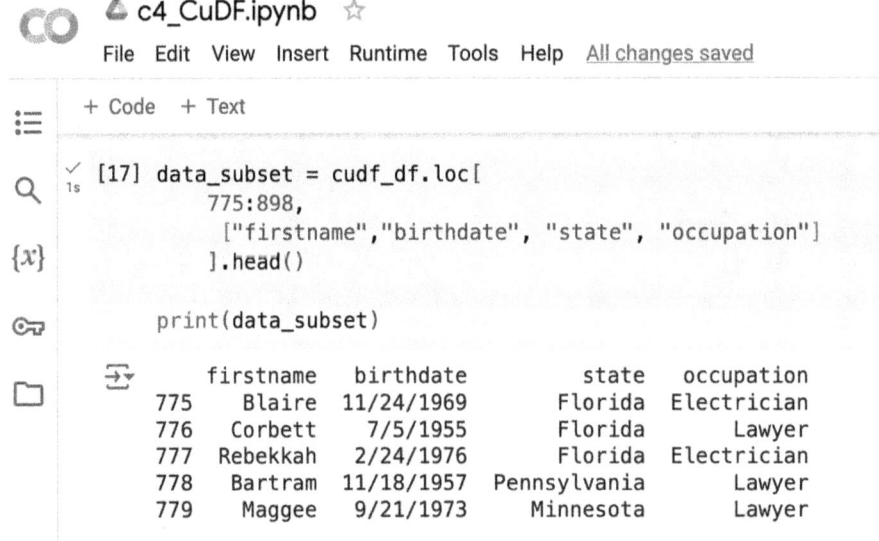

Figure 4-9. Examining a subset of data in CuDF

Sorting the Dataset

Let us try to sort the dataset by a specific column to either ascending or descending order:

```
data_subset_sort = cudf_df.loc[
    775:898,
    ["firstname","birthdate", "state", "occupation"]
    ] \
    .sort_values(
        "state",ascending=False
    ).head()

print(data_subset_sort)
```

would yield

```
    firstname   birthdate          state       occupation
811 Pierrette    7/4/1982      Wisconsin           Doctor
813 Florencia  12/27/1956  West Virginia           Doctor
825    Imelda   3/31/1975     Washington   IT Professional
859   Marlowe   5/20/1953     Washington           Lawyer
867   Atlante   2/12/1969       Virginia       Accountant
```

Here is how that may look on Google Colab:

Chapter 4 GPU Driven Data Wrangling Using CuDF

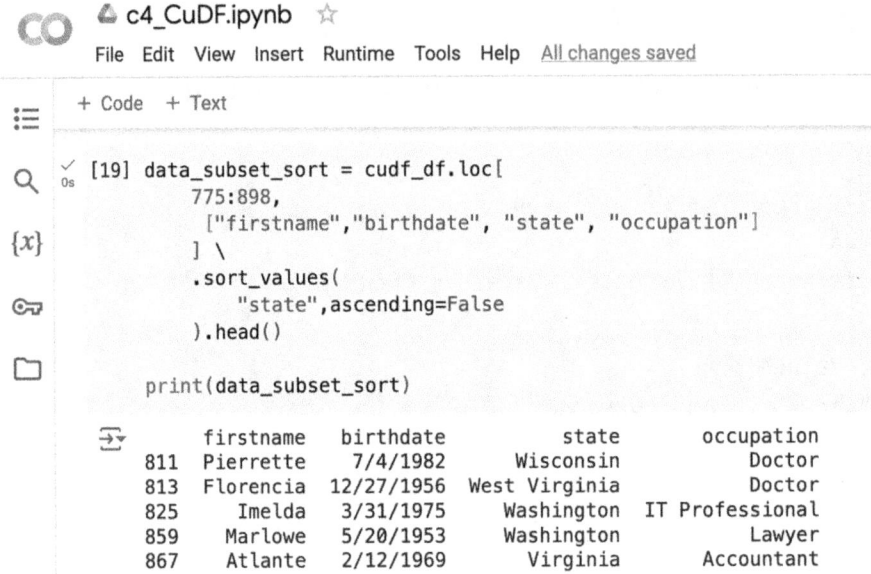

Figure 4-10. Sorting a subset of data on a specific column and returning first few rows

Combining Multiple CuDF Objects

CuDF supports database-like joins on data frames either by columns or by indexes. Let us look at them in detail.

Note CuDF does not support right joins.

Left Join

Left join is where everything is kept as is on the left table while looking for potential matches on the right table. We use `cudf.merge` to perform left joins on GPU DataFrames.

CHAPTER 4 GPU DRIVEN DATA WRANGLING USING CUDF

Let us look at it with an illustration:

```
import cudf

left_df1 = cudf.DataFrame(
    {
        'ID': [1, 2, 3, 4, 5],
        'Name': ['Alice', 'Bob', 'Charlie', 'David', 'John'],
        'DeptID': [101, 102, 103, 104, 103]
    }
)
right_df1 = cudf.DataFrame(
    {
        'DeptID': [101, 102, 103],
        'DeptName': ['HR', 'IT', 'Marketing']
    }
)
```

Now we are ready to perform the join:

```
leftjoin = left_df1.merge(
    right_df1,
    on=["DeptID"],
    how="left"
)
print(leftjoin)
```

would yield us

```
   ID     Name  DeptID   DeptName
0   1    Alice     101         HR
1   2      Bob     102         IT
2   3  Charlie     103  Marketing
3   5     John     103  Marketing
4   4    David     104       <NA>
```

Outer Join

Outer join is where the columns or indexes that are used to perform the join are unionized from both data frames. Here is how that looks:

```
outerjoin = left_df1.merge(
    right_df1,
    on=["DeptID"],
    how="outer"
)

print(outerjoin)
```

would yield us

	ID	Name	DeptID	DeptName
0	1	Alice	101	HR
1	2	Bob	102	IT
2	3	Charlie	103	Marketing
3	5	John	103	Marketing
4	4	David	104	<NA>

Inner Join

Inner join is the inverse of outer join. Inner join is where the intersection of columns or indexes is performed on both data frames.

Let us look at an example:

```
innerjoin = left_df1.merge(
    right_df1,
    on=["DeptID"],
    how="inner"
)

print(innerjoin)
```

would yield us

```
   ID    Name  DeptID   DeptName
0   1   Alice     101         HR
1   2     Bob     102         IT
2   3 Charlie     103  Marketing
3   5    John     103  Marketing
```

Left Semi Join

A left semi join is a type of join that returns only the rows from the left table that match with the right table; rows from the left table that do not seem to have a match on the right table are ignored.

Here is how that looks:

```
leftsemi = left_df1.merge(
    right_df1,
    on=["DeptID"],
    how="leftsemi"
)

print(leftsemi)
```

would yield us

```
   ID    Name  DeptID
0   1   Alice     101
1   2     Bob     102
2   3 Charlie     103
3   5    John     103
```

Left Anti Join

A left anti join can be seen as an inverse of a left semi join. Left anti joins would return the rows from the left table that do not have matches with the right table.

Here is an example to illustrate further:

```
leftanti = left_df1.merge(
    right_df1,
    on=["DeptID"],
    how="leftanti"
)
print(leftanti)
```

would yield us

```
   ID   Name   DeptID
0   4   David    104
```

Advanced Operations

Let us look at some advanced data wrangling operations using CuDF.

Group-By Function

A group-by function attempts to group values in a series or a data frame. The grouping of columns can occur either in a single column or more than one column as well. CuDF supports some of the most commonly used Pandas group-by functions. Let us look at some of them in detail:

```
import cudf

groupbydata = {
    'Person': ['Alice', 'Bob', 'Charlie', 'David'],
    'Dept': ['Marketing', 'Marketing', 'DigitalM', 'DigitalM'],
    'Sales': [20, 12, 35, 42]
}
gb_df = cudf.DataFrame(groupbydata)
print(gb_df)
```

would yield us

```
    Person       Dept  Sales
0    Alice  Marketing     20
1      Bob  Marketing     12
2  Charlie    DigitalM    35
3    David    DigitalM    42
```

Now let us try to get total sales by department. To accomplish this, we need to group the dataset by the department and obtain the sum of the sales numbers within that respective department.

Here is how one may achieve that:

```
totsale_dept = gb_df \
   .groupby('Dept')\
   .agg('sum')

print(totsale_dept)
```

would yield us

```
           Sales
Dept
Marketing     32
DigitalM      77
```

Transform Function

The transform() function is an important function in the context of data transformation. The transform() function performs aggregations and returns the result in the size that is of similar size to the data frame or series. This way the output may be added to the data frame.

Here is how that looks for the same dataset:

```
dept_sum = gb_df \
   .groupby('Dept', sort=True)\
   .transform('sum')

print(dept_sum)
```

would yield us

```
   Sales
0     32
1     32
2     77
3     77
```

Notice the size of the resultant dataset. It may be easy to add the same to the data frame as well.

```
gb_df['totalDeptSales'] = dept_sum
```

```
print(gb_df)
```

would yield us

```
   Person       Dept  Sales  dept_sales  totalDeptSales
0   Alice  Marketing     20          32              32
1     Bob  Marketing     12          32              32
2 Charlie   DigitalM     35          77              77
3   David   DigitalM     42          77              77
```

apply()

Though CuDF provides many of the commonly used methods and functions on GPU accelerator mode, it is certainly possible there may be cases where you have to write a custom function, also referred to as a user-defined function. These user-defined functions are often applied to a series or a data frame object on a given column or a set of columns. This is where the apply() method is beneficial. The apply() method enables to implement user-defined functions over a given column.

Here is an illustration of the apply() method. Let us define a series object:

```
ap1 = cudf.Series([1, cudf.NA, 3, 5, 6, 8])
```

```
print(ap1)
```

CHAPTER 4   GPU DRIVEN DATA WRANGLING USING CUDF

would yield us

```
0       1
1    <NA>
2       3
3       5
4       6
5       8
dtype: int64
```

We now create a custom method that is required to be applied on the CuDF series object:

```
def transformData(x):
    if x is cudf.NA:
        return 42
    else:
        return x + 100
```

Now, we are ready to apply this user-defined function in our series object:

```
ap1.apply(transformData)
```

would yield us

```
0    101
1     42
2    103
3    105
4    106
5    108
dtype: int64
```

Let us look at another application using our earlier dataset. We had

```
print(gb_df)
```

```
   Person       Dept  Sales  dept_sales  totalDeptSales
0   Alice  Marketing     20          32              32
1     Bob  Marketing     12          32              32
2 Charlie    DigitalM     35          77              77
3   David    DigitalM     42          77              77
```

155

Now, let us create a user-defined function for this dataset, where we assign ranks based on crossing a certain threshold of sales figure:

```
def salesRank(y):
  if y > 25:
    return 1
  else:
    return 2
```

Now, let us apply this function and add a new column to hold the ranks:

```
gb_df['SalesRank'] = gb_df['Sales'].apply(salesRank)
print(gb_df)
```

would yield us

	Person	Dept	Sales	dept_sales	totalDeptSales	SalesRank
0	Alice	Marketing	20	32	32	2
1	Bob	Marketing	12	32	32	2
2	Charlie	DigitalM	35	77	77	1
3	David	DigitalM	42	77	77	1

Cross Tabulation

Let us look at some advanced functions in CuDF. First, we will try to use the "crosstab" function. Crosstab stands for cross tabulation that displays the frequency distribution of variables in a table, where the row and column belong to a data point and the value in the table denotes how often the value occurs. Cross tabulation view is useful in exploring data, analyzing relationships, or even just looking at distribution of a variable across various categories.

Let us create a crosstab for two data points in CuDF:

```
from cudf import crosstab

data = cudf.DataFrame(
    {
        'A': [1, 2, 1, 2, 3, 2, 3, 3, 1, 2],
        'B': ['x', 'y', 'x', 'y', 'z', 'z', 'x', 'y', 'x', 'y']
    }
)
```

```
crosstab = crosstab(data['A'], data['B'])

print(crosstab)
```

would yield us

```
B  x  y  z
A
1  3  0  0
2  0  3  1
3  1  1  1
```

Feature Engineering Using cut()

The cut function is used for creating bins or buckets with discrete intervals and dropping the numerical values into appropriate buckets. It is commonly used to transform continuous data into categorical data. A common application of this function is to create categorical features from continuous variables.

Let us look at creating features using the cut function:

```
from cudf import cut

data = cudf.DataFrame(
    {
        'Age': [25, 30, 35, 40, 45, 50, 55, 60, 65, 86, 96, 97]
    }
)

bins = [0, 30, 50, 70, 100]

labels = ['<30', '30-50', '50-70', '>70']

data['AgeGroup'] = cut(data['Age'], bins=bins, labels=labels)

print(data)
```

would yield us

	Age	AgeGroup
0	25	<30
1	30	<30
2	35	30-50
3	40	30-50
4	45	30-50
5	50	30-50
6	55	50-70
7	60	50-70
8	65	50-70
9	86	>70
10	96	>70
11	97	>70

Factorize Function

The factorize function is another key operation in feature engineering. The factorize function is used to encode categorical variables into numerical values (but not continuous), so it is suitable for machine learning algorithms with numerical inputs. The way it works is that it assigns a unique value to each category so that we have a vector of numerical values that corresponds to a certain category. Here is how that works:

```
from cudf import factorize

fact_data = cudf.Series(
    [
        'apple', 'banana', 'apple',
        'orange', 'banana', 'apple',
        'banana', 'banana', 'orange'
    ]
)

fact_labels = factorize(fact_data)

print(fact_labels)
```

would yield us

```
(
    array([0, 1, 0, 2, 1, 0, 1, 1, 2], dtype=int8),
    StringIndex(['apple' 'banana' 'orange'], dtype='object')
)
```

Window Functions

A window function is another key analytical function, where operation is performed over a specific window of data, usually where they are placed. A good example is calculating moving averages on certain data. Let us look at an example:

```
ts_data = cudf.Series(
    [10, 20, 30, 40, 50, 55, 60, 65, 70, 72, 74, 76, 78, 80]
)

moving_avg = ts_data.rolling(window=3).mean()

print("Moving average:", moving_avg)
```

would yield

```
Moving average:
0            <NA>
1            <NA>
2            20.0
3            30.0
4            40.0
5     48.33333333
6            55.0
7            60.0
8            65.0
9            69.0
10           72.0
11           74.0
12           76.0
13           78.0
dtype: float64
```

CHAPTER 4   GPU DRIVEN DATA WRANGLING USING CUDF

CuDF Pandas

CuDF provides Pandas accelerator mode that supports all of the Pandas API functions in methods. This means that CuDF can accelerate Pandas code on GPU without having to entirely rewrite the code. In cases where the code may not be able to leverage GPU, this Pandas accelerator mode automatically reverts to executing the Pandas code on CPU.

It may be a good idea to recall that CuDF supports most of the commonly utilized Pandas functions and methods. If possible, it may be a better idea to rewrite the Pandas code into CuDF code. Moreover, if you are working with a large dataset with varying complex operations, it may be better to test the cudf.pandas with a subset and compare the same with Pandas.

cudf.pandas works well with Pandas, NumPy, scipy, TensorFlow, scikit-learn, and XGBoost packages so one can utilize GPU accelerator mode not just with data analysis but also with machine learning pipelines.

To enable cudf.pandas, simply add the following line in your Pandas code for the GPU accelerator mode:

```
%load_ext cudf.pandas
```

Note GPU acceleration means carrying out processing tasks on GPUs instead of CPUs.

Here is how that looks:

```
%load_ext cudf.pandas

import pandas as pd

df = pd.DataFrame(
    {
        'A': [1, 2, 3],
        'B': [4, 5, 6]
    }
)
```

```
def transform(data: pd.DataFrame) -> pd.DataFrame:
    data['C'] = data['A'] + data['B']
    data = data[data['C'] > 5]
    return data

result = transform(df)

print(result)
```

would yield

```
   A  B  C
1  2  5  7
2  3  6  9
```

The preceding code is purely written in Python. You may wish to replace the values in the data frame to something large enough that the CPU takes a minute to compute and try to run this. You will notice the CuDF Pandas accelerator mode in action. You can also enable GPU accelerator mode by executing the Python script in the following manner:

```
python3 yourprogram.py -m cudf.pandas
```

Conclusion

So far, we have gained an appreciation toward data processing in traditional CPU environments and GPU-accelerated environments. We looked at key concepts of GPU programming like kernels, threads, and memory management, which form the foundation of efficient GPU utilization. We also gained hands-on CuDF programming, performing advanced data manipulation operations, feature engineering, and complex joins. This chapter can serve as a foundation for setting up cost-effective and robust machine learning pipelines. We will see more of these in upcoming chapters.

CHAPTER 5

Getting Started with Data Validation using Pydantic and Pandera

Introduction

Data validation has been in existence for a long time. During the early adoption of relational database management systems (RDBMSs), basic validation rules were defined within a limited scope. As SQL evolved within the relational systems, it provided more opportunities with respect to specifying better validation rules by writing SQL code. In the modern days of big data and machine learning, data validation has occupied greater relevance. The quality of data directly affects the insights and intelligence derived from analytical models that are built using them. In this chapter we will explore two major data validation libraries, namely, Pydantic and Pandera, and delve into features, capabilities, and practical applications.

By the end of this chapter, you will

- Gain an appreciation to concepts and principles of data validation.
- Learn Pydantic, a data validation library, and its key concepts.
- Define data schemas using Pydantic models and explore validators and custom validation logic.
- Learn Pandera and define data schemas using Pandera's DataFrameSchema and Column classes.

- Explore checks in Pandera to specify data constraints and properties and perform statistical tests on the data.

- Learn lazy validation using Pandera and integrate Pandera in existing data engineering projects using decorators.

Introduction to Data Validation

As government organizations and enterprises adopted computer systems, software development gained more traction, leading to incorporating data validation specifications and rules within the software application development projects. These days, web applications and mobile applications increasingly dominate modern computing including but not limited to consuming and procuring services, communication, etc. And so, real-time client-side data validation has become common for ensuring data accuracy and integrity.

As we have now moved into fully mature machine learning and analytics development, advanced data validation techniques are now employed as part of data quality measures. Data quality is an important part of the data management process.

Need for Good Data

The entire machine learning modeling and data science analysis process runs on a few assumptions. Just like how generic linear models assume that the underlying data is normally distributed, the machine learning models and data science analysis are based on assumptions about data. Some of such assumptions in our projects may look like the following:

- Our data is current, that is, believed to represent the reality of the situation.

- If the model exceeds the accuracy threshold, then the model is good; otherwise, models and algorithms need to be retrained.

- Data from source systems or where we extract from should be accurate.

Data validation ensures these assumptions hold good.

Models are built and decisions are made based on the idea that underlying incoming data is good, consistent, and accurate. And so, any bottlenecks and inconsistencies may be attributed to the model itself and require further development cycles.

Here are some facts:

- Quality of data directly affects the insights and intelligence derived from the models; it affects accuracy, effectiveness, and trustworthiness of the insights gained from the models.

- Organizations increasingly have more data from a variety of data systems and external sources; it's essential to have a good data validation process to ensure quality and consistency.

- Data validation is an ongoing process, with periodic reviews on quality and consistency on various fields or data points across these sources, enabling a cohesive view or organization's operations and performance.

- Some industries have compliance and regulatory requirements with regard to data privacy, data integrity, and data accuracy; the data validation processes may assist in complying with various regulations.

Here are the most common pitfalls for data quality:

- Incomplete data
- Inconsistent data
- Outdated data

Definition

Data validation is the process of ensuring the data collected or entered meets certain predefined standards and rules.

The objective is to ensure accuracy, reliability, and integrity of the data coming from source systems.

Data validation processes usually operate by identifying and preventing errors, inconsistencies, or inaccuracies.

The data validation process may occur during data entry, data import, or data processing and involves checking data against predefined rules to ensure it meets the expected criteria.

Principles of Data Validation

We may look at these validation principles in detail when applied to a dataset; however, it is important to understand what they mean and the purpose they serve. Here are some basic principles of data validation.

Data Accuracy

Data accuracy is the process of verifying data is free from errors. The goal of data accuracy is to transform and perform sanitization operations to keep it acceptable and correct, as per the context of the project or business context. Lack of data accuracy may not necessarily affect the downstream data pipelines as long as they are of the same data type that the downstream processes are expecting. However, it may certainly have a major impact on the quality of results and intelligence that downstream processes may generate. Hence, it is important to keep the data accurate and current.

Data Uniqueness

Data uniqueness is when each row or a data point in either a row or column is not a copy of another data point and not a duplicate in any form. In many cases, a data point may repeat itself in several instances due to the nature and context (e.g., categorical bins in numerical format). However, in some cases, where the data point is the output of a transformation that involves more than one input data point, it may be beneficial to have the uniqueness tested and validated. Another application is that, when the data point is a floating point number with mandatory "n" number of digits after the decimal, it certainly helps to validate the uniqueness of the data point.

Data Completeness

Data completeness is a process of validating whether the data required as part of the dataset is actually present. In other words, data completeness is computing the missing values or partially available values, when a field or a data point is deemed mandatory or critical. A good example is the validation of email addresses; email addresses usually have a combination of strings, numbers, and acceptable special characters, followed by the "@" symbol and the domain name along with the .com. Any of these elements, if missing, can be referred to as data incomplete. Phone numbers are also another good example. Relevant area code, followed by the phone digits, would make a data complete. Also it should total to ten digits (in the case of the United States).

Data Range

Data range is the process of validating data falling between certain acceptable boundaries for a numerical data point or a set of predefined categories for a categorical data point. For instance, one cannot have age more than 110 or so (link to oldest person). Similarly, one cannot have negative values for credit scores and such (credit scores start from 300).

Data Consistency

Data consistency is the process of verifying and validating whether the value contained is consistent and aligns with other data within the same dataset. There may be some rare cases where the range of the data is well defined with lower and upper bounds. However, business rules may suggest that data should be between a subrange with upper and lower bounds that are well within a given range.

Data Format

Data format is the process of verifying whether the data follows different encoding than what is expected and whether the data types are in the appropriate format. Data format is an important part of validation. When data is transported between data systems, there is a greater likelihood that the one system may not support the native encoding of the other system. And so, the data types are often converted or imposed to data fields. A common issue is when date is read as a string. Another issue is when an identification number is casted as a string and gets auto-converted into an integer; problem is it may lose all the trailing zeroes, which is the very reason integers are casted as strings.

Referential Integrity

Referential integrity is the process of verifying and validating the data relationships that exist between tables. To illustrate this further, consider an example where a pipeline is writing relational tables from source to a target database system. It may be important to ensure that the referential integrity is maintained such that when in case of joining these tables with specific join conditions, they continue to produce similar results.

Data validation can be performed during the development phase, or you can set up an automated process of checking various data points. In this chapter, we are going to look at Pydantic library.

Introduction to Pydantic

Pydantic is the most popular Python library for performing data validations. The way it works is that Pydantic enforces data types and constraints while loading data into the Pydantic models. Some of the most effective applications of Pydantic are

- Validating JSON documents and datasets
- Loading configuration data from external files and environmental variables
- Validating input data from web APIs prior to inserting them in databases

Pydantic provides a relatively simple way to declare data schemas on incoming data and validate the Pydantic models. Please note that validation happens automatically when you load the data into a Pydantic model. Pydantic automatically catches errors and enforces constraints on the incoming data. Before we dive deep into the exciting world of Pydantic, let us refresh our memory on the concept of type annotations in Python.

Type Annotations Refresher

Type annotations or type hints, in Python programming practices, are an optional feature that specifies the type of input parameter or the type of variable that is returned in a given function. Type annotations are best practices that inform the developer what kind of data to pass to the function and what kind of data to expect when the same function returns a variable or value. Python type annotations specify only the expected data types for functions, variables, methods, etc. Python does not enforce them. Hence, type annotations, alone, cannot be used for type checking by default. Type annotations do improve the code readability and documentation.

Let's look at a simple example of annotating a variable in Python:

```
myVariable: int = 0
```

Now, this Python variable is mutable. And so, let's look at modifying this value:

```
myVariable = 1
print(myVariable)
```

The value of myVariable upon executing the print function would be

```
1
```

Now, let us go a step further:

```
myVariable = 9768.3491
print(myVariable)
```

The type of myVariable is now effectively float and not int; the value of myVariable has now changed as well. We get the following output:

```
9768.3491
```

Let's look at an example of a function:

```
def addition(a: int, b: int) -> float:
    c: int = 0
    c = a + b
    return c
```

And so, when we call the function

```
addition(33,44)
```

we would get

```
77
```

Please note that this is not a good programming practice to specify that this function addition would return a float type while explicitly defining an integer within the function. The idea is to show that Python does not enforce the data types.

Here is a better example:

```
def greet(name: str) -> str:
    return "Hello " + name
```

When we call this function

```
greet("your name")
```

it would return

```
Hello your name
```

Pydantic is the most widely used library for data validation in Python. Big tech like Apple and others along with large governmental organizations use Pydantic (and contribute too). Pydantic recently released a major upgrade called Pydantic V2. According to the release notes, Pydantic's validation and serialization logic has been written in Rust programming language, so the V2 will be tens of times faster than the earlier version.

Given Pydantic's ability to generate JSON schema, it would make it a default choice to integrate with other tools that support JSON schema. Pydantic offers a strict mode, where data coercion is made unavailable.

Note Data coercion is the process of converting data from one type to another, so the data is usable in a given context.

Setup and Installation

Here's how to install Pydantic from PyPI:

```
pip install pydantic
```

Pydantic is also available in conda environments.

CHAPTER 5   GETTING STARTED WITH DATA VALIDATION USING PYDANTIC AND PANDERA

Pydantic Models

Models are how you define schemas in Pydantic. Pydantic models are basically writing classes that inherit from BaseModel. Upon defining the class, you can begin to define the schemas for the dataset. Once the data is passed through the Pydantic model, Pydantic will make sure that every data field in the source data will conform to the schema defined in the Pydantic class.

Here is an example. Let's import a data file that is in comma-separated values format:

```
from pydantic import BaseModel
import pandas as pd

data = pd.read_csv("/home/parallels/Downloads/MOCK_DATA.csv")
```

Let's check the output to see if the data is loaded for our testing purposes:

```
print(data.head(2))
```

will yield

```
            name  age
0  Ermentrude Azema   22
1      Joey Morteo   56
```

Now let's define the schema for the Pydantic model class:

```
class TestPerson(BaseModel):
    name: str
    age: int
```

Now let's create a Pydantic model object:

```
for _, person in data.iterrows():
    person = person.to_dict()
    pydantic_model = TestPerson(**person)
    print(pydantic_model.name)
```

would yield

```
name='Ermentrude Azema' age=22
name='Joey Morteo' age=56
```

171

Nested Models

Pydantic helps define complex hierarchical data structures using Pydantic models as types. Here is an example.

Let's import the required libraries and data:

```
import pydantic, json
from typing import List

with open("/home/parallels/Documents/data2.json") as f:
    data = json.load(f)

print(data[0])
```

Here is how the first record in the dataset looks like:

```
{
    'first_name': 'Angil',
    'last_name': 'Boshers',
    'interests': {
        'first': 'sports',
        'second': 'mountains'
        }
}
```

Now let's define the Pydantic model:

```
class Interests(pydantic.BaseModel):
    first: str
    second: str

class Stu(pydantic.BaseModel):
    first_name: str
    last_name: str
    interests: List[Interests]
```

Both the parent and immediate child get their own Pydantic model class. The parent Pydantic model class references the immediate child model class using a List method.

Let's create the model object and print the model output:

```
for i in data:
    output = Stu(**i)
    print(output)
```

would yield

```
first_name='Angil' last_name='Boshers' interests=[Interests(first='sports', second='mountains')]
first_name='Aggi' last_name='Oswick' interests=[Interests(first='long-drives', second='beaches')]
```

Fields

The field function in Pydantic is used to specify a default value for a given field, in case it remains missing from source documents.

Here is an example to illustrate the field function. To begin with, let's import the libraries and define the Pydantic model:

```
import pydantic

class Person(pydantic.BaseModel):
    name: str = pydantic.Field(default='John Doe')
    age: int = pydantic.Field(default=999)
```

As you can see, the default value is mentioned as John Doe. Now let's define the Pydantic model object and observe the output:

```
newPerson = Person()
```

```
print(newPerson)
```

would yield

```
name='John Doe' age=999
```

CHAPTER 5 GETTING STARTED WITH DATA VALIDATION USING PYDANTIC AND PANDERA

JSON Schemas

Pydantic allows us to create JSON schemas from the Pydantic model class we created. As you may recall, a JSON schema is simply a JSON data structure. A JSON schema asserts how the JSON document will be like.

Let us revisit the earlier nested models example. We import the necessary libraries, create the Pydantic model class, create the Pydantic model object, and print the JSON schema:

```
import pydantic, json
from typing import List

with open("/home/parallels/Documents/data2.json") as f:
    data = json.load(f)

print(data[0])

class Interests(pydantic.BaseModel):
    first: str
    second: str

class Stu(pydantic.BaseModel):
    first_name: str
    last_name: str
    interests: List[Interests]

print(Stu.schema_json(indent=3))

# In case if the schema_json method in BaseModel class
# is deprecated, use the model_json_schema method instead:

#print(Stu.model_json_schema(indent=3))
```

would yield

```
{
   "title": "Stu",
   "type": "object",
   "properties": {
```

```
      "first_name": {
         "title": "First Name",
         "type": "string"
      },
      "last_name": {
         "title": "Last Name",
         "type": "string"
      },
      "interests": {
         "title": "Interests",
         "type": "array",
         "items": {
            "$ref": "#/definitions/Interests"
         }
      }
   },
   "required": [
      "first_name",
      "last_name",
      "interests"
   ],
   "definitions": {
      "Interests": {
         "title": "Interests",
         "type": "object",
         "properties": {
            "first": {
               "title": "First",
               "type": "string"
            },
            "second": {
               "title": "Second",
               "type": "string"
            }
         },
```

```
        "required": [
           "first",
           "second"
        ]
      }
   }
}
```

Constrained Types

A constrained type is a method offered by Pydantic that dictates the values each field can take on in the Pydantic model class. Pydantic offers the following types:

- For int, there is the conint method; conint can be used to set up upper and lower bounds for a continuous variable.

- For str, there is the constr method; constr can enforce uppercase or lowercase for a string, wherever required.

- For floats, there is a confloat method; once again, it helps to set lower and upper bounds for floats, decimals, etc.

```
from pydantic import BaseModel, validator, conint, confloat

class Person(BaseModel):
    name: str
    age: conint(ge=30)
    blood_sugar_level: confloat(gt=4, lt=10)
```

Validators in Pydantic

Pydantic provides a way to validate the field data using the validator method. You can even bind validation to a type instead of a model. It may be worthy to note that validators do not get executed if a default field value is specified. There are several types of validators:

- After validators run after Pydantic's internal parsing.

- Before validators run before Pydantic's internal parsing.

- Wrap validators run either before or after Pydantic's internal parsing.

A field validator enables validation of a specific field of the Pydantic model. Let us look at an example on the same concept.

Here, let us import necessary libraries and the data file (in this example, a flat file):

```
from pydantic import BaseModel, field_validator
import csv

data = csv.DictReader(open("/home/parallels/Downloads/MOCK_DATA.csv"))
```

The dataset is about persons and contains two variables, name and age. Name is a string and age is an integer. The idea is to validate how many of the persons are less than 21 years old. Now, let us define the Pydantic model class with a validator:

```
class Person(BaseModel):
    name: str
    age: int

    @field_validator('age')
    def agelimit(cls, value: int) -> int:
        if value < 21:
            raise ValueError("Less than 21 years old")
        return value
```

And so, when we run the model output this way

```
for person in data:
    pydantic_model = Person(**person)
    print(pydantic_model)
```

we get the following output:

```
name='Ermentrude Azema' age=22
name='Joey Morteo' age=56
name='Jordan Corp' age=56
name='Jehu Kettlesing' age=25
name='Adham Reeson' age=40
name='Ronda Colomb' age=32
name='Bond Dufton' age=47
name='Pearce Mantione' age=47
name='Reginald Goldman' age=39
```

CHAPTER 5   GETTING STARTED WITH DATA VALIDATION USING PYDANTIC AND PANDERA

```
name='Morris Dalrymple' age=32
name='Lemar Gariff' age=39
name='Leonhard Ruberti' age=63
name='Alyssa Rozier' age=63
name='Cassey Payze' age=22
name='Lonnie Mingaud' age=56
name='Town Quenby' age=39
Traceback (most recent call last):

  File ~/anaconda3/lib/python3.11/site-packages/spyder_kernels/py3compat.
py:356 in compat_exec
    exec(code, globals, locals)

  File ~/Documents/pyd_model.py:25
    pydantic_model = Person(**person)

  File pydantic/main.py:341 in pydantic.main.BaseModel.__init__

ValidationError: 1 validation error for Person
age
  Less than 21 years old (type=value_error)
```

Introduction to Pandera

Pandera is an MIT-licensed, open source Python library that performs data validation in the context of cleaning and preprocessing. Pandera works well with Pandas' data frames' data structure. Pandera enables declarative schema definition, as in you can define the expected data structure including data types, constraints, and other properties. Once a schema is defined, Pandera can validate whether the incoming Pandas data frame object adheres to the user-defined schema (including data types and constraints), checking whether the fields are present among other things. Furthermore, Pandera supports custom functions and user-defined validation checks, which may be very domain specific for the incoming dataset.

Pandera also supports statistical validation like hypothesis testing. Note that statistical validation is very different from data validation; statistical validation is where one hypothesizes that data should have a certain property or meet a condition, which is then validated by applying various statistical methods and formulas toward samples of

CHAPTER 5   GETTING STARTED WITH DATA VALIDATION USING PYDANTIC AND PANDERA

the population data. Pandera supports lazy validation; and so, when a data frame is set to be validated against a long list of data validation rules, Pandera will perform all the defined validation before raising an error.

Setup and Installation

Here's how to install Pandera in your system:

```
pip install pandera
```

Here is a simple example of data validation using Pandera. I have deliberately not used aliases in this code, as it may be confusing for someone who is looking at these for the first time.

Let's import the libraries:

```
import pandas
import pandera
```

As mentioned earlier, declare the schema definition:

```
panderaSchema = pandera.DataFrameSchema(
    {
      "id": pandera.Column(pandera.Int),
      "name": pandera.Column(pandera.String)
    }
)
```

Now, bring in the data:

```
data = {
        "id": [123,2435,234],
        "name": ["john","doe", None]
        }
pandas_df = pandas.DataFrame(data)
```

Let us now validate the "pandas" data frame against the "pandera" defined schema:

```
try:
    dataValidation = panderaSchema.validate(pandas_df)
    print(dataValidation)
```

CHAPTER 5 GETTING STARTED WITH DATA VALIDATION USING PYDANTIC AND PANDERA

```
except pandera.errors.SchemaError as e:
   print(e.failure_cases)  # dataframe of schema errors
   print("\n\ninvalid dataframe")
   print(e.data)  # invalid dataframe
```

would yield the following error:

```
   index failure_case
0    2        None
```

invalid dataframe

```
     id   name
0   123   john
1  2435    doe
2   234   None
```

Pandera provides a class-based API, similar to Pydantic. This means in Pandera, you can define a schema model like a Pydantic model class.

DataFrame Schema in Pandera

Pandera offers a data frame schema class that is used to define the specifications of the schema, which is then used to validate against columns (and index) of the data frame object provided by Pandas. Pandera's column class enables column checks, data type validation, and data coercion and specifies rules for handling missing values.

Here is an example:

Let's import the libraries, methods, and data:

```
import pandas as pd
from pandera import DataFrameSchema, Column, Check

data = pd.DataFrame({"series": [1, 2, 4, 7, None]})
```

Let's define the Pandera schema, which enforces strict conditions:

```
StrictlyNotNullSchema = DataFrameSchema(
   {
     "series": Column(float, Check(lambda x: x > 0))
   }
)
```

Let us validate the data frame against the schema:

```
try:
    strictNotNull = StrictlyNotNullSchema.validate(data)
    print(strictNotNull)
except errors.SchemaError as e:
    print(e.failure_cases)  # dataframe of schema errors
    print("\n\ninvalid dataframe")
    print(e.data)  # invalid dataframe
```

would yield

```
   index  failure_case
0      4           NaN
invalid dataframe
   series
0     1.0
1     2.0
2     4.0
3     7.0
4     NaN
```

As you can see, the schema enforces itself strictly onto the data frame. To accept null or missing values, one can specify an argument like nullable=True.

Here is how that looks:

```
NullAllowedSchema = DataFrameSchema(
    {
      "series": Column(float, Check(lambda x: x > 0),
                       nullable=True)
    }
)
```

And so, when we try to validate and print the results

```
try:
    nullAllowed = NullAllowedSchema.validate(data)
    print(nullAllowed)
```

CHAPTER 5 GETTING STARTED WITH DATA VALIDATION USING PYDANTIC AND PANDERA

```
except errors.SchemaError as e:
   print(e.failure_cases)  # dataframe of schema errors
   print("\n\ninvalid dataframe")
   print(e.data)  # invalid dataframe
```

it would yield

```
     series
0      1.0
1      2.0
2      4.0
3      7.0
4      NaN
```

Data Coercion in Pandera

Pandera can also perform data coercion in addition to validation. Let us talk about data coercion a little bit. Data coercion is the process of converting a given data column from one type to another user-specified type. The idea is to make the data accessible in a specific data format.

Data coercion needs to be handled with caution: in some cases, data coercion may hide errors, if an out-of-range value or invalid data is sourced but coerced anyways. Common examples include transforming a tuple to a list, an integer to a string, etc.

Here is how that looks.

Let us initialize a Pandas data frame:

```
import pandas as pd
from pandera import DataFrameSchema, Column

df = pd.DataFrame({"values": [1,4,5,7,9]})
```

Now, let us define the schema, where the incoming data must be a string. And so, we enable coerce=True:

```
schema = DataFrameSchema(
   {
     "values": Column(str, coerce=True)
   })

dataValidated = schema.validate(df)
```

We perform the validation and it looks good. Now let's examine the data type of the validated data here:

```
print(dataValidated.iloc[0].map(type))
```

would yield

```
values    <class 'str'>
Name: 0, dtype: object
```

And so, the integer value has been coerced into a string.

Checks in Pandera

Checks in Pandera are similar to constrained types and specifying conditions in the field in Pydantic. Checks specify properties about various objects like data frame, series, etc., which are validated against Pandera methods. For ease of reading the code, one can define checks early on to a variable and then call the same in the DataFrameSchema object.

Here is how that would look.

Let us define the check. In this example, let us check if the given number is odd or even. Let us define that in the check:

```
from pandera import (Check,
                     DataFrameSchema,
                     Column,
                     errors)
import pandas as pd

odd = Check.isin([1,3,5,7,9])
```

Let us define the Pandera schema and data:

```
schema = DataFrameSchema({
    "series": Column(int, odd)
    })

data = pd.DataFrame({"series": range(10)})
```

Let us attempt to validate the schema:

```
try:
    schema.validate(data)
except errors.SchemaError as e:
    print("\n\ndataframe of schema errors")
    print(e.failure_cases)  # dataframe of schema errors
    print("\n\ninvalid dataframe")
    print(e.data)  # invalid dataframe
```

would yield

```
dataframe of schema errors
   index  failure_case
0    0         0
1    2         2
2    4         4
3    6         6
4    8         8

invalid dataframe
   series
0    0
1    1
2    2
3    3
4    4
5    5
6    6
7    7
8    8
9    9
```

And so, both the index and element are given in the output.

Pandera, by default, performs these data validations by way of vectorized operations.

CHAPTER 5   GETTING STARTED WITH DATA VALIDATION USING PYDANTIC AND PANDERA

Note Vectorized operations are a way of performing operations on a vector or a matrix or an array of elements without having to loop through and iterate over each and every time.

You can also specify whether you need element-wise check operations. For our previous example, here is how that looks:

```
from pandera import (Check,
                     DataFrameSchema,
                     Column,
                     errors)
import pandas as pd

odd = Check(
    lambda x: x % 2 != 0,
    element_wise=True
)

data = pd.DataFrame(
    {
        "series": [1,3,5,7,9]
    }
)

schema = DataFrameSchema(
    {
        "series": Column(int, odd)
    }
)

try:
    schema.validate(data)
except errors.SchemaError as e:
    print("\n\ndataframe of schema errors")
    print(e.failure_cases)  # dataframe of schema errors
    print("\n\ninvalid dataframe")
    print(e.data)  # invalid dataframe
```

would yield

```
   series
0       1
1       3
2       5
3       7
4       9
```

Statistical Validation in Pandera

Pandera supports hypothesis testing, without having to explicitly load scipy or another statistical library. Let us cover a bit about hypothesis testing.

Hypothesis testing is primarily a complex statistical procedure that involves statistical analysis among other things. Hypothesis testing enables us to draw conclusions or inferences about a population from a sample data. It starts off by specifying the claim that one wishes to make. The rest of the statistical procedure is to design the analysis such that the results of the procedure would dictate whether the claim specified is true or false.

Now, let us learn some technical jargon (statistical words). As mentioned earlier, the procedure starts off with making a claim about the data. We call it the null hypothesis. The remaining procedure is about validating the claim. We then make an alternate hypothesis, which is the opposite of what we claimed.

We now begin to proceed with various complex formulas, essentially to compute whether the difference between the claimed metric and measured metric is due to chance or if there is really empirical evidence against our claim. Pandera offers two types of hypothesis tests, namely, one-sample t-test and two-sample t-test. One-sample t-test is used to determine whether a sample set with a sample mean indeed comes from the population. Two-sample t-tests are used to compare two groups and decide whether there is statistical evidence that a given group is really better than the other.

Let's look at an example.

I have generated a dataset that contains some kind of score for both iOS and Android operating systems. It is a randomly generated dataset. A sample output would look like

CHAPTER 5 GETTING STARTED WITH DATA VALIDATION USING PYDANTIC AND PANDERA

```
        os  hours
0      ios    4.5
1  Android    5.1
2      ios    5.1
3  Android    5.1
4      ios    5.7
```

Let us look at this in detail.

First, we import necessary libraries, methods, and data:

```
import pandas as pd
from pandera import DataFrameSchema, Column, Check, Hypothesis

df = pd.read_csv("/home/parallels/data.csv")

df[['hours','os']] = df['float\tname'].str.split('\t',expand=True)

data = df[['os','hours']]
```

Please note that the data is randomly generated and I needed to split the column into two columns, followed by rearranging the columns.

My claim is that iOS has a better score than Android. Once again, these are randomly generated floating point numbers and do not represent anything real here. Here is how they are interpreted in statistical language:

In statistical language

Null hypothesis: There is no significant difference in value scores between iOS and Android.

Alternate hypothesis: There is significant difference in value scores between iOS and Android.

It is essential to note that an alternate hypothesis does not necessarily mean the opposite of a given claim is true. It simply means that we do not have enough data to prove our claim is true.

CHAPTER 5 GETTING STARTED WITH DATA VALIDATION USING PYDANTIC AND PANDERA

Here is how the code looks:

```
schema = DataFrameSchema(
    {
      "hours" : Column(
          float, [
              Hypothesis.two_sample_ttest(
                  sample1="ios",
                  sample2="Android",
                  groupby="os",
                  relationship="greater_than",
                  alpha=0.02,
                  equal_var=True
                  )
              ], coerce=True
          ),
      "os" : Column(str)
      })
```

Let us try to run the validation:

```
try:
    schema.validate(data)
except errors.SchemaError as exc:
    print(exc)
```

We get

```
SchemaError: <Schema Column(name=hours, type=DataType(float64))> failed
series or dataframe validator 0:
<Check two_sample_ttest: failed two sample ttest between 'ios' and
'Android'>
```

As we discussed earlier, there is not enough data that is present to conclude whether our claim is correct, and we are about 98% accurate in our conclusion. The degree of confidence is specified by a parameter called "alpha."

Lazy Validation

So far we have seen that when we call the "validate" method, the SchemaError is raised immediately after one of the assumptions specified in the schema is falsified. And so, the program stops executing other instructions supplied. In many cases, it may be more useful to see all the errors raised on various columns and checks. Lazy validation is an option that enables just that. We have seen lazy evaluation of code in our earlier chapter. Lazy evaluation is when the code is executed only when it is actually needed.

Let's look at an example.

Let us validate a simple table with a record in it. First, import the libraries and method, followed by defining the Pandera schema:

```python
import pandas as pd
import json
from pandera import (Column,
                     Check,
                     DataFrameSchema,
                     DateTime,
                     errors)

# define the constraints

check_ge_3 = Check(lambda x: x > 3)
check_le_15 = Check(lambda x: x <= 15)
check_gpa_0 = Check(lambda x: x > 0.0)
check_gpa_4 = Check(lambda x: x <= 4.0)

# schema

schema = DataFrameSchema(
    columns={
        "StudentName" : Column(str, Check.equal_to("John")),
        "CreditsTaken" : Column(int, [check_ge_3, check_le_15]),
        "GPA" : Column(float, [check_gpa_0, check_gpa_4]),
        "Date" : Column(DateTime),
        },
    strict=True
)
```

CHAPTER 5   GETTING STARTED WITH DATA VALIDATION USING PYDANTIC AND PANDERA

Now, let us load the data:

```
df = pd.DataFrame(
    {
        "StudentName" : ["JohnDoe", "JaneDoe","DoeJohn"],
        "CreditsTaken" : [9,12,18],
        "GPA" : [3.7, 4.01, 3.5],
        "Date" : None,
    }
)
```

Let us perform the lazy validation:

```
try:
    schema.validate(df, lazy=True)
except errors.SchemaErrors as e:
    print(json.dumps(e.message, indent=2))
    print("dataframe of schema errors")
    print(e.failure_cases)  # dataframe of schema errors
    print("invalid dataframe")
    print(e.data)   # invalid dataframe
```

would yield

```
{
  "DATA": {
    "DATAFRAME_CHECK": [
      {
        "schema": null,
        "column": "StudentName",
        "check": "equal_to(John)",
        "error": "Column 'StudentName' failed element-wise validator number
        0: equal_to(John) failure cases: JohnDoe, JaneDoe, DoeJohn"
      },
      {
        "schema": null,
        "column": "CreditsTaken",
        "check": "<lambda>",
```

```
          "error": "Column 'CreditsTaken' failed element-wise validator
          number 1: <Check <lambda>> failure cases: 18"
        },
        {
          "schema": null,
          "column": "GPA",
          "check": "<lambda>",
          "error": "Column 'GPA' failed element-wise validator number 1:
          <Check <lambda>> failure cases: 4.01"
        }
      ]
    },
    "SCHEMA": {
      "SERIES_CONTAINS_NULLS": [
        {
          "schema": null,
          "column": "Date",
          "check": "not_nullable",
          "error": "non-nullable series 'Date' contains null
          values:0    None1    None2    NoneName: Date, dtype: object"
        }
      ],
      "WRONG_DATATYPE": [
        {
          "schema": null,
          "column": "Date",
          "check": "dtype('datetime64[ns]')",
          "error": "expected series 'Date' to have type datetime64[ns],
          got object"
        }
      ]
    }
}
dataframe of schema errors
```

	schema_context	column	check	check_number	failure_case	index
0	Column	StudentName	equal_to(John)	0	JohnDoe	0
1	Column	StudentName	equal_to(John)	0	JaneDoe	1
2	Column	StudentName	equal_to(John)	0	DoeJohn	2
3	Column	CreditsTaken	<lambda>	1	18	2
4	Column	GPA	<lambda>	1	4.01	1
5	Column	Date	not_nullable	None	None	0
6	Column	Date	not_nullable	None	None	1
7	Column	Date	not_nullable	None	None	2
8	Column	Date	dtype('datetime64[ns]')	None	object	None

invalid dataframe

	StudentName	CreditsTaken	GPA	Date
0	JohnDoe	9	3.70	None
1	JaneDoe	12	4.01	None
2	DoeJohn	18	3.50	None

Please note that in the preceding example, SchemaError is coming from the class errors. This exhaustive output of the "try..except" block is also a good way to see detailed validation output.

Pandera Decorators

Pandera can be integrated into your current production pipelines through the use of decorators.

Note Python decorators are a way to accentuate or modify the behavior of already written functions. Programmatically, a decorator is a function that takes a function as an input parameter and returns another function. When you use a decorator on an existing function, the decorator would extend the current function's ability to include the decorator function's ability as well.

Here is a simple example of decorators:

```
def decoratortest(func):
    def anotherfunc():
        func()
        print("the decorator function is now executing this line")
    return anotherfunc

@decoratortest
def hello_world():
    print("this is the main hello world function")

hello_world()
```

When you run this program, you would get

```
this is the main hello world function
the decorator function is now executing this line
```

Here, the original function "hello_world" is fitted with a decorator called "decoratortest." As you can see, the "decoratortest" is another function that takes a function as input and returns a different function.

When the decorator is applied to the hello_world() function, the hello_world() function becomes the input of the decorator function. Notice that it is the decorator function that wrapped the hello_world() function that got executed, not the hello_world() function by itself.

Pandera offers two decorators to perform validation against variables returned from functions. These two decorators serve the purpose of validating before and after data processing. They are called @check_input and @check_output, respectively. These decorators accept the DataFrameSchema object as a parameter.

Let us look at an example where we have a current Python code that adds two columns and stores the result in a third column, all within a Pandas data frame. We will leverage the idea of using decorators. Let's look at the code, without Pandera validation parameters:

```
import pandas as pd

data = pd.DataFrame({
    "a": [1,4,7,9,5],
```

```
    "b": [12,13,15,16,19],
})

def addition(dataframe):
    dataframe["c"] = dataframe["a"] + dataframe["b"]
    return dataframe

final_df = addition(data)

print(final_df)
```

It is a straightforward example where we add two columns and store their result in a third column. Now let's look at Pandera data validators along with decorator integration. First off, we import the necessary functions, data, and methods:

```
import pandas as pd
from pandera import DataFrameSchema, Column, Check, check_input

data = pd.DataFrame({
    "a": [1,4,7,9,5],
    "b": [12,13,15,16,19],
})
```

Now we define the Pandera data validation through DataFrameSchema. Please note the check is defined separately and called within the DataFrameSchema for purposes of readability. You can incorporate the check directly within the DataFrameSchema:

```
check_a = Check(lambda x: x <= 10)
check_b = Check(lambda x: x <= 20)

validateSchema = DataFrameSchema({
    "a": Column(int,
                check_a
                ),
    "b": Column(int,
                check_b
                ),
})
```

CHAPTER 5   GETTING STARTED WITH DATA VALIDATION USING PYDANTIC AND PANDERA

Now, let's utilize the Pandera decorators. We are going to use the @check_input, to validate the input columns. The DataFrameSchema is passed as parameter to the @check_input decorator:

```
@check_input(validateSchema)
def addition(dataframe):
    dataframe["c"] = dataframe["a"] + dataframe["b"]
    return dataframe

final_df = addition(data)

print(final_df)
```

Upon executing the code, we would obtain

```
   a   b   c
0  1  12  13
1  4  13  17
2  7  15  22
3  9  16  25
4  5  19  24
```

Now, let's say we change the validation condition of check_b. Currently it looks like

```
check_b = Check(lambda x: x <= 20)
```

The modified check would look like

```
check_b = Check(lambda x: x <= 18)
```

Upon executing the code, we obtain the following:

```
SchemaError: error in check_input decorator of function 'addition': <Schema Column(name=b, type=DataType(int64))> failed element-wise validator 0:
<Check <lambda>>
failure cases:
   index  failure_case
0      4            19
```

195

Conclusion

Data validation remains one of the key aspects of engineering data pipelines. We discussed various principles of data validation and the need for good data. We then looked at Pydantic, a data validation library that focuses on Python objects conforming to a certain data model. We then discussed Pandera, another data validation library that focuses on applying validation checks on data frame-like objects; these are very effective toward the data analysis and data preprocessing tasks. In our next chapter, we will discuss another data validation library that focuses heavily on data quality assurance and data validation in data pipelines.

CHAPTER 6

Data Validation using Great Expectations

Introduction

So far we looked at the need for and idea behind data validation, principles of data validation, and what happens when data isn't validated. We also looked at key Python data validation libraries Pydantic and Pandera. Pydantic is one of the most utilized libraries in Python. We looked at creating a model class and instantiating an object from the class. We also looked at Pandera, a specialized open source Python library, with advanced data validation options like checking conditions at column levels and so on. In this chapter, we will be looking at Great Expectations, an entire data validation framework that is designed for managing data validation and testing for several production pipelines.

By the end of this chapter, you will learn

- The building blocks of the Great Expectations framework
- Setup, installation, and troubleshooting
- The typical workflow of a Great Expectations data pipeline
- Developing expectations for data pipelines
- Creating an expectation suite and checkpoints

CHAPTER 6 DATA VALIDATION USING GREAT EXPECTATIONS

Introduction to Great Expectations

Great Expectations is a Python framework for testing and validation of a dataset in a data pipeline. It helps data professionals define testing and validation rules, develop documentation, and test the integrity of the data. The data professionals can set various expectations about the structure, data types, and various other properties of the datasets. Each and every testing and validation criteria (unit test) is defined as an expectation in this context. And so, we will be using the word "expectation" more often to define various testing and validation criteria for a given dataset.

Great Expectations can help you in defining what you wish to expect from the data that has been loaded and/or transformed. By employing these expectations, you can identify bottlenecks and issues within your data. In addition, Great Expectations generates documentation about data and reports about data quality assessment of the datasets.

Components of Great Expectations

Currently, Great Expectations is available in both on-premise and cloud environments. We will look at the on-premise option, where Great Expectations is set up locally on your computer. Internally, Great Expectations consists of four core components that enable its function. They are as follows.

Data Context

Data context is referred to as the starting point for the Great Expectations API; this component contains settings, metadata, configuration parameters, and output from validating data, for a given Great Expectations project. The metadata is usually stored in an object called Great Expectations stores. The Great Expectations API provides a way to configure and interact with methods using Python. You can access them using the Great Expectations public API. Great Expectations also generates data documentation automatically that contains where the data is coming from and the validations that the dataset is subjected to.

Data Sources

The data source component tells Great Expectations about how to connect to a given dataset. The dataset may be a flat file, JSON document, SQL database, Pandas data frame located locally, or remote machine or one in a cloud environment. The data source component contains one or more data assets. Data assets are collections of records from the source data. Data assets are stored in batches, where a given batch may contain an unique subset of records. Great Expectations submits a batch request internally to access the data to perform validation and testing. It may be interesting to note that a data asset in Great Expectations may contain data from more than one file, more than one database table, etc., whereas a given batch may contain data from exactly one data asset.

Expectations

Expectations would make the heart of this chapter! As mentioned earlier, an expectation is an assertion that needs to be verified against a data point. Expectations provide a descriptive way for describing the expected behaviors of a given data. These expectations apply to the data directly. And so, when Great Expectations applies a given expectation, it returns a report that contains whether the expectation is met or not. Great Expectations already supplies an exhaustive list of predefined expectations; however, you can also write custom expectations. Here is an example of an expectation:

```
my_custom_dataframe.expect_column_values_to_be_in_set (
"Hobbies",
["mountain hiking", "city surfing", "long drives"]
)
```

Checkpoints

A checkpoint is how the Great Expectations framework validates data in production pipelines. Checkpoints provide a framework to group together one or more batches of data and execute expectations against the said batched data. A checkpoint takes one or more batches, providing them to the validator module. The validator module generates results, which are then passed to a list of validation actions. The validation results tell you how and where the data stands with respect to the defined expectations. The final step is validation actions that can be completely configured to your needs. Actions include sending an email, updating data documentation, etc.

CHAPTER 6 DATA VALIDATION USING GREAT EXPECTATIONS

Setup and Installation

Great Expectations recommends the use of virtual environments for creating a project. And so, we will make use of a virtual environment setup.

First, let us make sure you have the latest version of Python, version 3.x.

Now let us create a virtual environment within your workspace:

```
python3 -m venv gework
```

This will create a new directory called "gework."

Now, activate the virtual environment by

```
source gework/bin/activate
```

If you are using a Windows PC, please activate the virtual environment using the following command:

```
gework/Scripts/activate
```

We use pip to install Python packages. Let us update pip to its latest version before installing Great Expectations:

```
python3 -m ensurepip –upgrade
```

To set up Great Expectations on your machine, please execute

```
pip3 install great_expectations
```

or

```
python3 -m pip install great_expectations
```

If you need to work with a SQL database as your data source, then you need the SQLAlchemy package as well. To install Great Expectations along with optional SQL dependencies, use this command:

```
pip install 'great_expectations[sqlalchemy]'
```

There are two ways of getting started with a new Great Expectations project. One is the legacy way of interacting with the command line interface, and the other way is through the modern fluent data source method.

CHAPTER 6 DATA VALIDATION USING GREAT EXPECTATIONS

Let's briefly look at the interaction with Great Expectations CLI:

```
great_expectations init
```

This will then prompt a screen like this:

```
    _____                _     _____                      _        _   _
  / ____|              | |   |  ____|                    | |      | | (_)
 | |  __ _ __ ___  __ _| |_  | |__  __  ___ __   ___  ___| |_ __ _| |_ _  ___  _ __  ___
 | | |_ | '__/ _ \/ _` | __| |  __| \ \/ / '_ \ / _ \/ __| __/ _` | __| |/ _ \| '_ \/ __|
 | |__| | | |  __/ (_| | |_  | |____ >  <| |_) |  __/ (__| || (_| | |_| | (_) | | | \__ \
  \_____|_|  \___|\__,_|\__| |_____/_/\_\ .__/ \___|\___|\__\__,_|\__|_|\___/|_| |_|___/
                                         | |
                                         |_|
           ~ Always know what to expect from your data ~

Let's create a new Data Context to hold your project configuration.

Great Expectations will create a new directory with the following structure:

    great_expectations
    |-- great_expectations.yml
    |-- expectations
    |-- checkpoints
    |-- plugins
    |-- .gitignore
    |-- uncommitted
        |-- config_variables.yml
        |-- data_docs
        |-- validations

OK to proceed? [Y/n]:
```

Figure 6-1. *Great Expectations prompt screen on a command line*

Great Expectations will create a project structure for your project. To proceed forward, type "y"; if not type "n" and restart the process.

Once the project structure creation has been completed, then you will see the following screen:

CHAPTER 6   DATA VALIDATION USING GREAT EXPECTATIONS

```
================================================================================
Congratulations! You are now ready to customize your Great Expectations configur
ation.

You can customize your configuration in many ways. Here are some examples:

  Use the CLI to:
    - Run `great_expectations datasource new` to connect to your data.
    - Run `great_expectations checkpoint new <checkpoint_name>` to bundle data w
ith Expectation Suite(s) in a Checkpoint for later re-validation.
    - Run `great_expectations suite --help` to create, edit, list, profile Expec
tation Suites.
    - Run `great_expectations docs --help` to build and manage Data Docs sites.

  Edit your configuration in great_expectations.yml to:
    - Move Stores to the cloud
    - Add Slack notifications, PagerDuty alerts, etc.
    - Customize your Data Docs

Please see our documentation for more configuration options!

(base) parallels@linux:~/Documents/ge/test1$
```

Figure 6-2. *Customizing the Great Expectations configuration page*

We can start to connect our data sources by issuing the following command:

`great_expectations datasource new`

```
(base) parallels@linux:~/Documents/ge/test1$ great_expectations datasource new
As of V0.16, the preferred method for adding a Datasource is using the fluent me
thod of configuration, and not using the CLI.
You can read more about this here: https://greatexpectations.io/blog/the-fluent-
way-to-connect-to-data-sources-in-gx
If you would like to proceed anyway, press Y. [Y/n]: y

What data would you like Great Expectations to connect to?
    1. Files on a filesystem (for processing with Pandas or Spark)
    2. Relational database (SQL)
: 1

What are you processing your files with?
1. Pandas
2. PySpark
: 1

Enter the path of the root directory where the data files are stored. If files a
re on local disk enter a path relative to your current working directory or an a
bsolute path.
: /home/myfolder/path-to-the-data-file
```

Figure 6-3. *Adding a new data source in Great Expectations*

Here is how the options look if we choose relational databases instead of files on a file system:

```
(base) parallels@linux:~/Documents/ge/test1$ great_expectations datasource new
As of V0.16, the preferred method for adding a Datasource is using the fluent me
thod of configuration, and not using the CLI.
You can read more about this here: https://greatexpectations.io/blog/the-fluent-
way-to-connect-to-data-sources-in-gx
If you would like to proceed anyway, press Y. [Y/n]: y

What data would you like Great Expectations to connect to?
    1. Files on a filesystem (for processing with Pandas or Spark)
    2. Relational database (SQL)
: 2

Which database backend are you using?
    1. MySQL
    2. Postgres
    3. Redshift
    4. Snowflake
    5. BigQuery
    6. Trino
    7. Athena
    8. Clickhouse
    9. other - Do you have a working SQLAlchemy connection string?
: 9
Because you requested to create a new Datasource, we'll open a notebook for you
now to complete it!
```

Figure 6-4. Choosing the appropriate option for a new data source in Great Expectations

This would then open up a Jupyter notebook, where you can complete the setup of your new data source.

Getting Started with Writing Expectations

Let us look at writing expectations. The objective of the example is to show the workings of expectations as a functionality. In production or development pipelines, it is required to initialize a data context, define a data asset, connect to an expectation store, write validations, and test validation against expectation among other steps.

Here we have a dataset that contains the gender of a subject and their respective blood sugar levels. We are going to write an expectation to see what types of gender are available.

CHAPTER 6 DATA VALIDATION USING GREAT EXPECTATIONS

Let's import the libraries and methods required:

```
import great_expectations as ge
import pandas as pd
import random
```

We are creating a dataset from scratch. And so the random library along with the following code:

```
gender = ["Male","Female","Others"]
random.seed(10)
blood_sugar_levels = [random.randint(70, 130) for i in range(20)]
gender = [random.choice(gender) for i in range(20)]
```

Now let us create a Pandas data frame:

```
data = {"Gender": gender, "BloodSugarLevels": blood_sugar_levels}
df = pd.DataFrame(data)
```

Now, import the Pandas data frame into a Great Expectations data frame:

```
df_ge = ge.from_pandas(df)
```

Great. Now, let us create the expectation. We expect that the dataset may contain only men and women. Here is the how that looks:

```
gender_exp = df_ge.expect_column_values_to_be_in_set(
    "Gender",
    ["Male", "Female"]
    )
print(gender_exp)
```

In the preceding code, we are mentioning both the header and the category that we expect to see. Here is the output:

```
{
  "success": false,
  "expectation_config": {
    "expectation_type": "expect_column_values_to_be_in_set",
```

```
    "kwargs": {
      "column": "Gender",
      "value_set": [
        "Male",
        "Female"
      ],
      "result_format": "BASIC"
    },
    "meta": {}
  },
  "result": {
    "element_count": 20,
    "missing_count": 0,
    "missing_percent": 0.0,
    "unexpected_count": 5,
    "unexpected_percent": 25.0,
    "unexpected_percent_total": 25.0,
    "unexpected_percent_nonmissing": 25.0,
    "partial_unexpected_list": [
      "Others",
      "Others",
      "Others",
      "Others",
      "Others"
    ]
  },
  "meta": {},
  "exception_info": {
    "raised_exception": false,
    "exception_traceback": null,
    "exception_message": null
  }
}
```

CHAPTER 6 DATA VALIDATION USING GREAT EXPECTATIONS

You can also explore the dataset's various categories by issuing the following code:

```
print(df_ge.Gender.value_counts())
```

In this case, we get

```
Gender
Female    10
Male       5
Others     5
Name: count, dtype: int64
```

Now, if we reprogram our expectation to include the "Others" category, we would get the following.

Here is the code:

```
gender_exp = df_ge.expect_column_values_to_be_in_set(
    "Gender",
    ["Male", "Female","Others"]
    )

print(gender_exp)
```

Here is the output:

```
{
  "success": true,
  "expectation_config": {
    "expectation_type": "expect_column_values_to_be_in_set",
    "kwargs": {
      "column": "Gender",
      "value_set": [
        "Male",
        "Female",
        "Others"
      ],
      "result_format": "BASIC"
    },
    "meta": {}
  },
```

```
  "result": {
    "element_count": 20,
    "missing_count": 0,
    "missing_percent": 0.0,
    "unexpected_count": 0,
    "unexpected_percent": 0.0,
    "unexpected_percent_total": 0.0,
    "unexpected_percent_nonmissing": 0.0,
    "partial_unexpected_list": []
  },
  "meta": {},
  "exception_info": {
    "raised_exception": false,
    "exception_traceback": null,
    "exception_message": null
  }
}
```

Data Validation Workflow in Great Expectations

So far, we have seen getting started with a command line interface, following various options and settings and a very simple "hello-world" like a great expectation project. The purpose of the previous section is to get you acquainted with the Great Expectations functionality and environment as Great Expectations is different from creating Pydantic model classes or Pandera schema validations.

In this section, we will look at a typical data validation workflow that is required to be followed for developing a data pipeline. Here are the following steps that Great Expectations has laid out:

1. Install Great Expectations on the workstation.

2. Create a data context, where you initialize the data context component to store metadata and other settings management.

3. Connect to data, where you list all your data sources from Pandas data frames to cloud-based databases, and perform necessary steps for each data source.

CHAPTER 6   DATA VALIDATION USING GREAT EXPECTATIONS

4. Create a validator, so you can validate an expectation suite against your data.

5. Create expectations, where you write assertions for your data.

6. Run a checkpoint, using a validator to run expectations against a batch of data.

7. View validation results, results that are generated after data is validated against an expectation or an expectation suite.

We have already seen installing Great Expectations, prior to using the CLI, in the "Setup and Installation" section. Upon successful installation of Great Expectations, it is time to set up the data context. The data context contains configurations for expectations, expectation stores, checkpoints, and data docs. Initializing a data context is quite straightforward, as importing the library and running the get_context() method:

```
import great_expectations as ge
context = ge.get_context()

print(context)
```

You will get something like this:

```
{
  "anonymous_usage_statistics": {
    "explicit_url": false,
    "explicit_id": true,
    "data_context_id": "14552788-844b-4270-86d5-9ab238c18e75",
    "usage_statistics_url": "https://stats.greatexpectations.io/great_expectations/v1/usage_statistics",
    "enabled": true
  },
  "checkpoint_store_name": "checkpoint_store",
  "config_version": 3,
  "data_docs_sites": {
    "local_site": {
      "class_name": "SiteBuilder",
      "show_how_to_buttons": true,
```

```
      "store_backend": {
        "class_name": "TupleFilesystemStoreBackend",
        "base_directory": "/tmp/tmp_oj70d5v"
      },
      "site_index_builder": {
        "class_name": "DefaultSiteIndexBuilder"
      }
    }
  },
  "datasources": {},
  "evaluation_parameter_store_name": "evaluation_parameter_store",
  "expectations_store_name": "expectations_store",
  "fluent_datasources": {},
  "include_rendered_content": {
    "expectation_validation_result": false,
    "expectation_suite": false,
    "globally": false
  },
  "profiler_store_name": "profiler_store",
  "stores": {
    "expectations_store": {
      "class_name": "ExpectationsStore",
      "store_backend": {
        "class_name": "InMemoryStoreBackend"
      }
    },
    "validations_store": {
      "class_name": "ValidationsStore",
      "store_backend": {
        "class_name": "InMemoryStoreBackend"
      }
    },
    "evaluation_parameter_store": {
      "class_name": "EvaluationParameterStore"
    },
```

CHAPTER 6   DATA VALIDATION USING GREAT EXPECTATIONS

```
    "checkpoint_store": {
      "class_name": "CheckpointStore",
      "store_backend": {
        "class_name": "InMemoryStoreBackend"
      }
    },
    "profiler_store": {
      "class_name": "ProfilerStore",
      "store_backend": {
        "class_name": "InMemoryStoreBackend"
      }
    }
  },
  "validations_store_name": "validations_store"
}
```

In the recent versions of Great Expectations, all expectations get stored as expectation suites. An expectation suite is a collection of one or more expectations or assertions about your data, stored in JSON format. And so, we will have to configure how to connect to the expectation suite. This is where expectation stores come in.

The next step in the process is to connect to data assets. Data assets are simply data in their original form. A data source is where these data assets are stored. Since we are operating Great Expectations locally on our file system, let us specify the flat file locally stored in the file system, to be read into a data asset:

```
import great_expectations as ge

context = ge.get_context()

validator = context.sources.pandas_default.read_csv(
  "/home/parallels/Downloads/mockaroo_health_data.csv"
)
```

In the preceding code, we are writing a data asset directly into a validator. The next step is to create a data source.

First, create a data source:

```
datasource_name = "test2"
ge.datasource = context.sources.add_pandas(datasource_name)
```

CHAPTER 6 DATA VALIDATION USING GREAT EXPECTATIONS

Then, create a data asset:

```
asset_name = "asset1"
path_to_data = "/home/parallels/Downloads/mockaroo_health_data.csv"

asset = ge.datasource.add_csv_asset(asset_name, filepath_or_buffer=path_to_data)
```

Now, let's build a batch request:

```
batch_request = asset.build_batch_request()
```

Now, let us create the expectation suite:

```
context.add_or_update_expectation_suite("my_expectation_suite")
```

Now let us create a validator.

The validator needs to know the batch that contains the data, used to validate the expectations and the expectation suite, which contains the combined list of expectations you create:

```
validator = context.get_validator(
    batch_request=batch_request,
    expectation_suite_name="my_expectation_suite",
)

print(validator.head())
```

We will get

```
Calculating Metrics:    0%|              | 0/1 [00:00<?, ?it/s]
   id             name  age state  gender exercise_daily? copay_paid
0   1  Allayne Moffett  NaN    SC    Male             NaN    $206.76
1   2 Kerby Benjafield  NaN    NM    Male            True     $21.58
2   3  Raina Vallentin  NaN    MI  Female            True    $125.18
3   4   Kaela Trodden   NaN    OH  Female           False     $86.20
4   5  Faber Kloisner   NaN    MN    Male           False    $219.38
```

CHAPTER 6 DATA VALIDATION USING GREAT EXPECTATIONS

Note The dataset used in the preceding example is randomly generated with missing values among other things.

Now, let us run the validator to create and run an expectation. Let us look at the missing values for one of the columns, say "age." The expectation we use for this operation is

expect_column_values_to_not_be_null

Here is how the code looks:

```
expectation_notnull = validator.expect_column_values_to_not_be_null(
    column="age"
)
print(expectation_notnull)
```

This would yield

```
Calculating Metrics:   0%|            | 0/6 [00:00<?, ?it/s]
{
  "success": false,
  "expectation_config": {
    "expectation_type": "expect_column_values_to_not_be_null",
    "kwargs": {
      "column": "age",
      "batch_id": "test2-asset1"
    },
    "meta": {}
  },
  "result": {
    "element_count": 1000,
    "unexpected_count": 1000,
    "unexpected_percent": 100.0,
    "partial_unexpected_list": [
      null,
      null,
      null,
```

```
        null,
        null,
        null,
        null,
        null,
        null,
        null,
        null,
        null,
        null,
        null,
        null,
        null,
        null,
        null,
        null,
        null
      ]
  },
  "meta": {},
  "exception_info": {
    "raised_exception": false,
    "exception_traceback": null,
    "exception_message": null
  }
}
```

By running the validator, the expectation has been recorded in the expectation suite. Now let us run the expectation related to missing values for another column called "copay_paid."

Here is how the code looks:

```
expectation_notnull = validator.expect_column_values_to_not_be_null(
    column="copay_paid"
)
print(expectation_notnull)
```

CHAPTER 6   DATA VALIDATION USING GREAT EXPECTATIONS

This would return

```
Calculating Metrics:    0%|             | 0/6 [00:00<?, ?it/s]
{
  "success": false,
  "expectation_config": {
    "expectation_type": "expect_column_values_to_not_be_null",
    "kwargs": {
      "column": "copay_paid",
      "batch_id": "test2-asset1"
    },
    "meta": {}
  },
  "result": {
    "element_count": 1000,
    "unexpected_count": 126,
    "unexpected_percent": 12.6,
    "partial_unexpected_list": [
      null,
      null,
      null,
      null,
      null,
      null,
      null,
      null,
      null,
      null,
      null,
      null,
      null,
      null,
      null,
      null,
      null,
      null,
```

```
      null,
      null,
      null
    ]
  },
  "meta": {},
  "exception_info": {
    "raised_exception": false,
    "exception_traceback": null,
    "exception_message": null
  }
}
```

As you can see, there are about 126 records where the copay_paid value is missing.

So far, we have run two expectations. While the expectations are saved in the expectation suite on the validator object, validators do not persist outside the current Python session. To reuse the current expectations, it is necessary to save them. Here is how that looks:

```
validator.save_expectation_suite("/home/parallels/Documents/my_expectation_suite.json", discard_failed_expectations=False)
```

Upon running the program, Great Expectations would create an expectation suite in the form of a JSON document.

For our current program, here is how that JSON document looks like:

```
{
  "data_asset_type": null,
  "expectation_suite_name": "my_expectation_suite",
  "expectations": [
    {
      "expectation_type": "expect_column_values_to_not_be_null",
      "kwargs": {
        "column": "age"
      },
      "meta": {}
    },
```

```
    {
      "expectation_type": "expect_column_values_to_not_be_null",
      "kwargs": {
        "column": "copay_paid"
      },
      "meta": {}
    }
  ],
  "ge_cloud_id": null,
  "meta": {
    "great_expectations_version": "0.18.4"
  }
}
```

Creating a Checkpoint

A checkpoint is a means of validating data in production pipelines using Great Expectations. Checkpoints allow you to link an expectation suite and data; the idea is to validate expectations against the data.

Here is how to create a checkpoint:

```
checkpoint = context.add_or_update_checkpoint(
    name="my_checkpoint",
    validations=[
        {
            "batch_request": batch_request,
            "expectation_suite_name": "my_expectation_suite",
        },
    ],
)
```

You can return the checkpoint results by using the following code:

```
checkpoint_result = checkpoint.run()
print(checkpoint_result)
```

CHAPTER 6   DATA VALIDATION USING GREAT EXPECTATIONS

This produces the following output

```
Calculating Metrics: 0it [00:00, ?it/s]
{
  "run_id": {
    "run_name": null,
    "run_time": "2023-12-15T13:40:18.348374+05:30"
  },
  "run_results": {
    "ValidationResultIdentifier::my_expectation_suite/__
    none__/20231215T081018.348374Z/test2-asset1": {
      "validation_result": {
        "success": true,
        "results": [],
        "evaluation_parameters": {},
        "statistics": {
          "evaluated_expectations": 0,
          "successful_expectations": 0,
          "unsuccessful_expectations": 0,
          "success_percent": null
        },
        "meta": {
          "great_expectations_version": "0.18.4",
          "expectation_suite_name": "my_expectation_suite",
          "run_id": {
            "run_name": null,
            "run_time": "2023-12-15T13:40:18.348374+05:30"
          },
          "batch_spec": {
            "reader_method": "read_csv",
            "reader_options": {
              "filepath_or_buffer": "/home/parallels/Downloads/mockaroo_
              health_data.csv"
            }
          },
```

217

```
          "batch_markers": {
            "ge_load_time": "20231215T081018.350419Z",
            "pandas_data_fingerprint": "9c19ba3fc9d218a2c39f81794f4f37fe"
          },
          "active_batch_definition": {
            "datasource_name": "test2",
            "data_connector_name": "fluent",
            "data_asset_name": "asset1",
            "batch_identifiers": {}
          },
          "validation_time": "20231215T081018.363378Z",
          "checkpoint_name": "my_checkpoint",
          "validation_id": null,
          "checkpoint_id": null
        }
      },
      "actions_results": {
        "store_validation_result": {
          "class": "StoreValidationResultAction"
        },
        "store_evaluation_params": {
          "class": "StoreEvaluationParametersAction"
        },
        "update_data_docs": {
          "local_site": "file:///tmp/tmpy1si_l4d/validations/my_
          expectation_suite/__none__/20231215T081018.348374Z/test2-
          asset1.html",
          "class": "UpdateDataDocsAction"
        }
      }
    }
  },
  "checkpoint_config": {
    "config_version": 1.0,
    "notify_on": null,
```

```
  "default_validation_id": null,
  "evaluation_parameters": {},
  "expectation_suite_ge_cloud_id": null,
  "runtime_configuration": {},
  "module_name": "great_expectations.checkpoint",
  "template_name": null,
  "slack_webhook": null,
  "class_name": "Checkpoint",
  "validations": [
    {
      "name": null,
      "expectation_suite_name": "my_expectation_suite",
      "expectation_suite_ge_cloud_id": null,
      "id": null,
      "batch_request": {
        "datasource_name": "test2",
        "data_asset_name": "asset1",
        "options": {},
        "batch_slice": null
      }
    }
  ],
  "action_list": [
    {
      "name": "store_validation_result",
      "action": {
        "class_name": "StoreValidationResultAction"
      }
    },
    {
      "name": "store_evaluation_params",
      "action": {
        "class_name": "StoreEvaluationParametersAction"
      }
    },
```

```
      {
        "name": "update_data_docs",
        "action": {
          "class_name": "UpdateDataDocsAction"
        }
      }
    ],
    "site_names": null,
    "notify_with": null,
    "batch_request": {},
    "profilers": [],
    "run_name_template": null,
    "name": "my_checkpoint",
    "expectation_suite_name": null,
    "ge_cloud_id": null
  },
  "success": true
}
```

You can add more validations to the current checkpoint as well:

```
checkpoint = context.add_or_update_checkpoint(
    name="my_test_checkpoint", validations=more_validations
)
```

Data Documentation

Data documentation, or simply data doc, provides interactive documentation of the data validation results that Great Expectations processed.

Here is how we can generate the data documentation in Great Expectations:

```
context.build_data_docs()
```

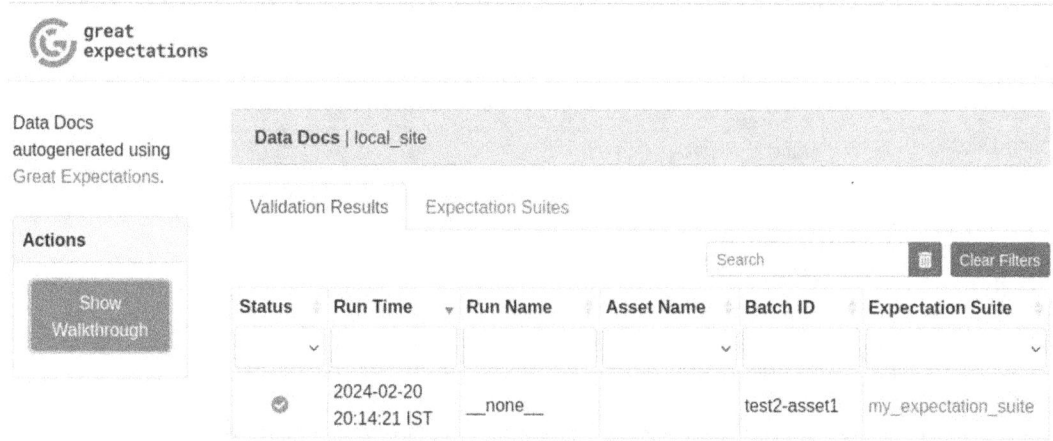

Figure 6-5. *Data documentation in Great Expectations*

Great Expectations will generate a full web-based documentation along with other metadata. You can also edit the documentation. Here is how the documentation may look for you.

When you click the expectation suite, you will be able to see more metadata. Here is how that may look for you:

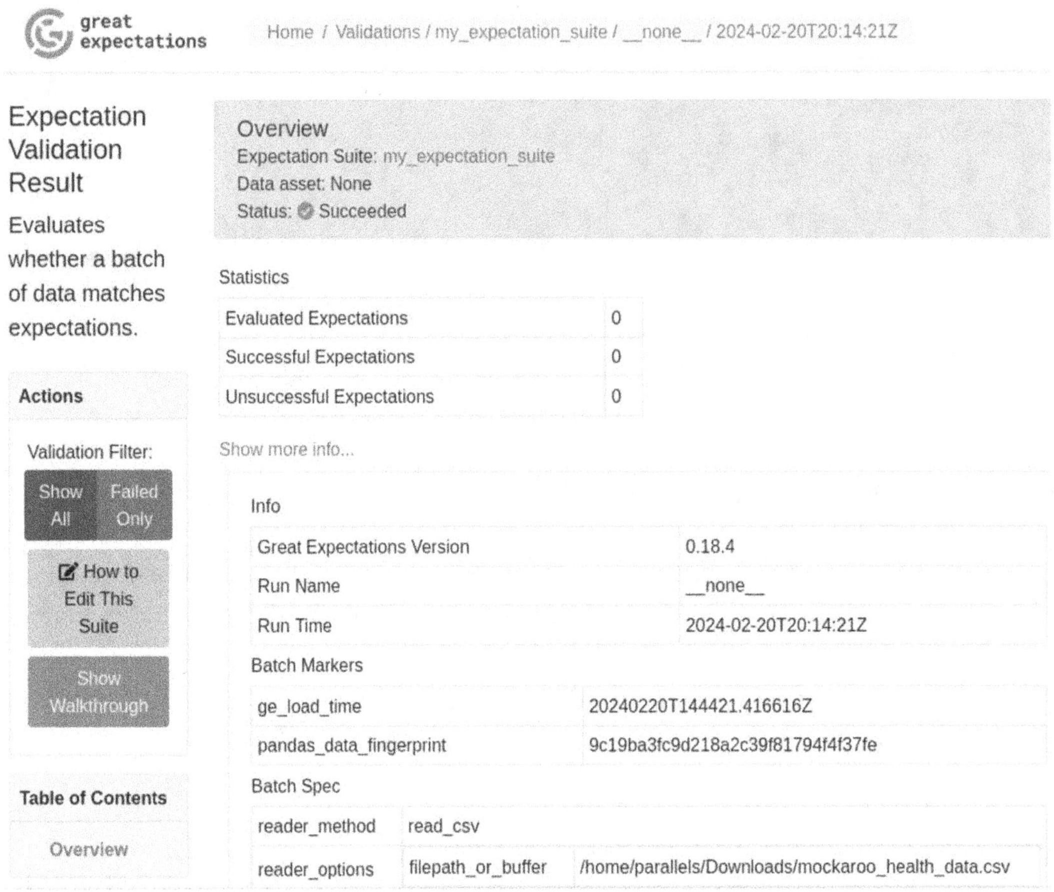

CHAPTER 6   DATA VALIDATION USING GREAT EXPECTATIONS

Figure 6-6. Expectation validation result in Great Expectations

Expectation Store

So far we have seen connecting to data assets, building batch requests, creating validators, running expectations on validators, and saving the expectations on a JSON document. Now, let us see how to manage the expectations we have built in an expectation store. An expectation store is a connector to a store and that helps retrieve collections of expectations about data.

CHAPTER 6 DATA VALIDATION USING GREAT EXPECTATIONS

The process is relatively simple, where you create a folder within your Great Expectations folder, move all your JSON documents into that folder, and update your configuration file to include a store for expectation results:

```
mkdir shared_expectations
mv path-to/my_expectation_suite.json destination-path/shared_expectations/
```

And update the "base directory" of the configuration file. Here is a screenshot of how that looks:

```
stores:
# Stores are configurable places to store things like Expectations, Validations
# Data Docs, and more. These are for advanced users only - most users can simply
# leave this section alone.
#
# Three stores are required: expectations, validations, and
# evaluation_parameters, and must exist with a valid store entry. Additional
# stores can be configured for uses such as data_docs, etc.
  expectations_store:
    class_name: ExpectationsStore
    store_backend:
      class_name: TupleFilesystemStoreBackend
      base_directory: expectations/
```

Figure 6-7. Expectation store in Great Expectations

Conclusion

We have continued to discuss more on data validations. Great Expectations is one of the innovative data validation libraries, which comes with predefined expectations. The number of predefined expectations continues to grow, which is also a good resource for exploring various ways of performing data validation on your data. You can also create a template with predefined expectations called an expectation store. The Great Expectations community appears to be evolving rapidly in my opinion as during the process of authoring the book, I have passed two different versions. By the time you are reading this book, you will be witnessing a major 1.0 release.

CHAPTER 7

Introduction to Concurrency Programming and Dask

Introduction

We have discussed earlier that processing large datasets effectively and efficiently is important. As datasets continue to grow over time, traditional single-thread processing can be seen as a limitation. Python, by default, has a global interpreter lock (GIL) that allows only one thread to hold the interpreter at a given point in time. While this design ensures the integrity of computations submitted, it would be much more effective to use many threads within a multi-core CPU that is commonly available today. In this chapter, you will learn concepts of parallel programming, concurrent programming, and Dask, a Python library that supports distributed processing and works around the global interpreter lock limitation by using multiple processes. Dask also supports various data processing libraries that we have seen in earlier chapters.

By the end of this chapter, you will learn

- Parallel processing concepts and how they differ from sequential processing
- Fundamentals of concurrent processing
- Dask
- Dask architecture and concepts

CHAPTER 7   INTRODUCTION TO CONCURRENCY PROGRAMMING AND DASK

- Core data structures of Dask and how they enable distributed computing
- Optimizing Dask computations

Introduction to Parallel and Concurrent Processing

Parallel processing, in general, is a process of partitioning a large task into several smaller tasks and executing these smaller tasks side by side so that the time taken to complete the task has now been reduced significantly while at the same time all the resources were utilized optimally. In computing, it is the same case. It is a process of breaking down a large programming task into several smaller tasks, executing the same simultaneously, and bringing together all the results of smaller tasks to obtain the result. It may not look that evident from a programming point of view.

History

Parallel processing gained momentum during the time when a CPU started to contain more than one processing unit within its microprocessor architecture. In general, the electrons carrying the digital signals in the microprocessor travel very fast; there are upper limits on how quickly a single processor can operate. Such limitations along with the need for minimizing heat generation, among other things, paved the way for multi-core processors. Today, we have several processing units (we call them cores) in the microprocessor embedded in our laptops, phones, computers, etc.

Python and the Global Interpreter Lock

The Python programming language that we use is implemented in the C programming language. There are other implementations based on Java and .NET for specialized purposes; however, Python for data processing, engineering, and machine learning is based out of C programming language. One of several mechanisms that are available in core Python interpreters is called the global interpreter lock, or GIL. The purpose of GIL is to simplify memory management. So when we have more than one Python byte code (for instance), each one of them would proceed by acquiring the GIL and execute the

Python byte code instruction. Once that process is completed, the GIL gets released so that the next available Python byte code can acquire it for its execution. While it ensures thread-safe computing, it also disables parallel processing through the method of multi-threading. However, there are other ways to achieve parallel processing, and some file input/output operations do not fall under the GIL, which is good news.

Concepts of Parallel Processing

When we talk about a process, at any given point in time, there are more than one process that are running simultaneously in a device. A process is basically a program (code) that is currently running. Any given process can have more than one thread.

A thread is like a subprocess, so to speak. A process may contain one or more threads. More than one thread can execute at the same time (or not in some cases). From a CPU point of view, a thread is not something that is physically present on a microprocessor; instead, it is a logical execution unit.

A core is a physical execution unit that can execute computational operations. A core can execute a thread. A microprocessor can have one or more cores.

In short

A process is an operating system–level abstraction.

A thread is a software-level abstraction.

A core is a physical-level abstraction.

There is a Python library called "multiprocessing" that bypasses the global interpreter lock in Python. Instead of leveraging the idea of multiple threads, the Python library uses the idea of subprocesses running on more than one core at the same time; and so, the way it works is that the "multiprocessing" library provides an inter-process communication system that provides mechanism for effective collaboration of parallel processing by using multiple cores. The number of cores that can be used by "multiprocessing" libraries is limited by how many of them are in your system.

Identifying CPU Cores

You can write simple Python code to determine how many cores are available on your computer.

Here is how that looks:

```python
from multiprocessing import cpu_count

numberOfCores = cpu_count()

print(f"There are {numberOfCores} cores in your CPU")
```

would return a number that denotes the number of cores that are on the microprocessor of your computer.

Here is a simple example of parallel processing using the "multiprocessing" library:

```python
from multiprocessing import Pool

def cube(a):
    return a * a * a

inputs = [5,12,17,31,43,59,67]

with Pool(processes=4) as parallelprocess:
    cubevalue = parallelprocess.map(cube, inputs)

print(cubevalue)
```

The preceding code will display the cube values for the input supplied. Here is how the output may look for you:

```
[125, 1728, 4913, 29791, 79507, 205379, 300763]
```

In the code, you can see the Pool method and its parameter called processes.

Concurrent Processing

Concurrent processing is similar to the idea of parallel processing as far as dealing with simultaneous tasks are concerned. However, the method of execution is different from what we have earlier. Concurrent processing refers to two or more tasks getting executed at the same point in time. However, they may not execute at the exact same time.

For instance, within a single core, one task may start and perform operations for a bit of time. Let's say that the first task may need to wait for some other process to perform another operation. And so it waits; during this time one or more tasks might start and may even finish, depending upon their compute, memory, and storage requirements and what is available to them.

CHAPTER 7   INTRODUCTION TO CONCURRENCY PROGRAMMING AND DASK

Rather than waiting for full completion of a task, concurrency enables dynamic sharing and allocation of memory, compute, and storage during runtime. It is a way of using resources seamlessly.

Note   Imagine you are a chef at a restaurant. There are five orders of food that arrived at the exact same time. You know that not all items in the menu take equal amounts of time. You have one stove and one microwave for five orders. For the sake of simplicity, assume you have one deep fryer, grill, or something like that. Now let us illustrate concurrency and parallelism.

Concurrency

Imagine starting off a slow-cooking dish. While that is cooking on simmer, prep vegetables for another dish. As the vegetables are chopped and prepped, periodically check the slow-cooking dish, and stir maybe a bit. Transfer the slow-cooking dish to the oven while placing the prepped vegetables on the stove. As the vegetables start cooking, prepare eggs for another dish. Check the vegetables and the oven periodically. You are switching between tasks, making progress on three orders even though you are working alone. This is concurrency.

Parallelism

Imagine five chefs for five orders. Each chef takes care of one order with their own complete set of equipment (stove, oven, etc.) and tools (knife, spoon, etc.). They work on their own dish, with their own equipment and focusing on their own task. This is parallelism.

Concurrency is when one CPU core deals with multiple tasks, using shared resources, and rapidly switches between tasks. **Parallelism** is when "n" CPU cores deal with "n" tasks using dedicated resources to each core.

Introduction to Dask

Dask is a specialized Python library for enabling parallel and distributed computing for data analysis and machine learning tasks in Python. Dask is a Python native library and does not depend on non-Python programming languages for its implementation.

Dask can work well with major Python libraries like Pandas, Polars, CuDF, scikit-learn, and XGBoost, to name a few. If you are working with data processing and machine learning pipelines with data that exceeds the compute and memory requirements of a single machine, then you may benefit from adopting Dask in your technology stack.

Dask has a relatively quicker learning curve, and it is relatively easy to integrate Dask into your production data pipelines. Dask is used by big banks and major healthcare and national laboratories and organizations.

Dask supports many of the common data formats that are prevalent in big data pipelines such as Avro, HDF5, Parquet, etc. while also supporting traditional data formats like CSV, JSON, etc.

Dask primarily leverages CPUs for performing its computations and operations. However, Dask can also work with GPUs. Dask and the CuDF Python library interoperate well with each other. Let us dive in detail on this topic in the coming sections.

Setup and Installation

It is relatively simple to set up Dask in the development environment. You can pip or conda install the Dask package from your command line:

```
sudo apt install graphviz
```

and

```
python3 -m pip install "dask[complete]"
```

or

```
pip3 install dask dask distributed bokeh --upgrade
```

or

```
conda install dask
```

One can also install Dask in a distributed computing cluster. Dask works well with Hadoop clusters; you can deploy Dask with YARN, a resource manager for Hadoop that deploys various services within a given Hadoop cluster. You can also create a cloud computing Dask cluster on AWS or GCP.

Here is a simple hello world–like implementation using Dask:

```
import dask
import dask.array as da

arr = da.from_array(
   [1,2,3,4,5,6,7,8,9,10]
   , chunks=2
)

squared = arr ** 2

result = squared.compute()

print(result)
```

In the preceding example, we started off by importing the necessary class and method. We then created a Dask array with chunks of 2, meaning create two different chunks of this array. We then created a new function called squared that would take the square of each value in the array. This is followed by execution of that function by using a method called "compute()," which triggers the actual computation and returns the result. This is due to the lazy evaluation nature of Dask.

When we print the result, it will return

```
[  1   4   9  16  25  36  49  64  81 100]
```

Features of Dask

Tasks and Graphs

In Dask, computation is represented as a collection of tasks as nodes that are connected using edges. These nodes and edges are organized in a directed acyclic graph (DAG) manner. The tasks that are needed to be accomplished are represented as nodes. The data dependencies between various tasks (or no data dependency) are represented as edges. The directed acyclic graph representation enables Dask to effectively execute the tasks and their underlying dependencies in multiple threads and processes or even across multiple CPUs.

Lazy Evaluation

We have seen the concept of lazy evaluation in previous chapters. To refresh your memory, lazy evaluation is where the computation does not happen immediately. The execution of a given set of code happens only if it is actually needed. Until then, this code may be delayed and sometimes it may not be executed at all. As we already know, Dask creates a directed acyclic graph to represent its execution flow. But it never actually executes every node, unless it is needed. This way, Dask optimizes the computation.

Partitioning and Chunking

Partitioning means splitting a very large dataset into multiple smaller or manageable portions of datasets. These smaller datasets can be processed independently. By leveraging the idea of partitioning, Dask can handle datasets that are larger than its available memory. Chunking is similar to the concept of partitioning except it is used in smaller scope. The term "partitioning" is used heavily in the context of databases and data tables, where a given table is split into multiple smaller tables. These tables, upon processing, may be regrouped or joined back into its original structure (similar to the concept of MapReduce programming).

On the other hand, "chunking" is greatly associated with splitting arrays into smaller chunks. You can specify the size of the chunks during the creation of Dask arrays. In the Dask context, it may be easier to conclude that the term "partitioning" can be applied in the case of databases, tables and data frames, while the term "chunking" can be applied in the case of one-dimension or n-dimension arrays.

Serialization and Pickling

Serialization means converting a given dataset into a format that can be stored or transmitted in an efficient manner. When required, this format can be reconstructed back to the original data format. Dask uses serialization to convert the dataset so it can efficiently transport the partitions or chunks of serialized data into various worker nodes for processing. Dask determines which serialization techniques to implement when dealing with large data frames, n-dimensional arrays, etc.

Pickling is one of the serialization methods available exclusively in Python programming. Pickling is a way to serialize Python data objects into byte streams that can be stored or transmitted to various worker nodes. This can be then reconstructed back into datasets in their original format. Dask is a general-purpose computing framework that consists of several serialization techniques. However, it is only fair to acknowledge the essential role pickling serialization plays in the world of machine learning modeling within the Python ecosystem.

Dask-CuDF

In a previous chapter, we have discussed GPU computing and CuDF in detail. As we know already, CuDF is a GPU computation framework, which enables data frame processing across one or more GPUs. CuDF provides various methods and functions that can import, export, and perform complex data transformations, which can run strictly on GPUs.

Dask, on the other hand, is a general-purpose parallel and distributed computing library. We can use Dask on top of Pandas and other leading data analysis libraries to scale out our workflow, across multiple CPUs.

Dask-CuDF is a specialized Python package that enables processing data frame partitions using CuDF instead of Pandas. For instance, if you call Dask-CuDF to process a data frame, it can efficiently call CuDF's method to process the data (by sending the same to GPU). If you have very large data for your data processing pipeline, larger than the limits of your machine, then you may benefit from using Dask-CuDF.

Dask Architecture

Let us review the various components in Dask. The architecture of Dask is flexible in that it allows single-instance computations, single-instance with multi-core computations, and distributed computations.

CHAPTER 7 INTRODUCTION TO CONCURRENCY PROGRAMMING AND DASK

Core Library

At the heart of the Dask system, we have the Dask collections that consist of various data structures. Some of them are Dask arrays, which resemble NumPy arrays; Dask data frames, which resemble Pandas' data frames; and Dask bags, which resemble Python lists and iterators, respectively. These collections enable you to perform parallel and distributed computations natively. Just by calling Dask's collections, you will be able to parallelize or asynchronously execute the computation.

Schedulers

The Dask scheduler is a major component of the Dask execution mechanism. The Dask scheduler sits in the heart of the cluster, managing the computation, execution of tasks among workers, and communication to workers and clients. The scheduler component itself is asynchronous, as it can track progress of multiple workers and respond to various requests from more than one client. The scheduler tracks work as a constantly changing directed acyclic graph of tasks. In these directed acyclic graphs, each node can be a task and each arc (edge between nodes) can be a dependency between tasks.

A task is a unit of computation in Dask that is basically a Python function or an execution on a Python object. There are different types of schedulers available in Dask. There is a single-thread scheduler that executes tasks on a single thread on a single machine; there is also a multi-thread scheduler that executes tasks using multiple threads of a single machine; there is also a multi-core scheduler that executes tasks using multiple cores of a processor; and there is a distributed scheduler where tasks are executed using multiple worker nodes in a given cluster.

Client

Dask's client is the primary mode of communication with Dask. You communicate with Dask using a client. The client is the default scheduler and worker; the client manages and coordinates communication between schedulers. From the computation point of view, the client manages communication with the scheduler process, and the scheduler coordinates and assigns work to worker processes; these worker processes would perform the necessary computation and return the result back to the client. Dask's client communicates with the scheduler and collects results from the workers as well. Here is how you may instantiate the Dask client:

CHAPTER 7   INTRODUCTION TO CONCURRENCY PROGRAMMING AND DASK

```
from dask.distributed import (
            Client,
            progress
        )
client = Client(
        threads_per_worker=4
        , n_workers=2
    )
client
```

Here is the response you may get:

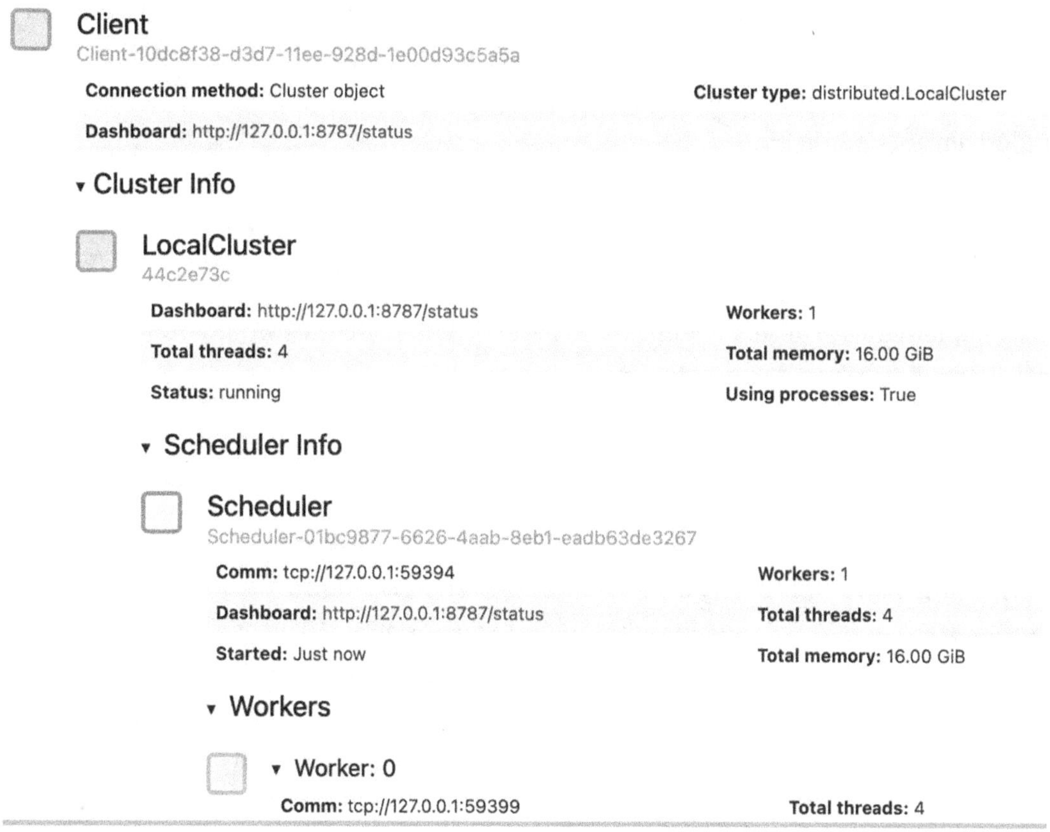

Client
Client-10dc8f38-d3d7-11ee-928d-1e00d93c5a5a

Connection method: Cluster object   **Cluster type:** distributed.LocalCluster
Dashboard: http://127.0.0.1:8787/status

▼ **Cluster Info**

LocalCluster
44c2e73c

Dashboard: http://127.0.0.1:8787/status   **Workers:** 1
Total threads: 4   **Total memory:** 16.00 GiB
Status: running   **Using processes:** True

▼ **Scheduler Info**

Scheduler
Scheduler-01bc9877-6626-4aab-8eb1-eadb63de3267

Comm: tcp://127.0.0.1:59394   **Workers:** 1
Dashboard: http://127.0.0.1:8787/status   **Total threads:** 4
Started: Just now   **Total memory:** 16.00 GiB

▼ **Workers**

▼ **Worker: 0**
Comm: tcp://127.0.0.1:59399   **Total threads:** 4

Figure 7-1. Initialization of the Dask client

Workers

Workers are end computation machines that perform the computation that is assigned to them by the scheduler. A worker can be a single thread or a separate process or a separate machine in a cluster; each worker has access to compute, memory, and storage. Workers communicate with the schedulers to receive tasks from them, execute the tasks, and report the results.

Task Graphs

As discussed in the earlier section, the computation in a task is represented as a constantly changing directed acyclic graph of tasks where the edges may be dependency between tasks. Task graphs define the dependencies between various tasks. When you write Dask code, you are implicitly creating task graphs by performing operations on a Dask array or Dask DataFrame or Dask bag. The Dask system automatically designs the underlying task graph.

Dask Data Structures and Concepts

Dask implements a subset of NumPy and Pandas data structures while supporting parallel and distributed computations on the same.

Dask Arrays

A Dask array is an implementation of NumPy's ndarray. Dask arrays are made up of many NumPy arrays, and they are split into smaller and uniform sizes of arrays called chunks. Dask uses a blocked algorithm to perform this action. You can either specify how many chunks you would like the Dask to split the array or ask Dask to automatically split the array for computations and calculations. These chunks are then processed independently, thereby allowing a distributed and parallel execution of data processing across cores or workers within a cluster. A Dask array supports parallelism and multi-core processing out of the box so that you do not have to write additional code or add a decorator to enable concurrency.

Let us create a simple Dask array:

```
from dask import array

var1 = array.random.random(
        (10,10),
        chunks=(2,2)
    )

print(var1)
```

would return

```
dask.array<random_sample, shape=(10, 10), dtype=float64, chunksize=(2, 2), chunktype=numpy.ndarray>
```

Here, we have a random array of (10 × 10) dimensions that is broken into smaller chunks of (2 × 2) dimensions each.

When we add "T.compute()" to our variable, we get a transposed random matrix. Here is how that looks:

```
print(var1.T.compute())
```

```
>>> var1.T.compute()
array([[0.67307784, 0.33297361, 0.43449907, 0.87693594, 0.11630913,
        0.32511344, 0.25753008, 0.63318751, 0.49384467, 0.06931872],
       [0.46455715, 0.10148165, 0.1096482 , 0.36356486, 0.01606927,
        0.49648592, 0.35737028, 0.88264313, 0.38212025, 0.30888879],
       [0.95134913, 0.84782068, 0.72521106, 0.08989213, 0.89135709,
        0.34014486, 0.27589355, 0.82262447, 0.56631172, 0.55527946],
       [0.53861738, 0.79816048, 0.77497625, 0.63111302, 0.96465897,
        0.75253021, 0.12548712, 0.29267116, 0.97756565, 0.91079222],
       [0.23970689, 0.30043349, 0.10452291, 0.85425859, 0.26423908,
        0.05007975, 0.66632572, 0.66853232, 0.14236124, 0.70064729],
       [0.50350135, 0.67990453, 0.13675301, 0.36314849, 0.77022817,
        0.80155082, 0.43052538, 0.96108671, 0.94222871, 0.43986962],
       [0.00517627, 0.50852426, 0.63408599, 0.84626007, 0.9335204 ,
        0.52350993, 0.68395926, 0.32672695, 0.2340683 , 0.84824245],
       [0.50096949, 0.1123642 , 0.54798468, 0.96876626, 0.53769494,
        0.59995999, 0.2153332 , 0.79808185, 0.14999928, 0.58387984],
       [0.50779711, 0.21795679, 0.19489835, 0.22923678, 0.43063149,
        0.62818265, 0.48885064, 0.34103434, 0.08419418, 0.22683588],
       [0.93634719, 0.98852583, 0.94796399, 0.82796294, 0.36243628,
        0.92536089, 0.41533759, 0.87309056, 0.87972462, 0.82257482]])
```

Figure 7-2. Retrieving the results of a Dask array

CHAPTER 7   INTRODUCTION TO CONCURRENCY PROGRAMMING AND DASK

The following is something that is not to be tried at home, but instead in a cloud account. Assuming you have adequate memory for your dataset, you can try to persist the data in memory.

Here is how we can do that:

```
var2 = var1.persist()
```

You can also re-chunk your Dask array. Here is how we can do that:

```
var1 = var1.rechunk((5,5))

print(var1)
```

would return

```
dask.array<rechunk-merge, shape=(10, 10), dtype=float64, chunksize=(5, 5), chunktype=numpy.ndarray>
```

We can export this Dask array in various formats. Let us look at HDF5 format, which is meant to store open source large complex data. To do so, we need to install the Python library "h5py" as a starting point:

```
pip install h5py
```

Let us create a slightly bigger random array:

```
import dask.array as da
import h5py

var3 = da.random.random(
    (10000,10000),
    chunks=(100,100)
)

da.to_hdf5(
    'daskarray.hdf5',
    '/dask',
    var3
)
```

We have imported the necessary libraries and created the Dask array. We are using the "to_hdf5" method to export the Dask array as a HDF5 file, where we also specify the path, where we want to store the file, and the variable that we are exporting as the HDF5 file.

With the Dask array, you can perform many NumPy operations. Here is how you can calculate the mean() function of the array:

```
dask_mean = da.array(var3) \
    .mean() \
    .compute()

print(dask_mean)
```

would yield

```
0.5000587201947724
```

Dask Bags

Dask bags are another type of data structure, designed to handle and process lists like objects. In a way they are quite similar to Python lists or iterators, with the added feature that they are capable of parallel and distributed computations. The word "bag" is actually a mathematical term. Let us quickly recall what a set is. It is an unordered list with no element repeating itself. Bags are basically sets where elements can repeat themselves. It is also known as multisets.

> **Note** A list is a collection of items that can repeat itself.
>
> A set is a collection of items that are unique with no duplicates involved.
>
> A multiset is a collection of items that can repeat itself.

Here's how to create a new Dask bag:

```
from dask import bag

new_bag = bag.from_sequence(
        [1,2,3,4,5,6,7,8,9,10]
        ,npartitions=2
    )
```

```
print(new_bag)

print(new_bag.compute())
```

would yield

```
dask.bag<from_sequence, npartitions=2>
[1, 2, 3, 4, 5, 6, 7, 8, 9, 10]
```

Here's how to import a file as a Dask bag:

```
import json
from dask import bag

file = "/content/dask_data1.json"
bags = bag \
    .read_text(file) \
    .map(json.dumps) \
    .map(json.loads)

print(bags.take(1))
```

would yield

```
('[{"id":1,"name":"Godard Bloomer","language":"Greek","gender":"Male","occupation":"Financial Analyst"},\n',)
```

Here is another illustration of reading CSV files into a Dask bag:

```
from dask import bag
import csv

file = "/your-folder-path-here/mockdata-0.csv"

bags = bag \
   .read_text(file) \
   .map(
       lambda x: next(csv.reader([x]))
   )

print(bags.take(5))
```

would yield

```
(['name', 'age'], ['Ermentrude Azema', '22'], ['Joey Morteo', '56'],
['Jordan Corp', '56'], ['Jehu Kettlesing', '25'])

print(bags2.count().compute())
```

would yield

15

Dask DataFrames

Dask data frames are implementation of Pandas' data frame objects except that Dask's data frames can parallelize and perform distributed computations across multiple cores of the same processor in a system or across multiple systems in a given cluster. Dask data frames also support several of native Python functions that can be executed in a parallelized manner. Let us look at some illustrations of Dask data frames:

```
!pip install -q dask-expr

import dask
import dask.dataframe as dd
import pandas as pd
import numpy as np

dask.config.set(
    {
        'dataframe.query-planning': True
    }
)

df = dd.from_pandas(
        pd.DataFrame(
            np.random.randn(10,3),
            columns=['A','B','C']
            ),
        npartitions=3)
print(df)
```

CHAPTER 7   INTRODUCTION TO CONCURRENCY PROGRAMMING AND DASK

would yield

```
Dask DataFrame Structure:
                    A        B        C
npartitions=3
0              float64  float64  float64
4                 ...      ...      ...
7                 ...      ...      ...
9                 ...      ...      ...
Dask Name: frompandas, 1 expression
Expr=df
```

The "dask-expr" is a Dask expressions library that provides query optimization features on top of Dask data frames. As for the "query-planning" property set to "True," this may be enabled by default in up-and-coming Dask versions. This "query-planning" property provides better optimization planning for execution of Dask data frames.

To convert the lazy Dask collection to in-memory data, we can use the compute() function. The caveat is that the entire data should fit into the memory:

```
print(df.compute())
```

would yield

```
          A         B         C
0  -0.580477  0.299928  1.396136
1   0.076835 -0.343578 -0.381932
2  -0.092484 -2.525399  1.520019
3  -0.631179  0.215883  0.409627
4  -0.655274  0.117466 -0.525950
5  -0.003654 -0.215435 -0.522669
6   0.947839 -1.978668  0.641037
7  -0.755312  0.247508 -1.881955
8   1.267564  1.342117  0.601445
9  -1.395537  0.413508 -0.896942
```

Internally, a Dask DataFrame manages several smaller Pandas data frames arranged on an index. You will see the true performance of a Dask DataFrame when you have a large dataset that requires execution on multiple cores or even multiple machines.

CHAPTER 7 INTRODUCTION TO CONCURRENCY PROGRAMMING AND DASK

You cannot truly appreciate the value of Dask if you are dealing with smaller data frames, whose computational times are less than a few seconds on a single core.

By default, a data frame in Dask is split into one partition. If you specify the number of partitions to a number, then Dask will partition the dataset accordingly. In our preceding example, we have set the value to 3. This would enable Dask to create three Pandas data frames, split vertically across the index. The way we define our partitions heavily influences the performance of our data munging script. For instance, if we needed aggregations of a specific month, say January, that fell on two partitions, then Dask would have to process two of these partitions. On the other hand, if our partitions were made by month, then the queries would execute faster and more efficiently.

As with loading data, you can load data from any number of sources. Let us look at an example:

```
import dask
import dask.dataframe as dd
import pandas as pd
import numpy as np

dask.config.set(
    {
        'dataframe.query-planning': True
    }
)

csvfile = "/content/mock_health_data.csv"

df1 = dd.read_csv(csvfile)

print(df1)
```

may provide something like

```
Dask DataFrame Structure:
                id     name      age    state   gender  exercise_daily?
copay_paid
npartitions=1
             int64   string   float64   string   string    string           string
              ...      ...       ...      ...      ...       ...              ...
Dask Name: to_pyarrow_string, 2 graph layers
```

Chapter 7 Introduction to Concurrency Programming and Dask

Let's try to run the following command:

```
print(df1.compute())
```

would yield

	id	name	age	state	gender	exercise_daily?	copay_paid
0	1	Allayne Moffett	NaN	SC	Male	<NA>	$206.76
1	2	Kerby Benjafield	NaN	NM	Male	True	$21.58
2	3	Raina Vallentin	NaN	MI	Female	True	$125.18
3	4	Kaela Trodden	NaN	OH	Female	False	$86.20
4	5	Faber Kloisner	NaN	MN	Male	False	$219.38
..	...	...	...	...	...	...	...
995	996	Rani Wiskar	NaN	CA	Female	True	$122.31
996	997	Nona McMaster	NaN	<NA>	Female	False	$198.55
997	998	Eduard Trail	NaN	MI	Male	<NA>	$258.98
998	999	Shaylynn Hutchins	NaN	NY	Female	False	$227.44
999	1000	Gelya Seabert	NaN	FL	Female	False	$116.78

[1000 rows x 7 columns]

In the preceding example, the function "compute()" is applied in order to execute the computation of the group-by function on the Dask data frame. As mentioned earlier, Dask has lazy evaluation enabled by default, and so you need to call the compute() function explicitly to begin the execution of the function.

Dask Delayed

There are cases where the problems do not easily fit into the abstraction of series, arrays, or data frames. This is where the Dask delayed feature comes in. Using Dask delayed, you can parallelize the execution of your own algorithms. Let us consider the following example:

```
def add(x):
    return x + 1

def square(x):
    return x * x

def square_add(a,b):
    return a+b
```

```
data = [1, 2, 3, 4, 5]
output = []
for x in data:
    a = add(x)
    b = square(x)
    c = square_add(a,b)
    output.append(c)

[print(i) for i in output]
```

In the preceding example, we have functions that do not depend on each other; and so, they can be executed in parallel. But this is not an easy task, or we simply cannot try to fit this into a series, array, or data frame. Now let us use the Dask delayed feature on this problem:

```
import dask

output = []
for x in data:
    a = dask.delayed(add)(x)
    b = dask.delayed(square)(x)
    c = dask.delayed(square_add)(a, b)
    d = c.compute()
    output.append(d)
```

In this code, we have decorated our functions with Dask delayed features. Dask delayed will enable the parallel computation of these functions. However, notice that these functions do not execute. They simply get added in the task graph that represents the computations performed in a specific order. In order to execute the function, we need to add the function "compute()." So when you run this code, you will see

```
> python3 dd.py
3
7
13
21
31
```

The following way of enabling Dask delayed is also possible:

```python
@dask.delayed
def add(x):
    return x + 1

@dask.delayed
def square(x):
    return x * x

@dask.delayed
def square_add(a,b):
    return a+b

output = []
for x in data:
    a = add(x)
    b = square(x)
    c = square_add(a,b)
    output.append(c)

results = dask.compute(*output)
```

Dask Futures

A future, simply, in Python, represents an operation that is yet to be executed. It is basically a placeholder for a result that has not yet been computed, but it may happen at some point. In Python, the concurrent.future module provides asynchronous execution of functions. This module provides a way to execute the tasks and obtain the results concurrently. Dask supports this concurrent.future module to scale your current data pipeline across multiple systems in a given cluster, with minimal code changes.

Dask futures form a part of Dask's greater asynchronous execution model, which is within Dask's distributed scheduler. So when you submit a task to the Dask scheduler, it returns a future representing that given task's result. After that, the scheduler may execute the task in the background by using the available worker (can be within the system or another system in the cluster). It may be a good time for us to recall that the future allows for concurrent execution and can enable parallel computations of code.

CHAPTER 7 INTRODUCTION TO CONCURRENCY PROGRAMMING AND DASK

Let us also revisit the concept of eager and lazy evaluation and where futures stand in this point of view. Eager evaluation executes code immediately regardless of whether a given function or set of code is relevant to the result or not. Lazy evaluation does not execute anything unless you explicitly specify to execute, upon which it will execute only the necessary code for the result. You have already seen the "compute()" function that performs lazy evaluation on Dask. Dask futures can be seen as somewhere in between.

Dask futures are not lazy evaluation; when you submit a future for processing, it will begin the task in the background. Dask futures are not eager evaluation either, because futures enable parallel and distributed computation of tasks and functions, natively. One can naively (with caution exercised) view the Dask futures as eager evaluation with added features of asynchronous execution and parallelism in a distributed environment, error handling, and task scheduling.

Internally Dask futures are, by default, asynchronous. When you submit a computation, it returns a future object immediately. Using the future, you can check the status of the computation at a given point in time or just wait it out till it completes in the background using the workers. Dask futures are beneficial in designing data pipelines, where you may need finer-level control over task scheduling.

To work with the Dask futures, you need to instantiate the Dask client. You will also be able to view results in a dashboard:

```
from dask.distributed import Client
import dask

client = Client() # Client(processes=False)
```

When you instantiate the client without any arguments, it starts local workers as processes, and when you provide the argument "processes=False", it starts local workers as threads. Processes offer better performance when it comes to fault tolerance over threads. Here is how you can submit a task to the Dask scheduler:

```
def square(x):
    return x ** 2

a = client.submit(square, 10)
c = client.submit(square, 20)
```

In the preceding example, the submit method that submits the task returns a future. The result of this task may not yet be completed and be completed at some point using a process. The result of this function will stay in that specific process unless you ask for it explicitly. Let's look at the contents of the variable "a":

```
print(a)
```

would return

```
<Future: finished, type: int, key: square-aecf50e0301cc2c62c9310c396ebcbee>
```

You have to explicitly ask it to return the result. Here is how that looks:

```
print(a.result())
```

would return

```
100
```

There is also the cancel() method that would stop the execution of a future if it is already scheduled to run; if it has already run, then the cancel() method would delete it. Here is how that looks:

```
a.cancel()
```

You can wait on a future or a set of futures using the `wait()` function. Once you ask a future to wait, then it will wait until all the other tasks have been completed or finished execution successfully. Here is how that looks:

```
wait(a)
```

In computing, we have this term called "fire and forget" to describe a process or a task that is executed without waiting for the response or the outcome. Dask also provides a "fire and forget" feature to tasks, where the tasks are submitted to be executed without the need for checking their result or the tasks to even complete. Once the task is fired, the system is told to forget about it, and so the system no longer monitors that task or even waits for the task to finish:

```
client.fire_and_forget(a)
```

or simply

```
fire_and_forget(a)
```

Optimizing Dask Computations

In Dask distributed computing, the scheduler performs dynamic task scheduling, a process of building a diagonal acyclic graph, where the nodes change over time. By dynamic, we are talking about making decisions as we go along, as tasks are getting completed and new tasks are becoming ready to execute. The idea is to execute the correct order of tasks while consuming the least amount of time. Let us look in detail how each and every task passes through various stages in Dask computations.

Let us look at the following example:

```
def square(x):
    return x ** 2

a = client.submit(square, 10)
c = client.submit(square, 20)

print(a.result())
print(c.result())
```

When we invoke the client.submit function, the Dask client sends this function to the Dask scheduler. As we know already, the Dask client would create a future and return the future object to the developer. This happens before the scheduler receives the function.

The scheduler then receives this function and updates its state on how to compute the a or c (whichever comes in first). In addition, it also updates several other states. Once the scheduler gathers all the required information in the memory, it would then proceed to assign this to a worker. All of these happen in a few milliseconds.

The scheduler now goes by its playbook that defines a set of criteria for identifying which worker to assign this work. It would start off to see if there are already workers who have this algorithm saved in their local memory. If none exists, it would select the least number of bytes a worker might require to process this function. If there are multiple workers who share similar least values, it would select the worker with least workload.

The scheduler has now placed this computation at the top of the stack. A message is generated with this function, worker assigned and other metadata; the scheduler serializes this message and sends it over to the worker via TCP socket.

The worker receives this message, deserializes the message, and understands what is required from the task point of view. The worker starts the execution of this computation. It looks to see whether it requires values that may be with other workers by investigating the "who_has" dictionary. The who_has dictionary is basically a list of

CHAPTER 7 INTRODUCTION TO CONCURRENCY PROGRAMMING AND DASK

all the metadata about workers and their tasks and contains other metadata that would enable the workers to find each other. After this step, the worker launches the actual computation and completes the task. The worker sends the output to the scheduler.

The scheduler receives this message and repeats the process of sending another task to the worker that is available. The scheduler now communicates with the client that the computation is ready, and so the future object that is in waiting mode should wake up and receive the result.

At this point, the Python garbage collection process gets triggered. If there are no further computations that are based out of the current task, it removes the key of this computation from the local scheduler and assigns to a list that gets deleted periodically.

Without delving deeply into the lower-level details, this captures a typical process of task scheduling. Optimizing the task scheduling process may guarantee efficient execution of all tasks within the DAG.

Data Locality

Moving data between workers or unnecessary data transfers can limit the performance of the data pipeline in a distributed system environment. Data locality is an important term in the broader context of distributed computing. As we mentioned earlier, schedulers intelligently assign to workers that either have the data already or are located close to the data transfer over the network. Let's look at a scenario:

```
from dask.distributed import Client
import dask.dataframe as dd

client = Client()

df = dd.read_csv("/home/parallels/dask/mockdata-*.csv")

result = df["age"].mean()

final_result = result.compute()
```

In this example, we are loading multiple CSV files stored in a given folder. Dask would intelligently partition this dataset across various workers available in the given cluster. Upon loading the dataset, the next line performs a group-by computation and sum operations on top of that. Dask will intelligently identify the workers who already have copies of the dataset available. In this case, the workers would be the ones that are already holding partitions of this dataset. Dask will utilize these workers to minimize the network transfers and increase the magnitude of the computation.

As we discussed in earlier sections, in certain cases (like custom algorithms), you may require more control over which worker to process what type of data. Certain data might require GPU-heavy workers, and so you can specify that in a following manner:

```
f_result = client.submit(result, *args, workers= IP address/Hostname)
```

```
f_result = client.submit(square, 3453453, workers = [192.168.100.1, 192.168.100.2, 192.168.100.3]
```

For instance:

```
workers = ['hostname1','hostname2', etc...]
Workers [IPaddress1, IPaddress2]
```

Prioritizing Work

When there are many items in the queue for the Dask cluster to process, Dask has to decide which task to process first and the ones to prioritize over others. Dask prefers to process the tasks that were submitted early. Dask's policy is first in, first out. You can custom define priority keywords, where tasks with higher priorities will run earlier when compared with ones with lower priority:

```
a = client.submit(square, 5, priority=10)
b = client.submit(square, 7, priority=-1)
```

The operation with priority set to 10 carries the top priority, whereas the operation with priority set to –1 carries relatively lower priority.

Work Stealing

In Dask distributed computing, work stealing is another strategy that is employed as part of performance optimization. A task may prefer to run on a specific worker because of the nature of processing or task dependencies that exist in that specific worker. Stealing strategy may be good in cases where the time taken to perform a computation is longer than time taken to communicate with the scheduler. Other cases where stealing is a good choice would be to obtain workload from an overworked worker and replicating data that is used more often. In Dask, work stealing is enabled by default with the idea of minimizing computational time while allowing for optimal usage of workers.

Conclusion

We have seen a lot of concepts in this chapter; we learned how breaking down a large operation into several smaller operations and running them either parallelly or concurrently utilizes lesser resources than running everything on a single thread. We looked at Dask, a Python distributed computing library that can perform concurrent tasks and overcome traditional Python limitations. We also looked at various data structures of Dask and various optimization concepts. Dask also integrates well with the modern data stack. The tools and concepts you have learned in this chapter lay a good foundation on parallel and concurrent data processing.

CHAPTER 8

Engineering Machine Learning Pipelines using DaskML

Introduction

Machine learning is a branch of artificial intelligence and computer science. A machine learning model is an intelligent system that can learn from data in order to make decisions. Organizations of all nature have gained significant value by leveraging machine learning models in their processes. Training a machine learning model is done using algorithms, which can be computationally expensive. As datasets grow larger, both in terms of volume and level of complexity of the data points, scalable machine learning solutions are highly sought after. In this chapter we will look at Dask-ML, a library that runs ML algorithms in a distributed computing environment and integrates well with existing modern data science libraries.

By the end of this chapter, you will learn

- The workflow of a typical machine learning pipeline
- Dask-ML
- Dask integration with current machine learning libraries
- Setup and installation
- Data preprocessing using Dask-ML

CHAPTER 8   ENGINEERING MACHINE LEARNING PIPELINES USING DASKML

- Cross validation, hyperparameter tuning, and statistical imputation in Dask-ML

- Machine learning pipelines using Dask-ML

Machine Learning Data Pipeline Workflow

Data engineering tasks for development of machine learning models involve a series of interconnected steps that form an organized workflow for application development and deployment. The following stages provide you an outline for building and implementing machine learning solutions. Please note that these steps are iterative in nature; they often require revisiting a step at an appropriate time. These are not in any way transient or linear. Let us explore these steps in greater detail.

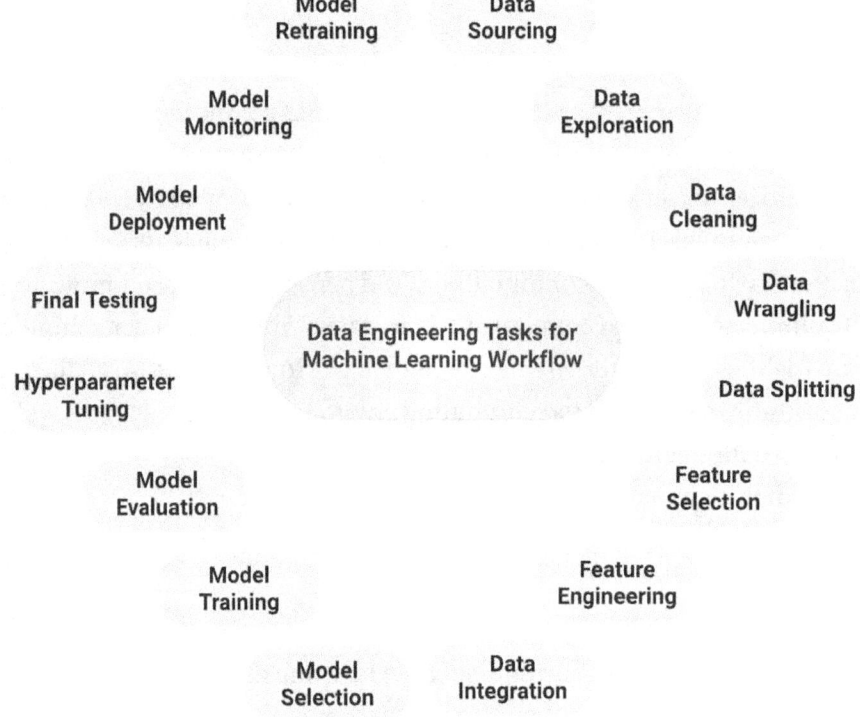

Figure 8-1. Data engineering tasks for machine learning

> **Note** This represents best practices derived from my experience in the field. I have found it to be effective to categorize project tasks based on these stages. This approach has helped me manage complex projects and believe following this approach can be beneficial in your work.

Data Sourcing

Data sourcing is the determination and gathering of data from various sources. Even prior to that is to identify the various sources and how often new data is generated. The data sources can also involve various data types including but not limited to text, video data, audio data, pictures, and so on. The amount of data that is coming in, how often new data gets generated, and the various types of data that is required for processing set the stage for data sourcing and ingestion tools and strategies.

Data Exploration

Data exploration is all about getting to know a dataset. A dataset has its own nature and characteristics, depending upon the nature, type, and context of data points. Data exploration involves understanding distribution of each data point, understanding correlation between one or more data points, looking at various patterns, identifying whether missing values exist, and gaining an initial assessment of how accurate or complete and consistent the dataset is (even if it is approximate). Data exploration helps identify critical insights about data points, minimize potential biased results (if one or more data points represent the same thing; for instance, length of a flower petal in millimeters, in centimeters, etc.), and most importantly reduce aimless data analytics work that leads to nowhere.

Data Cleaning

Data cleaning is caring for the hygiene of the dataset. For lack of a better term, clean data helps with effective data analysis. Data cleaning involves cleaning up errors and outliers in data points that may negatively affect and possibly skew the dataset into bias. Data cleaning also is about handling those missing values that were previously "discovered," deciding whether to impute the value that is missing or remove that row completely. Data cleaning can also involve handling data duplication, which would most likely be removing it. Clean data leads to accurate analysis and accurate regression models and makes the judgment more reliant.

Data Wrangling

Data wrangling is a continuous process of transforming data to be able to use them for analytics and model building purposes. The most common data wrangling technique is often referred to as data standardization, where the data is transformed into a range of values that can be measured in common scale. Other data wrangling tools include feature scaling and variable encoding where data is normalized and categorical variables are converted into numerical values in a specific format that can be easily analyzed by a mathematical model. While every step in this model building process is equally important, data wrangling directly deals with improving the accuracy of machine learning models.

Data Integration

Data integration is the idea of combining one or more datasets from one or more data sources into a single dataset that can provide a comprehensive view. Data integration would also involve further identifying and removing duplicates that may arise from combining several data sources. Data integration involves schema alignment, where data from one or more sources are compatible and aligned. Data integration also deals with data quality, where one would develop validations and checks to ensure data quality is good after the data is integrated from several sources. Data integration helps expand the scope of the entire project. Data integration also helps improve accuracy and integrity of the dataset.

Feature Engineering

Once we have the dataset, then we can proceed to the next step, that is, feature engineering. By this time, the dataset may contain several data points. In addition to the existing data points in the dataset, feature engineering involves creating new data points based on the existing data points. Some of the simplest data points are calculating the age of a person either at birth or at a specific date, calculating Fahrenheit from Celsius, or calculating distance between two points based on a specific distance measure, to name a few. Sometimes feature engineering could involve complex computations like Fourier transformations or other complex methods that are computationally expensive. The idea is to come up with new data points that have higher correlation with the data point that the model is looking to predict or classify.

Feature Selection

Now that you have the dataset you need for the project and engineered new features based on the data points of the existing dataset, it is time to select which of these data points enter into the model. There are many ways of selecting the features for a given machine learning model. Tests like the chi-square test help determine the relationship between categorical variables, while metrics like the Fisher score help generate ranks for various data points. There are approaches like forward selection or backward selection that iteratively generate various feature combinations and respective accuracy scores; however, they can be expensive as data and features grow beyond a lab experiment. Other techniques include dimensionality reduction using tools like principal component analysis and autoencoders that encode data into lower dimensions for processing and decode back to the original shape for output.

Data Splitting

Data splitting is the process of partitioning the entire dataset into smaller chunks for training a machine learning model. Depending upon the training strategy, the approach of splitting the data may differ. The most common approach of splitting the data is the idea of holding out training, validation, and test datasets. The training set is usually allotted for the model to learn from the data, the validation set is primarily designed for hyperparameter tuning purposes, and the test set is to evaluate the performance of the model with chosen hyperparameters. There is also k-fold cross validation where the dataset is partitioned into k chunks and iteratively trained by keeping k − 1 as training and the remaining as the test set. One can even go further by leveraging advanced statistical methods like stratified sampling to obtain partitions of similar characteristics and distributions.

Model Selection

So we have a good understanding of the dataset, performed various operations, integrated from multiple sources, generated features, and looked at data distributions. The next step is to start looking at selecting the right machine learning model. It is also a good stage to decide how important it is to tell a story of target variables based on variables entering into the model. In case of predicting a customer churn, it may be relevant to tell a story that is based on few variables like demographics, preferences, etc., whereas in case of image classification, one is more focused on whether the picture is a cat or a dog and less concerned about how the picture came about to be the one or the other. In cases of supervised learning, one would usually start off with logistic regression as a base point and move toward ensemble methods like bootstrapping or stacking methods.

Note Bootstrapping is basically taking several samples of your dataset, building machine learning models, and combining the results by voting or computing their mean. Stacking is training multiple machine learning models on the same population and training a meta model based on the several models and obtaining the result.

Model Training

Based on the characteristics of data and nature and context of the problem we are trying to solve using the machine learning model, the training strategy and algorithm vary. The common training strategies are supervised and unsupervised learning. Supervised learning is the process of training a model with a labeled dataset, where data points have specific names and usually the ones with higher correlations with the decision variables are picked to train. Common examples include regression models and classification models.

Unsupervised learning is the process of training with an unlabeled dataset. A good example is the clustering algorithm that groups similar data points based on features like geographical proximity and so on. Several other strategies exist; for instance, there is the incremental learning, where the model is trained continuously as new data is available. There is also ensemble modeling, which involves combining multiple machine learning models in order to improve the predictive performance of the model.

Model Evaluation

Once we have a machine learning model trained, the next step is model validation. Sometimes testing alone may suffice, and in other cases testing and validation are required. The idea is to evaluate the accuracy of a model by making a decision on a completely new set of data and evaluate its accuracy statistically by looking at precision metrics like r-squared, F1-score, precision, recall, etc.

Hyperparameter Tuning

Hyperparameters are parameters that control the training process. These hyperparameters are set prior to the model training stage. Hyperparameters are technical parameters that guide an algorithm mathematically. In an n-dimensional vector space, these hyperparameters dictate which direction to search for the best value and how far to go in that direction at each iteration.

Imagine a scenario where you are tasked to find a football in a large football field. You may have to walk in various directions and in each direction decide how many steps to take and look around to see if you can spot the football in your region. This is basically how the internals of training algorithms work. Hyperparameter tuning is the process of finding the optimal parameters for a machine learning model. Imagine if I say, "Turn 45 degrees toward your left, take 10 steps, turn right, and take 5 steps. You will find the football." Imagine a process that will guide you to figure out these steps. This is hyperparameter tuning.

Final Testing

We have performed various checks and validations in the process of building a machine learning model up to the process of selecting the best hyperparameters for our machine learning model. Final testing is basically testing the final chosen model on a held-out test set. This process exists in order to obtain realistic expectations when the model is deployed in production. The main goal behind this step is to get an idea of how this model may perform when exposed to a new set of data. Once we test out the final model, establish an understanding of strengths and weaknesses of the model and document the results for future reference and maintaining audit trails.

Model Deployment

Once we have the training model, the model is validated against test sets, and hyperparameters are tuned, we now have a machine learning model that is ready to be deployed. The concept of deployment changes with respect to the nature and context of the business and problem statement in hand. The conceptual definition is that deployment means the usage of a model in everyday scenarios, where you are leveraging the model to inform better decisions.

Model Monitoring

Once the model is deployed, then the model needs to be monitored in frequent time intervals to see if the model is performing appropriately or not. The model also requires frequent maintenance every so often. The maintenance activity could involve various strategies ranging from simple steps like incremental training or adjusting the coefficients to tiny bit to adding/modifying features and even to complete model overhaul. This will be the focus of the next step.

Model Retraining

Model retraining is a crucial step in the machine learning model process. This step involves sourcing the new data that has been collected since the last time and performing various data wrangling and feature engineering processes. Based on the dataset that is generated, it will then be decided as to whether incremental training may suffice or a complete model overhaul may be required. Based on the appropriate training strategy, the model is retrained, followed by employing testing and validation. The model is then redeployed to the production environment. This process is repeated subsequently to keep the model accurate and up to date.

Dask-ML Integration with Other ML Libraries

Dask's machine learning library provides scalable machine learning solutions using Dask and collaborating with other popular machine learning libraries like scikit-learn and others. Dask-ML's ability to extend parallel computing to machine learning algorithms enables faster computations and makes it possible to train models with large amounts of data. As we discuss the features and abilities of Dask-ML in this chapter, let us briefly look at various popular machine learning libraries that Dask-ML currently integrates with.

scikit-learn

scikit-learn is a popular machine learning and data processing library in Python. scikit-learn library is widely adopted by both academia and industry for building machine learning applications. The library supports many supervised and unsupervised learning algorithms and also has prebuilt methods for feature engineering, model selection, and cross validation strategies.

Many of the algorithms in scikit-learn are developed for parallel execution by specifying a "joblib" parameter. This "joblib" parameter provides thread-based and process-based parallelism for some of the machine learning algorithms. The Dask-ML library can scale these "joblib"-based algorithms to execute over multiple machines in a given cluster.

XGBoost

XGBoost is another popular library in the machine learning space for gradient-boosted trees. Gradient-boosted trees are a machine learning technique that is used for both continuous and categorical decision variables (regression and classification, respectively). It belongs to the ensemble learning paradigm, where multiple decision tree models minimize the error and combine to make the final prediction model. Dask integrates well with the XGBoost model, where Dask sets up the master process of XGBoost onto the Dask scheduler and the worker process of XGBoost to the workers. Then it moves the Dask data frames to XGBoost and let it train. When XGBoost finishes the task, Dask would clean up XGBoost's infrastructure and return to its original state.

PyTorch

PyTorch is an open source machine learning library primarily used for natural language processing, computer vision, and neural networks among other models. PyTorch supports tensor computation, where tensors are GPU-efficient multidimensional arrays, made for high-performance computing tasks. PyTorch integrates well with scikit-learn through a library called Skorch. This Skorch library brings a scikit-learn wrapper to

PyTorch methods. This means that Skorch-wrapped PyTorch models can be used with the Dask-ML API.

Other Libraries

Dask-ML also supports Keras and TensorFlow libraries. Keras is an open source Python library for developing deep learning models. TensorFlow is an end-to-end machine learning library, developed by Google. The scikit-learn package has an API wrapper to Keras, which allows Dask-ML to be utilized with Keras models.

Dask-ML Setup and Installation

You can obtain the latest wheel from PyPI:

```
pip install dask-ml
```

You can also install XGBoost and Dask-XGBoost using this command:

```
pip install dask-ml[xgboost]
```

You can also install all the dependencies:

```
pip install "dask-ml[complete]"
```

Similar to what we discussed earlier, Dask partitions the dataset into multiple smaller chunks and assigns these chunks to various workers in a given computing resource. The machine learning model is trained in parallel on these workers, and the results are aggregated upon training. Dask-ML offers parallel processing of several tasks including data preprocessing tasks like encoding and estimators like principle component analysis. Furthermore, Dask-ML also provides effective cross validation methods and hyperparameter tuning methods that would use all the workers in the given computing environment to perform computations for obtaining better parameters. Let us discuss these features.

CHAPTER 8 ENGINEERING MACHINE LEARNING PIPELINES USING DASKML

Dask-ML Data Preprocessing

Dask-ML provides various data preprocessing techniques that can perform large-scale data processing across multiple cores or multiple machines in a given cluster. These methods can be further incorporated into scikit-learn's pipeline methods.

RobustScaler( )

The robust scaler method enables data to be resistant against outliers. It employs a two-step approach of centering and scaling, where it subtracts the median and divides the data point with the quartile range.

Let us look at this with an illustration:

```
import dask.dataframe as dd
from dask_ml.preprocessing import RobustScaler
from sklearn.datasets import load_iris

iris = load_iris()

X = iris.data

df = dd.from_array(X)

RScaler = RobustScaler() \
         .fit_transform(df) \
         .compute()

print(RScaler.head())
```

would yield

```
Dask DataFrame Structure:
                    0        1        2        3
npartitions=1
0               float64  float64  float64  float64
149                ...      ...      ...      ...
Dask Name: from_array, 1 graph layer
>>> RobustScaler().fit_transform(df).compute()
           0     1         2         3
0  -0.538462   1.0 -0.842857 -0.733333
```

263

```
1    -0.692308   0.0  -0.842857  -0.733333
2    -0.846154   0.4  -0.871429  -0.733333
3    -0.923077   0.2  -0.814286  -0.733333
4    -0.615385   1.2  -0.842857  -0.733333
```

MinMaxScaler( )

MinMaxScaler is another useful scaling function provided by the Dask-ML preprocessing methods. For each value in a given column, MinMaxScaler subtracts the minimum value of the feature and divides by the range. The idea behind MinMaxScaler is to bring all the values within the range of [0,1].

It may be good to note that if the column had negative values, then it would attempt to fit the data in the range of [-1,1]. MinMaxScaler is a good fit when you are attempting to fit a neural network with a Tanh or Sigmoid activation function. Please note that MinMaxScaler may not play well with outliers. Either you remove the outliers or use the RobustScaler function, instead.

Let us look at an example:

```
import dask.dataframe as dd
from dask_ml.preprocessing import MinMaxScaler
from sklearn.datasets import load_iris

# load the dataset from sklearn datasets library
iris = load_iris()

X = iris.data

# initialize the Dask data frame
df = dd.from_array(X)

MMScaler = MinMaxScaler() \
          .fit_transform(df) \
          .compute()

print(MMScaler.head())
```

would return

```
         0         1         2         3
```

0	0.222222	0.625000	0.067797	0.041667
1	0.166667	0.416667	0.067797	0.041667
2	0.111111	0.500000	0.050847	0.041667
3	0.083333	0.458333	0.084746	0.041667
4	0.194444	0.666667	0.067797	0.041667

One Hot Encoding

One hot encoding is a technique that converts a categorical variable with "n" categories into "n" number of columns where each category is switched on if the value is present or switched off if the value is not present. This can be seen as a better option when compared with directly assigning numerical labels; the numerical labels may create a bias given that the algorithm may interpret it for the value it presents and not as a categorical variable. Let us look at an example:

```
from dask_ml.preprocessing import OneHotEncoder
import dask.array as da
import numpy as np

encoder = OneHotEncoder()

var1 = da.from_array(
    np.array(
        [['Apples'], ['Melons'], ['Melons'], ['Oranges']]
    ),
    chunks=2
)

# encoder = encoder.fit(var1)
result = encoder.fit_transform(var1)

print(result)
```

would return

```
dask.array<_check_and_search_block, shape=(4, 3), dtype=float64,
chunksize=(2, 3), chunktype=scipy.csr_matrix>

print(result.compute())
```

would return

```
<4x3 sparse matrix of type '<class 'numpy.float64'>'
    with 4 stored elements in Compressed Sparse Row format>
```

Cross Validation

As we have seen in a previous section, cross validation is a machine learning technique that helps assess a recently trained machine learning model; you may be able to understand how well the model adapts to non-training data using cross validation methods. The method usually partitions the dataset into "k" different chunks. The model is trained "k" times where at each time one of the k chunks acts as a test set and the rest all helps with the training/validation. This process is repeated k times where in each iteration a unique chunk serves as the test set.

For instance, let us say we have five partitions of a dataset, calling them d1, d2, d3, d4, and d5:

- During the first iteration, d1 will be the test set and d2, d3, d4, and d5 will be training sets.

- During the second iteration, d2 will be the test set and d1, d3, d4, and d5 will be training sets.

- During the third iteration, d3 will be the test set and d1, d2, d4, and d5 will be training sets.

And so on ... At each iteration, the performance metrics are gathered, and at the end of all "k" iterations, the performance metrics are aggregated, and it is used to assess the model's performance. What we have seen here is k-fold cross validation.

Let us look at an example of performing fivefold cross validation using Dask-ML:

```
import dask.array as da
from dask_ml.model_selection import KFold
from dask_ml.datasets import make_regression
from dask_ml.linear_model import LinearRegression
from statistics import mean

X, y = make_regression(n_samples=200, # choosing number of observations
            n_features=5, # number of features
```

CHAPTER 8 ENGINEERING MACHINE LEARNING PIPELINES USING DASKML

```
            random_state=0, # random seed
            chunks=20) # partitions to be made

model = LinearRegression()

# The Dask kFold method splits the data into k consecutive subsets of data
# Here we specify k to be 5, hence, 5-fold cross validation

kf = KFold(n_splits=5)

train_scores: list[int] = []
test_scores: list[int] = []

for i, j in kf.split(X):
    X_train, X_test = X[i], X[j]
    y_train, y_test = y[i], y[j]

    model.fit(
        X_train,
        y_train
    )

    train_score = model.score(
                    X_train,
                    y_train
                )
    test_score = model.score(
                    X_test,
                    y_test
                )

    train_scores.append(train_score)
    test_scores.append(test_score)

print("mean training score:", mean(train_scores))
print("mean testing score:", mean(train_scores))
```

You may wish to include your own example in the preceding code and work with tenfold cross validation and examine the output as an exercise.

CHAPTER 8 ENGINEERING MACHINE LEARNING PIPELINES USING DASKML

Dask-ML also provides a method to split one or more Dask arrays. Let us look at the following code:

```
import dask.array as da
from dask_ml.datasets import make_regression
from dask_ml.model_selection import train_test_split

X, y = make_regression(
        n_samples=200,
        n_features=5,
        random_state=0,
        chunks=20
    )

# notice the use of method from Dask ML library which resembles similar to Sklearn

X_train, X_test, y_train, y_test = train_test_split(X, y)

print(X)
```

would return

```
dask.array<normal, shape=(200, 5), dtype=float64, chunksize=(20, 5), chunktype=numpy.ndarray>
```

After splitting

```
print(X_train)
```

would return

```
dask.array<concatenate, shape=(180, 5), dtype=float64, chunksize=(18, 5), chunktype=numpy.ndarray>
```

```
print(X_train.compute()[:5])
```

results may look like

```
array(
    [[ 0.77523133, -0.92051055, -1.62842789, -0.11035289,  0.68137204],
     [ 0.02295746,  1.05580299, -0.24068297,  0.33288785,  2.14109094],
     [-0.40522438, -0.59160913,  0.0704159 , -1.35938371, -0.70304657],
```

```
    [-1.0048465 ,  0.02732673,  0.02413585,  0.48857595, -1.22351207],
    [-1.64604251,  1.44954215, -0.9111602 , -2.27627772, -0.02227428]]
)
```

Hyperparameter Tuning Using Dask-ML

Hyperparameters are parameters that govern the model training process. Before we begin training the model, we need to supply these parameters for the algorithm to search for the optimal solution iteratively. If we input values to hyperparameters that are slightly off (so to speak), then it might not converge toward the optimal solution. And so, the concept of finding good values for hyperparameters is an essential part of machine learning modeling.

When you are expecting a good model for production, it might take several hours to compute the best hyperparameters for a given problem. We do encounter challenges when we scale the hyperparameter searching effort. The most common issues that we may come across are compute constraint and memory constraint.

Memory constraint is when the data a model needs to be trained on is too large to fit into the physical memory available. Even when the data fits into the memory (by throwing more physical memory onto it), we may come across another issue called compute constraint, where the computation task takes too long to return the results (even when the data fits into the memory). Usually, this happens when the algorithm is more expensive.

Memory constraint and compute constraint are mutually exclusive. When the data presented as the training set does not pose either of these issues (memory or compute constraint), we could utilize the grid search and random search methods.

Grid Search

You define a grid of values (a matrix of values). The grid search would exhaustively search through all the combinations of hyperparameters. It would evaluate the performance of every possible combination of hyperparameters within the grid using a cross validation method. While it will deterministically search and provide the best possible hyperparameters for your model, it gets expensive as the model grows and a lot depends on the grid you have defined.

Random Search

You define the distribution for each hyperparameter. Random search randomly samples hyperparameter values from predefined distributions. Once again, this method evaluates the performance of the model using cross validation. While this might sound like a more efficient approach, there is likelihood that statistical distribution may not assist with finding the global optimal values for these hyperparameters.

When you are facing memory constraints but your computer is enough to execute the training model, you can use an incremental search method.

Incremental Search

Incremental search starts off with predefined values for hyperparameters and adjusts them based on the performance over every iteration. It is an efficient method when incremental adjustments are feasible; it may even converge to optimal values for the hyperparameters. But the search strategy requires work and expects the hyperparameters to be continuous.

When compute and memory are both constrained and posing a challenge to obtain good hyperparameters for your machine learning model, you may choose to explore Hyperband algorithms for hyperparameter tuning.

The way I understand it is that the Hyperband algorithm is basically a grid search where if a combination takes too much time to converge, then the algorithm dynamically cuts off the compute and memory resources allocated to it and focuses on the ones that yielded the hyperparameters while covering faster than the others. It has a more aggressive early stopping criteria at each iteration. Once all predefined iterations are completed, the hyperparameters that converged with least computing resources will be returned.

Let us look at an example, where we are creating a stochastic gradient descent classifier and initializing the hyperparameters to be in log uniform and uniform distribution, respectively. The Hyperband algorithm computes the hyperparameters and you can see that at the end. Here is how it looks:

```
from dask.distributed import Client
from dask_ml.datasets import make_classification
from dask_ml.model_selection import train_test_split
from sklearn.linear_model import SGDClassifier
```

```python
from scipy.stats import uniform, loguniform
from dask_ml.model_selection import HyperbandSearchCV

client = Client()

X, y = make_classification(
        chunks=10,
        random_state=0
    )

X_train, X_test, y_train, y_test = train_test_split(X, y)

clf = SGDClassifier(
        tol=1e-3,
        penalty='elasticnet', # regularization term
        random_state=0
)

params = {
        'alpha': loguniform(1e-2, 1e0),
        'l1_ratio': uniform(0, 1)
    }

search = HyperbandSearchCV(
        clf,
        params,
        max_iter=80,
        random_state=0
    )

search.fit(
    X_train,
    y_train,
    classes=[0, 1]
    )

print(search.best_params_)

print(search.best_score_)
```

```
print(search.score(X_test, y_test))
```

would yield

```
HyperbandSearchCV(
    estimator=SGDClassifier(penalty='elasticnet', random_state=0),
    parameters={
        'alpha': <scipy.stats._distn_infrastructure.rv_continuous_frozen
        object at 0x7fb9e4c47d90>,
        'l1_ratio': <scipy.stats._distn_infrastructure.rv_continuous_
        frozen object at 0x7fb9e43cfb80>
},
    random_state=0
)
print(search.best_params_)
{
    'alpha': 0.057240478677855075,
    'l1_ratio': 0.40506608504946273
}
print(search.best_score_)
0.6
print(search.score(X_test, y_test))
0.8
```

Statistical Imputation with Dask-ML

Imputation is the process of replacing missing data with appropriate values so you can retain the majority of the dataset. Dask-ML provides a method called SimpleImputer that we will use to impute a missing value. Furthermore, we will also use logistic regression to perform imputation for another value. Let us look at these in greater detail.

First, we will import the necessary classes and methods for our application:

```
import dask.dataframe as dd
```

```
import pandas as pd
from dask_ml.preprocessing import OneHotEncoder
from dask_ml.impute import SimpleImputer
from dask_ml.model_selection import train_test_split
from dask_ml.linear_model import LogisticRegression
from dask_ml.metrics import accuracy_score
```

Now let us create a sample dataset for us to work with the Dask-ML methods. We are keeping it a simple n-dimensional array that can be converted into a data frame for downstream processing purposes:

```
data = {
    'color':
        ['red', 'orange', 'green', None, 'yellow', 'green'],
    'weight':
        [150, 180, 200, 160, None, 220],
    'taste':
        ['sweet', 'sweet', 'sour', 'sweet', 'sweet', 'sour'],
    'fruit':
        ['apple', 'orange', 'apple', 'apple', 'orange', 'melon']
}
```

Now, let us create the Dask data frame for the preceding data with two partitions:

```
df = dd.from_pandas(
        pd.DataFrame(data),
        npartitions=2
    )
```

Let us initialize the SimpleImputer. You can specify the strategy that can be used to fill the missing value. The most common approach for a categorical variable is to fill in with a category that is most commonly occurring. For a continuous variable, you can choose the option of taking the statistical average and fill in the missing value:

```
imputer1 = SimpleImputer(strategy='most_frequent')

df_fit = imputer1.fit(df)

df[['color']] = imputer1.transform(df[['color']])
```

Let us perform the imputation for the weight of the fruit:

```
imputer2 = SimpleImputer(strategy='mean')
ddf[["weight"]] = imputer2.fit_transform(df[['weight']])
print(ddf.compute())
```

would yield

```
   color  weight  taste   fruit
0    red   150.0  sweet   apple
1 orange   180.0  sweet  orange
2  green   200.0   sour   apple
3  green   160.0  sweet   apple
4 yellow   182.0  sweet  orange
5  green   220.0   sour   melon
```

Notice the first imputer assigned the most commonly occurring category to the categorical variable. The second imputer assigned the statistical mean of the column to the missing value. As it so happens the statistical average of the missing orange fruit is close to the value found for orange.

Let us look at another machine learning model in Dask-ML where we fit a linear regression model to a set of data. Linear processes are primarily concerned with predicting a value of a continuous variable. For instance, forecasting the demand for a specific item given inventory levels at a given time can be a good fit for linear regression. Let us look at the code in detail.

First, let us import all the necessary classes and methods:

```
import dask.dataframe as dd
import numpy as np
import pandas as pd
from dask_ml.linear_model import LinearRegression
from dask_ml.datasets import make_regression
from dask_ml.model_selection import train_test_split
```

Let us generate some sample data for our machine learning model:

```
number_of_samples = 1000
number_of_features = 4
```

```python
X, y = make_regression(
    n_samples= number_of_samples,
    n_features= number_of_features,
    noise=0.1,
    chunks=4
)

y = np.exp(y / 200).astype(int)

# Split the data into training and testing sets
X_train, X_test, y_train, y_test = train_test_split(
                                    X,
                                    y,
                                    test_size=0.2
                                    )

# Initialize and fit the linear regressor
linear_model = LinearRegression()

linear_model.fit(
        X_train,
        y_train
    )

# Score the model on the test set
score = linear_model.score(
        X_test,
        y_test
    )

print("Model score:", score)
```

would yield

0.9999

```python
print(linear_model.get_params())
```

{'C': 1.0, 'class_weight': None, 'dual': False, 'fit_intercept': True, 'intercept_scaling': 1.0, 'max_iter': 100,

```
'multi_class': 'ovr', 'n_jobs': 1, 'penalty': 'l2', 'random_state': None,
'solver': 'admm',
 'solver_kwargs': None, 'tol': 0.0001, 'verbose': 0, 'warm_start': False}
```

Conclusion

The idea of utilizing distributed computing to run machine learning algorithms in the Python-based ecosystem can be seen as achieving a significant milestone in the applied research. Dask-ML integrates well with scikit-learn and other leading libraries. So far we looked at various preprocessing and tuning applications using Dask-ML. By knowing Dask-ML and being able to use it in a production environment, organizations can leverage gaining high-value insights and significant savings, enabling them to push the boundaries to the next level in data-driven decision making.

CHAPTER 9
Engineering Real-time Data Pipelines using Apache Kafka

Introduction

So far we looked at batch processing pipelines. In this chapter, we will be looking at real-time streaming processing pipelines and a library that is incredibly rich in functionalities. We are going to discuss Apache Kafka. Apache Kafka is a distributed and fault-tolerant streaming and messaging platform. Kafka helps build event streaming pipelines that can capture creation of new data and modification of existing data in real time and route it appropriately for various downstream consumption purposes. In this chapter, we will look at Kafka, its architecture, and how Kafka can help build real-time data pipelines.

By the end of this chapter, you will learn

- Fundamentals of distributed computing and its connection with real-time data pipelines
- Apache Kafka and its core components
- Setup and installation of Kafka
- Kafka architecture and writing producers and consumers
- ksqlDB

CHAPTER 9   ENGINEERING REAL-TIME DATA PIPELINES USING APACHE KAFKA

Introduction to Distributed Computing

Now is a good time to briefly touch on the evolution of distributed computing. Traditionally, computing helped in developing software applications, programming heavy computations that utilized data. Here, the data is stored in relational tables, while the application and programming are stored in a production computer, which may be bigger when compared with end user computers. As adoption of computing increased, the amount of data generated at a given point in time also increased. In addition, the type of data and the speed in which the data was generated were also on the rise. Given the need for more computing, distributed systems and the concept of distributed computing came into play.

Apache Hadoop, a successful distributed systems project, saw a rise in its adoption; their commercial spin-off partners witnessed a huge demand for distributed system processing. Several other tools and frameworks to address challenges and opportunities in distributed data processing complemented Hadoop. The term "big data" evolved. Every trend and best practice that was adopted by computing was challenged, and new frameworks and ideas evolved. For instance, relational SQL witnessed the adoption of NoSQL, which in turn forced relational SQL vendors to develop and fine tune more features.

One of the big data evolutions is called event processing or streaming data. The idea behind event processing is that the data is captured at the time of event and written to a data store as events. The relational data models and data-capturing methods focus on capturing data as entities. Entities can be seen as nouns; it could be a person, an organization, a project, an experiment, a vehicle, or even a light bulb (by definition of entities). On the other hand, streaming data processing is basically an action or an event that happened in a given time interval. They can denote anything.

Event data processing or streaming data processing is data generated on a massive scale. These are data that are arriving to a system at much higher speeds when compared with relational data arriving from, say, a point-of-sale system or number of items produced in a given day, to name a few applications. One of the notable projects that evolved in this streaming data processing is Kafka. Apache Kafka originated at LinkedIn, which helped solve LinkedIn's data movement problems. In this chapter, we will talk about Apache Kafka and ideas behind the streaming data processing and gain further ground on Kafka development.

Introduction to Kafka

Apache Kafka is a distributed, scalable, fault-tolerant, elastic, and secure streaming and messaging platform. Written in Java and Scala, it is used to build real-time data pipelines and streaming pipelines as well. Event streaming pipelines help capture data in real time (or near real time) from sources like databases, sensors from devices (like IoT devices and others), and other systems. These captured data are stored in a database for further analysis, data processing, and routing to other downstream software applications. As per the official documentation, Apache Kafka provides three core functionalities:

- To process streams of event data as they occur real-time
- To store these streams of event data
- To read and write streams of event data and continuously import or export data from or to other systems

As mentioned, Kafka is a distributed, scalable, fault-tolerant, elastic, and secure platform. Let's learn some software architecture terms:

"Distributed" means the ability to run on one or more computers; the software would come with an abstraction layer that would sit on top of one or more commercial-grade computers so that the abstraction would mimic one single computer, where the resources (computing, memory, storage) are shared and utilized as and when they are needed.

"Scalable" means the ability of the system to handle an increasing amount of work and the system's ability to accommodate its increased usage. Traditional computers may not be seen as scalable; they are limited by the number of cores, total storage, and memory present in the motherboard. When something is scaled up, it means more resources are added to the system, whereas scale-down means that resources are minimized for the system.

"Elastic" means the ability of the system to dynamically allocate hardware resources to meet the growing demand for computing. It is the automatic ability to scale up or scale down given the demand for computing resources. There may be some peak hours, where more resources are required for computing, and there may be other instances where the system might just function with minimal resources. Elasticity also means optimization and efficient resource utilization.

"Fault tolerant" means the ability of the system to continuously function in case of failure. Fault tolerance is usually achieved by the idea of replication. In the case of data processing in a distributed environment, data that are partitioned into chards are usually replicated; this means that every subset of data would have more than one copy in the system. These copies would be spread all across the cluster. And so, when one node containing one copy of the chard fails, the system would automatically locate the other copy of the same chard, in a different node.

"Secure" means that the data, systems, and resources are protected from unauthorized access, damages, and attacks from external entities. A secure system may contain access control, two-factor authorization, encryption in motion and at rest (disk encryption and transporting data in secure tunnels), and regular security updates to protect the system from potential threats and other vulnerabilities that may exist.

Note

Encryption at Rest

Encryption at rest means the idea of disk encryption. Disk encryption makes sure that when an unauthorized entity gains access to disk and attempts to read, it won't be able to access the data. The data is encrypted and it requires a "key" to decrypt.

Encryption in Motion

Encryption in motion means the idea of encrypting the data in transit. The transport layer encryption refers to securing data while it is in transit over a given network. There are network protocols (industry-approved instructions) that implement encryption in motion. Some of them are HTTPS, TLS, and SSL protocols. And so, when the data is intercepted in between transit, the information does not make sense (requires some sort of a key to decrypt).

Kafka supports three major messaging systems architectures, namely, streaming applications, pub–sub model, and messaging queue application. For reference, the most basic messaging system is where there is a sender, a receiver, and a message or messages. At the risk of stating the obvious, the sender sends a message or messages to the receiver, and the receiver would receive the message sent by the sender. There are also message brokers, which act as an intermediary layer between various messaging systems architectures, performing validation and routing.

Let us talk about the messaging queue systems architecture. These are where the messages are kept in a queue to be processed. The processing starts by consuming these messages in an asynchronous manner. Each message is consumed by only one consumer. Once a message is consumed, then that message is removed from the queue.

Now, let us talk about the pub-sub messaging system. Pub-sub stands for publisher and subscriber, respectively. In this architecture, the messages are "published" into what is called a topic. The topic can then be read by another program, called "subscriber." These subscribers can receive messages from the topic. You can have multiple subscribers receiving messages from topics; and so, these messages are broadcasted to multiple subscribers.

Kafka Architecture

Kafka can be deployed in virtual machines, cloud environments, containers, and even bare metal appliances. Kafka consists of servers and clients within its distributed system. Kafka servers actively provide brokerage and Kafka Connect services. The brokers form the storage layer, whereas Kafka Connect provides continuous import or export of data as event streams. The Kafka client provides you with libraries and APIs to write distributed applications and microservices that can process streams of events, function in parallel, scale up where needed, and be fault tolerant. Let us look at some of the components of Kafka.

Events

An event is simply something that is taking place or happened at a given point in time. Of course, every day there are lots of events happening in our world; in the Kafka context, it would specifically refer to the events for the use case. For instance, in a bank, you have lots of events happening at the same time—customer withdrawing, customer depositing money, etc. The anatomy of an event is such that it may contain key, value, timestamp, and other metadata relating to that event, for instance:

```
Key: John
Value: John withdrew $100 cash
Timestamp: July 14, 2023, 10:10:33 am
```

Topics

Topics are where events are organized and stored securely. The concept is similar to that of files and folders. A folder is where one or more files are stored; similarly, one or more events are stored in topics. In the preceding example, John withdrawing a sum of $100 is considered an event. If similar events were grouped, then it would be referred to as the "Cash withdrawal" topic. Events can be read as many times as needed and do not get deleted after consumption. You can define how long to retain these events. These retention periods can be modified topic to topic.

Partitions

A partition is the process of dividing a large data block into multiple smaller ones so that it helps with reading and access. Topics are partitioned. Any given topic is partitioned, as spread over a number of buckets. Events with the same event key (in the preceding example, John is an event key) will be written to the same partition. These partitions can also be replicated such that in case of a given partition failure, there is always another partition that contains the same topic and event. The replication is performed at the topic partition level. For instance, if the replication factor is 3 and you have four partitions for one given topic, then there will be 12 different partitions (as in three copies of each partition per topic).

Broker

Brokers in Kafka are storage services provided by the Kafka server. The abovementioned topic partitions are stored across a number of brokers. This way, when one broker fails, Kafka will locate the other broker that contains the similar topic partition and continue with the processing. Brokers communicate with each other for data replication, and they try to ensure consistency of data across the entire Kafka cluster.

Replication

Kafka automatically replicates the topic partition, and each partition gets stored in more than one broker, depending upon the replication factor set by the Kafka administrator. This is to enable fault tolerance so that even when a cluster fails, the messages continue to remain available. The main partition is called the leader replica, and the copies of topic partitions are stored at a place called follower replicas. Every partition that gets replicated has one leader and n – 1 follower replicas. When you are reading and writing data, you are always interacting with the leader replica. The leader and followers work together to figure out the replication. The whole thing is automated and happens in the background.

Producers

Producers are part of the Kafka client, enabling the process of publishing events to Kafka topics. Typically Kafka producers send a message to the topic. These messages are partitioned based on a mechanism (e.g., key hashing). For a message to be successfully written in a topic, the producer must specify a level of acknowledgment (ack). Some examples of a Kafka producer include web servers, IoT devices, monitoring and logging applications, etc. Kafka may serialize the data that is coming in using any of Java's built-in serializers or using Apache Avro, Protobuf, etc.

> **Note** Key hashing is a process of determining the mapping of a key to a partition.

You connect to the cluster using the Kafka producer, which is supplied with configuration parameters including brokers' addresses in the cluster, among other things.

Consumers

Once producers have started sending events to the Kafka cluster, we need to get the events back out into other parts of the system. Consumers are those that subscribe to (read and write) these events. They read these events from producers and generate a response with regard to the event. Note that consumers only read the topics and not "consume" in any other ways; these topics are still available in the Kafka cluster. The data that is coming from producers needs to be accurately deserialized in order to be consumed. Consumers read topics from oldest to newest.

Schema Registry

Now that we have producers and consumers ready, the producers will start publishing messages into topics, while the consumers will begin to subscribe to these same messages that the producers publish. The consumers will start consuming only if they identify and understand the structure of the data the producers publish. The schema registry maintains a database of all schemas that producers have published into topics. The schema registry is a standalone process that is external to Kafka brokers. By utilizing the schema registry, you are ensuring data integrity among other great benefits.

With Avro

Avro is a serializer for structured data.

With Protobuf

Protobuf stands for Protocol Buffers. It is a way of serializing structured data and comes with its own schema definition language. Protobuf is developed by Google; it is text based and readable when compared with JSON format. Protobuf is binary based, language neutral, and efficient in terms of speed.

Kafka Connect

Kafka Connect is a data integration mechanism, offered by Kafka to connect and integrate with various enterprise systems. Kafka Connect provides necessary methods and functions through an API, to connect both from and to various relational databases, NoSQL databases, file systems, etc. You can use Kafka to connect to both source data from these systems to publish them as topics and consume topics from a Kafka producer and load them to these various databases and file systems. Kafka Connect can function in standalone mode for development and testing and also in a distributed model for large-scale production deployments.

Kafka Streams and ksqlDB

Kafka Streams is a service that enables the idea of streaming data pipelines. A streaming data pipeline is a such that processes data as it is being generated or as an event is happening. Streaming data pipelines do not necessarily extract data; rather, they just receive data from a perennial flow of data coming from a given system. The easiest way to visualize is the idea of weather forecasting systems. These systems receive atmospheric data from a satellite continuously. And so, there is no "extraction" happening here. Instead, they is configured to receive the same.

Kafka Streams is scalable and fault tolerant and can be deployed in leading operating systems, bare metal appliances, containers, etc. Kafka Streams primarily caters to Java programming language to be able to write streaming pipelines and also to enrich current Java applications with stream processing functionalities. For other programming languages, we have another service called ksqlDB that enables writing Kafka stream functionalities.

ksqlDB is a database that integrates with Kafka, enabling you to build streaming data pipelines, similar to building traditional data applications. More specifically, ksqlDB is a computational layer that uses SQL-like syntax to interact with the storage layer, which is Kafka. You can interact with ksql using web, command line, and REST APIs. To use ksqlDB with Python REST API, you can install

```
pip install ksql
```

You can create a new table or stream within the ksqlDB and stream your existing data in relational tables or other file systems into topics, for instance:

```
CREATE TABLE campaign(
     AdvertiserID FOREIGN KEY referencing table,
     CampaignID INT PRIMARY KEY,
     CampaignName STRING,

     StartDate DATE,
     EndDate DATE,
     Budget FLOAT
)
WITH (
     Kafka_topic = "Campaign",
     Partitions = 3,
     Value_format = "Protobuf"
);
```

CHAPTER 9   ENGINEERING REAL-TIME DATA PIPELINES USING APACHE KAFKA

If you have existing Kafka topics, you can use ksqlDB to create a stream or table and then begin streaming data into ksqlDB:

```
CREATE STREAM ads
WITH (
    kafka_topic="campaign",
    value_format= "protobuf"
);
```

Just like relational SQL, you can perform lookups, joins, filtering, transformations, and aggregations to derive real-time analytic insights from data.

Kafka Admin Client

The Kafka admin client allows you to manage resources on your Kafka cluster.

Here is a basic illustration of Kafka's architecture:

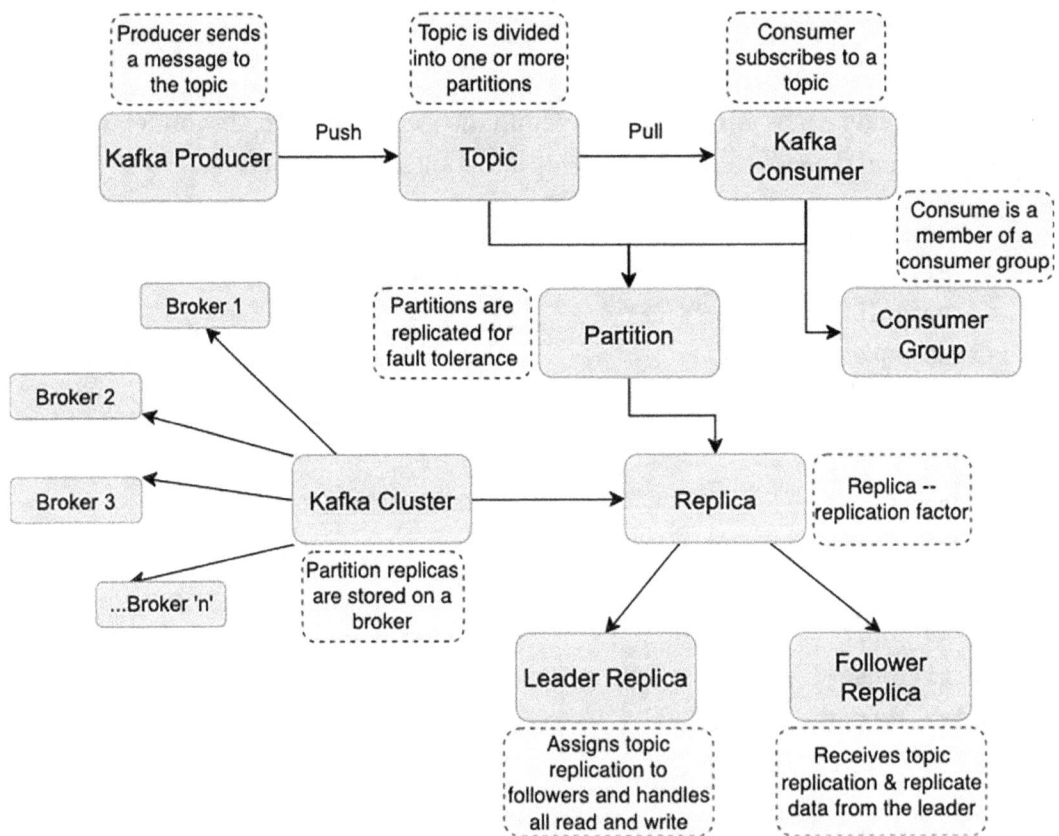

Figure 9-1. *Kafka architecture*

Setup and Development

There are three major flavors of Kafka distribution that are available. The first one is Apache Kafka distribution that is open source and Apache licensed. The second is the commercial distribution of Kafka, provided by Confluent Inc., a vendor. The last one is cloud-based distribution provided by major cloud computing vendors like AWS and others. It would be likely you use a secure and well-managed Kafka distribution for production, and so it makes sense to explore the commercially available distribution provided by Confluent, Inc. Besides, they also offer a great deal of documentation, which can be helpful.

To get started, visit Confluent's Kafka website, `www.confluent.io/get-started/`, and follow the instructions to set up a new account. You will be offered a choice of a cloud vendor to host your Kafka cluster. From a learner's perspective, it may not matter as much as to which vendor you choose to host your Kafka cluster. However, if your organization is adopting a specific cloud computing platform, it may help to adopt the same vendor for hosting your Kafka cluster. Here, we will look at registering an individual user for learning purposes. First, register yourself on the Confluent website. Click Environments and proceed to create a cluster.

Here is how that looks:

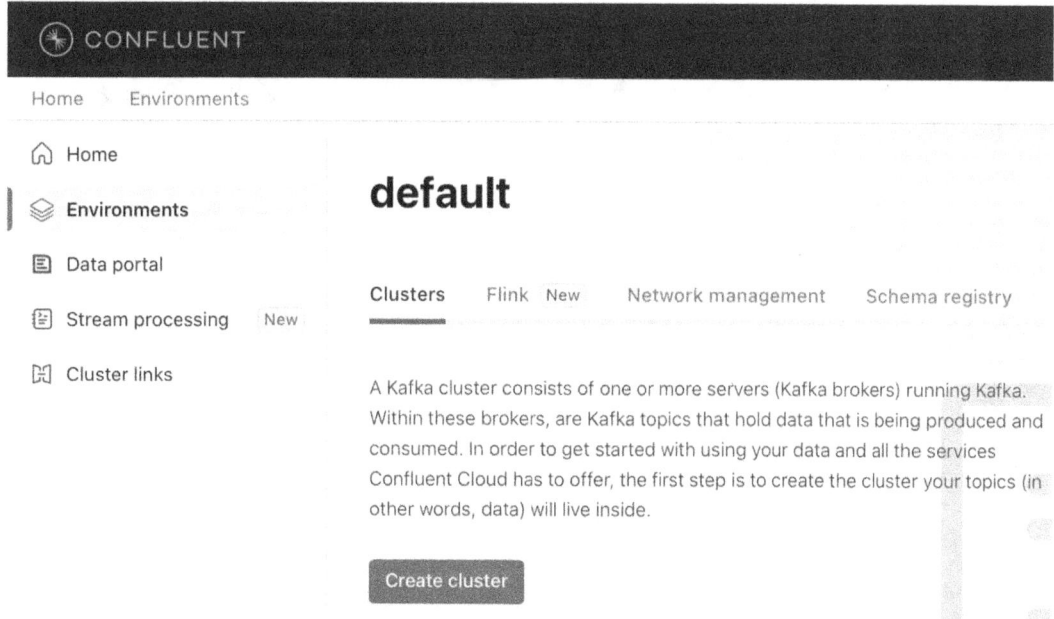

Figure 9-2. Home screen of the Confluent Kafka environment

CHAPTER 9   ENGINEERING REAL-TIME DATA PIPELINES USING APACHE KAFKA

Choose the appropriate cluster type. For learning purposes, the basic option should suffice. Here is how that may look:

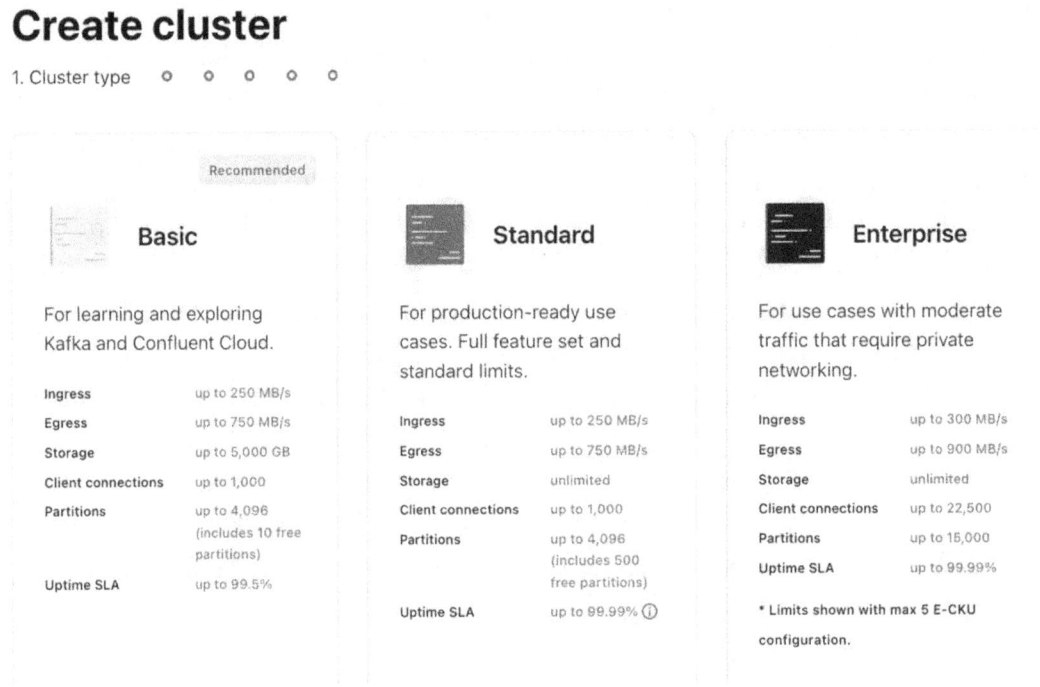

Figure 9-3. Choosing the cluster with appropriate bandwidth

CHAPTER 9 ENGINEERING REAL-TIME DATA PIPELINES USING APACHE KAFKA

You will have the option to specify where your cluster should be hosted. Please feel free to select your favorite cloud service provider and a region closest to where you are located. You will see the estimated costs at the bottom. Here is how this may look for you:

Figure 9-4. *Choosing the appropriate region*

This will be followed by a payment page. If you are registering with Confluent for the first time, you may get free credits; so you can skip the payments and proceed to launch the cluster.

Here is how the cluster may look for you:

CHAPTER 9 ENGINEERING REAL-TIME DATA PIPELINES USING APACHE KAFKA

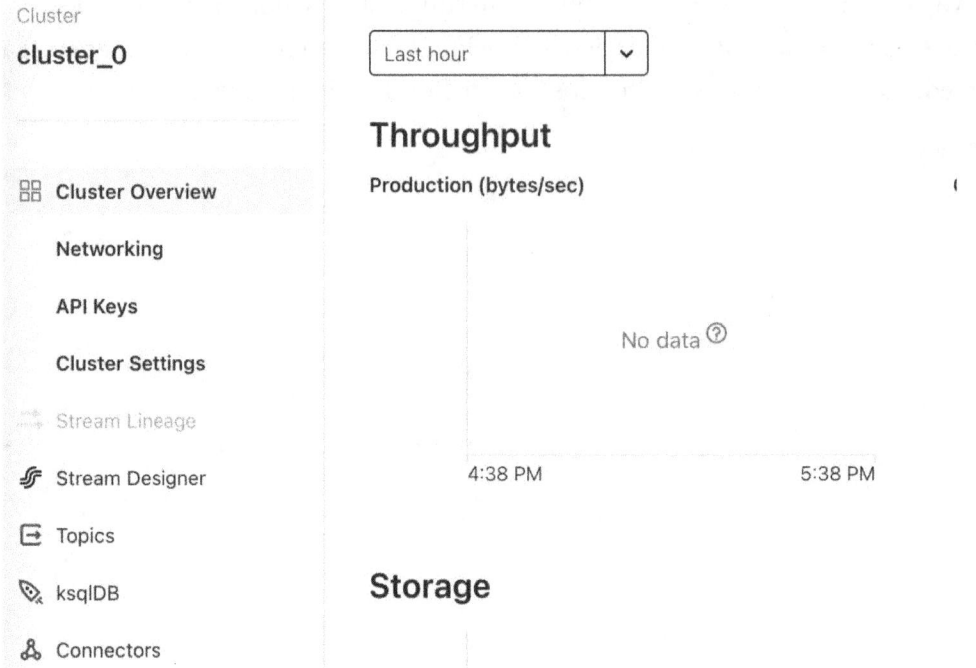

Figure 9-5. *Cluster dashboard within Confluent Kafka*

Once you have a Kafka cluster ready, proceed to create a topic within that cluster. A topic is an append-only log register that registers messages.

At this time, navigate to your cluster within the Kafka dashboard, locate the "Connect to your systems" section, choose Python programming language, and create a cluster API key and a schema registry API key.

Here is how that may look:

CHAPTER 9 ENGINEERING REAL-TIME DATA PIPELINES USING APACHE KAFKA

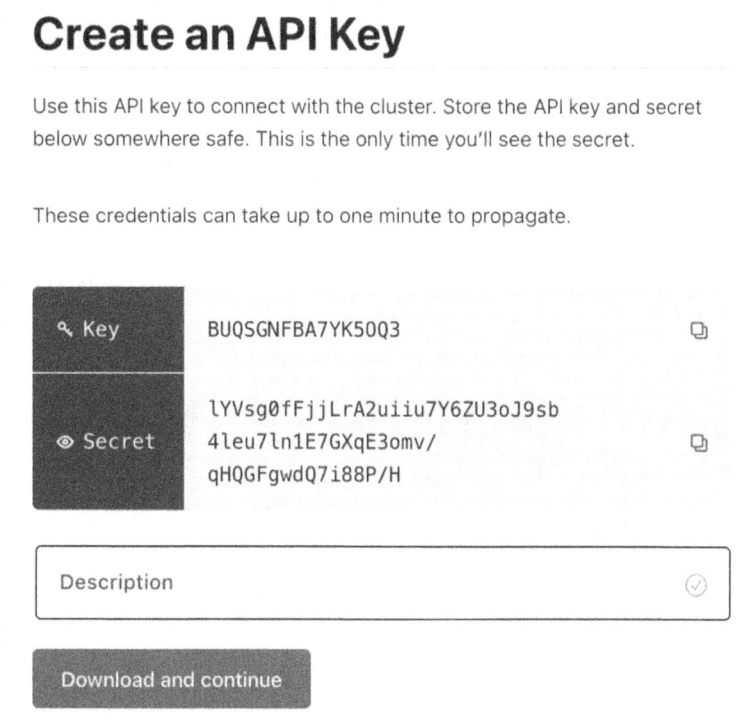

Figure 9-6. *API key and secrets for the Kafka cluster*

For development and sandbox purposes, create a config.py file within your development environment and store the following:

```
config = {
    'bootstrap.servers': '<bootstrap-server-endpoint>',
    'security.protocol': 'SASL_SSL',
    'sasl.mechanisms': 'PLAIN',
    'sasl.username': '<CLUSTER_API_KEY>',
    'sasl.password': '<CLUSTER_API_SECRET>'}

sr_config = {
    'url': '<schema.registry.url>',
    'basic.auth.user.info':'<SR_API_KEY>:<SR_API_SECRET>'
}
```

Make sure to replace the endpoints, key, and secret with appropriate data that is obtained during the process. This is essential for your local Kafka code to interact with your Kafka cluster in the cloud.

Now, let us set up the Kafka development environment in your local workstation. We will start off by revising the creation of the virtual environment:

```
pip3 install venv
```

```
python3 -m venv kafkadev
```

```
source kafkadev/bin/activate
```

If you are on MS Windows operating system, use the following command to activate your Python virtual environment:

```
kafkadev\Scripts\activate
```

To deactivate this environment, go to the command line and enter

```
deactivate
```

Let us now set up the Kafka development environment, from the command line:

```
pip3 install confluent-kafka config
```

Now let us create a simple Kafka publisher–subscriber program. To get started, create a work directory:

```
mkdir Kafka-dev
```

```
cd Kafka-dev
```

Create a new Python script named "config.py". Copy the earlier configuration into the script and comment out the schema registry section that is not needed for our program.

It would look like this:

```python
# python kafka cluster configuration
config = {
    'bootstrap.servers':
        'abcdef-4r47.us-north2.gcp.confluent.cloud:9092',
    'security.protocol':
        'SASL_SSL',
    'sasl.mechanisms':
```

```
            'PLAIN',
        'sasl.username':
            'MG2334H787GH097FRT',
        'sasl.password':
            'vYGUYTGuyGVJHGVUYtt76&*^YUG+ArDmnoHxENzv**o#$%^ijkjkn+buNjCbOW5E'
        }
"""

sr_config = {
    'url': '<schema.registry.url>',
    'basic.auth.user.info':'<SR_API_KEY>:<SR_API_SECRET>'
}
"""
```

Create a new file named producer.py and enter the following:

```
from confluent_kafka import Producer
from config import config

p = Producer(config)

def callback(error, event):
    if error:
        print(f"producer topic {event.topic()} failed for the event
        {event.key}")
    else:
        val = event.value().decode('utf-8')
        print(f"value is sent to the partition {event.partition}")

greetings = ['alice','bob','jenny','rhonda','steve','miller']

for g in greetings:
    p.produce("hello_topic",
            f"hello {g}",
            g,
            on_delivery=callback
            )

p.flush()
```

Now let us examine the code. We start off by importing the necessary libraries, namely, Confluent distribution of Kafka and config. We then instantiate the producer class with the config. The most important methods of the producer class are produce and flush.

The produce method would contain

- The topic to produce the messages to (contains a string)
- The message itself (payload)
- The message key

In the preceding example, we also have an "on_delivery" parameter that takes in a callback function as an argument. This callback function contains information to print upon delivery of messages to the appropriate topic in the Kafka cluster.

When you run this code

```
python3 producer.py
```

the program will connect with the Kafka cluster in the cloud, locate the appropriate topic, and write out the messages to the topic. From the program point of view, it will print the following:

```
value is sent to the partition <built-in method partition of cimpl.Message object at 0x104fbaec0>
value is sent to the partition <built-in method partition of cimpl.Message object at 0x104fbaec0>
value is sent to the partition <built-in method partition of cimpl.Message object at 0x104fbaec0>
value is sent to the partition <built-in method partition of cimpl.Message object at 0x104fbaec0>
value is sent to the partition <built-in method partition of cimpl.Message object at 0x104fbaec0>
value is sent to the partition <built-in method partition of cimpl.Message object at 0x104fbaec0>
```

Let us now create a Kafka consumer that would read the messages from the topic within the Kafka cluster. Please note that the Kafka consumer only tries to read the messages and not in any way alter the form or perform operations other than read. These messages will continue to persist in the topic within the Kafka cluster.

Create a new Python script named "consumer.py" and enter the following code:

```python
from config import config
from confluent_kafka import Consumer

def consumer_config():
    config['group.id'] = 'hello_group'
    config['auto.offset.reset'] = 'earliest'
    config['enable.auto.commit'] = False # type: ignore

consumer_config()

subscriber = Consumer(config)
subscriber.subscribe(["hello_topic"])

while True:
    event = subscriber.poll(1.0)

    if event is None:
        continue

    if event.error():
        print(event.error())

    val = event.value().decode('utf8')
    partition = event.partition()
    print(f'Received: {val} from partition {partition}')

subscriber.close()
```

We start off, once again, by importing the necessary classes. In addition to the configuration supplied by config.py, we also need to mention the consumer group id, whether to start consuming messages that arrived earliest or latest, etc. A consumer class is instantiated in the name of the subscriber, and the same is subscribed to consume messages from the given topic. Finally, the subscriber will operate on an infinite while loop to output the messages, unless the user interrupts with a key combination. As a consumer program, the idea here is to read the messages from the topic and, if possible, print the same to the console screen.

Kafka Application with the Schema Registry

As you can see producers and consumers operate independently. Both these programs have no prior knowledge of each other and do not interact directly during runtime. This method of program design constitutes a loosely coupled architecture. While this is a good method of designing programs, it is critical for both producers and consumers to concur on the data structure to ensure data integrity. In Kafka, this can be achieved using the schema registry. Producers would serialize the data into bytes before sending it to the Kafka cluster. Consumers, on the other hand, would need to deserialize the bytes into data when they read the messages in the topic within the Kafka cluster.

Let us look at an example. Here we have student data, and we are using JSONSerializer to serialize and deserialize our data in publisher and subscriber scripts, respectively.

Let us create a publisher.py. Here is how that looks:

```
from confluent_kafka import Producer
from confluent_kafka.serialization import (SerializationContext,
                                            MessageField)
from confluent_kafka.schema_registry import SchemaRegistryClient
from confluent_kafka.schema_registry.json_schema import JSONSerializer
from config import (config,
                    sr_config)

class Student(object):
    def __init__(self, student_id, student_name, course):
        self.student_id = student_id
        self.student_name = student_name
        self.course = course

schema_str = """{
    "$schema": "https://json-schema.org/draft/2020-12/schema",
    "title": "Student",
    "description": "Student course data",
    "type": "object",
    "properties": {
      "student_id": {
        "description": "id of the student",
        "type": "number"
```

```python
      },
      "student_name": {
        "description": "name of the student",
        "type": "string"
      },
      "course": {
        "description": "course of the student",
        "type": "string"
      }
    }
  }
"""

def student_to_dict(stud, ctx):
    return {
        "student_id": stud.student_id,
        "student_name": stud.student_name,
        "course": stud.course
    }

student_data = [
    Student(55052,"Johndoe","Chem101" ),
    Student(55053,"TestStudent","Physics"),
    Student(55054,"JaneDoe","Biology")
]

def delivery_report(err, event):
    if err is not None:
        print(f'Delivery failed on reading for {event.key().
        decode("utf8")}: {err}')
    else:
        print(f'Student reading for {event.key().decode("utf8")} produced
        to {event.topic()}')

if __name__ == '__main__':
    topic = 'student_data'
    schema_registry_client = SchemaRegistryClient(sr_config)
```

```
json_serializer = JSONSerializer(schema_str,
                                 schema_registry_client,
                                 student_to_dict)

producer = Producer(config)
for i in student_data:
    producer.produce(topic=topic,
                     key=str(i.student_id),
                     value=json_serializer(i,
                                           SerializationContext(
                                           topic,
                                           MessageField.VALUE
                                           )),
                     on_delivery=delivery_report)

producer.flush()
```

We start off with loading the various classes and methods required for this publisher script. We also have loaded a data class and a JSON schema for demonstrating the student data example. We have defined a function that converts student data to a dictionary. The remaining portion of the program is quite familiar to you, as we are instantiating the producer class, schema registry, and JSON serializer, supplying various parameters required for the producer to serialize the data and iterate over the list to write to the topic within the Kafka cluster.

The JSON serializer accepts the schema both in string format and as a schema instance. The student_to_dict function converts the Python object (in this case a list) to a dictionary. The "ctx" parameter provides metadata relevant to serialization.

When this program is run

```
python3 publisher.py
```

we would get

```
Student reading for 55054 produced to student_data
Student reading for 55052 produced to student_data
Student reading for 55053 produced to student_data
```

CHAPTER 9 ENGINEERING REAL-TIME DATA PIPELINES USING APACHE KAFKA

Now let us focus on the consumer portion of this project. Create a file, called consumer.py, and load the following code into the file:

```python
from confluent_kafka import Consumer
from confluent_kafka.serialization import SerializationContext, MessageField
from confluent_kafka.schema_registry.json_schema import JSONDeserializer
from config import config

class Student(object):
    def __init__(self, student_id, student_name, course):
        self.student_id = student_id
        self.student_name = student_name
        self.course = course

schema_str = """{
    "$schema": "https://json-schema.org/draft/2020-12/schema",
    "title": "Student",
    "description": "Student course data",
    "type": "object",
    "properties": {
      "student_id": {
        "description": "id of the student",
        "type": "number"
      },
      "student_name": {
        "description": "name of the student",
        "type": "string"
      },
      "course": {
        "description": "course of the student",
        "type": "string"
      }
    }
  }
"""
```

299

```python
def set_consumer_configs():
    config['group.id'] = 'student_group'
    config['auto.offset.reset'] = 'earliest'

def dict_to_student(dict, ctx):
    return Student(dict['student_id'], dict['student_name'], dict['course'])

if __name__ == '__main__':
    topic = 'student_data'
    print(type(schema_str))
    json_deserializer = JSONDeserializer(schema_str, from_dict=dict_to_student)

    set_consumer_configs()
    consumer = Consumer(config)
    consumer.subscribe([topic])

    while True:
        try:
            event = consumer.poll(1.0)
            if event is None:
                continue
            student = json_deserializer(event.value(),
                SerializationContext(topic, MessageField.VALUE))
            if student is not None:
                print(f'Student with id {student.student_id} is {student.student_name}.')

        except KeyboardInterrupt:
            break

    consumer.close()
```

Once again, the program loads all the required classes and methods to the file, followed by the data class for the student data schema and JSON configuration defining the schema string. We then add the consumer configuration required in addition to the existing configuration and a function that would convert the dictionary to a list.

From here on, we proceed to instantiate our consumer class, mention the specific topic, instantiate the JSON deserializer, and create a loop that would print the student data unless interrupted by keyboard keys.

When you run this program

```
python3 consumer.py
```

it would yield

```
Student with id 55054 is JaneDoe.
Student with id 55053 is TestStudent.
Student with id 55052 is Johndoe.
```

Protobuf Serializer

As we saw earlier, Protobuf stands for Protocol Buffers. Protobuf is developed by Google and considered to be more efficient in terms of speed and size of serialized data. To be able to use the Protobuf serializer, you need to follow the mentioned guidelines as follows:

- First off, define your Protobuf schema in your schema definition file. You need to define the schema and save the file with ".proto" file format. These proto schema files are very much human readable and contain information in much simplified manner when compared with JSON documents.

- Second, compile the ".proto" file into your programming language, in our case a Python script. As mentioned earlier, Protobuf is language agnostic and supports many programming languages.

- Finally, you have to import the generated file into your main code. In our case, we would obtain a Python file. This Python file needs to be imported in our main publisher or subscriber program.

CHAPTER 9   ENGINEERING REAL-TIME DATA PIPELINES USING APACHE KAFKA

Here is an example of schema definition mentioned in a ".proto" file:

```
/*
Protobuf person model
---------------------
*/

syntax = "proto3";

package com.mykafka.personmodel;

message Person {
    string name = 1;
    int32 age = 2;
    float temperature = 3;
    bool needsParkingPass = 4;
    optional string email = 5;
}
```

As you can see, the schema reads much easier when compared with JSON. You can define the data types and name of the fields and also specify whether a specific item requires to be filled, by using the keyword "optional." The numbers that are assigned at the end of each line are field numbers.

To investigate further, you can visit https://protobuf.dev. If you experience issues with installing protoc in your workstation, I suggest obtaining the binaries from the git repository and copying the OS-appropriate executable in a separate folder. Then, open a terminal and use the following command within that folder.

```
./protoc --python_out=. <your-schema-file-here.proto>
```

It is essential to keep the proto file within the same folder. You will get a Python script, carrying the same model name suffixed with "pb2" like the following:

```
myschemafile_pb2.py
```

This file will be imported in our Kafka code. Let us look at an illustrated example using the Protobuf serializer. Here we shall utilize the same student model that we saw previously. Here is how the Protobuf model may look like:

```
/*
Protobuf student model
----------------------
student model to interact with confluence kafka cluster
*/

syntax = "proto3";

package com.kafka.studentmodel;

message student {
    int32   student_id = 1;
    string  student_name = 2;
    string  course = 3;
}

message student_data {
    repeated student student_records = 1;
}
```

"Repeated" is a Protobuf syntax meaning that the student_data can contain multiple "student" entries.

Now, let us compile this proto model using the Protobuf compiler:

```
./protoc --python_out=. model.proto
```

This would yield us

```
model_pb2.py
```

Let us examine the contents of this generated Python file:

```
# -*- coding: utf-8 -*-
# Generated by the protocol buffer compiler.  DO NOT EDIT!
# source: model.proto
# Protobuf Python Version: 4.25.2
"""Generated protocol buffer code."""
from google.protobuf import descriptor as _descriptor
```

```
from google.protobuf import descriptor_pool as _descriptor_pool
from google.protobuf import symbol_database as _symbol_database
from google.protobuf.internal import builder as _builder
# @@protoc_insertion_point(imports)

_sym_db = _symbol_database.Default()

DESCRIPTOR = _descriptor_pool.Default().AddSerializedFile(b'\n\x0bmodel.
proto\x12\x16\x63om.kafka.studentmodel\"C\n\x07student\x12\x12\n\nstudent_
id\x18\x01 \x01(\x05\x12\x14\n\x0cstudent_name\x18\x02 \x01(\t\x12\x0e\n\
x06\x63ourse\x18\x03 \x01(\t\"H\n\x0cstudent_data\x12\x38\n\x0fstudent_
records\x18\x01 \x03(\x0b\x32\x1f.com.kafka.studentmodel.studentb\
x06proto3')

_globals = globals()
_builder.BuildMessageAndEnumDescriptors(DESCRIPTOR, _globals)
_builder.BuildTopDescriptorsAndMessages(DESCRIPTOR, 'model_pb2', _globals)
if _descriptor._USE_C_DESCRIPTORS == False:
  DESCRIPTOR._options = None
  _globals['_STUDENT']._serialized_start=39
  _globals['_STUDENT']._serialized_end=106
  _globals['_STUDENT_DATA']._serialized_start=108
  _globals['_STUDENT_DATA']._serialized_end=180
# @@protoc_insertion_point(module_scope)
```

Now, let us look at our producer script:

```
from confluent_kafka import Producer
import model_pb2
from config import config

class student:
    def __init__(self, student_id, student_name, course):
        self.student_id = student_id
        self.student_name = student_name
        self.course = course

def callback(error, msg):
    if error:
```

```
        print(f"Message delivery did not succeed due to {error}")
    else:
        print(f"Messages of {msg.topic()} delivered at {msg.partition()}")
producer = Producer(config)

student_data = model_pb2.student_data()

data = [
    student(55052,"Johndoe","Chem101" ),
    student(55053,"TestStudent","Physics"),
    student(55054,"JaneDoe","Biology")
]

for i in data:
    student_record = student_data.student_records.add()
    student_record.student_id = i.student_id
    student_record.student_name = i.student_name
    student_record.course = i.course

serialized_data = student_data.SerializeToString()

# Send serialized data
producer.produce('student_data', serialized_data, callback=callback)
producer.flush()
```

We have imported the necessary classes and methods, including the Python file generated by the Protobuf compiler. Then we create a student data class, followed by a callback function that would write out messages to console in case of success or failure of publishing messages to the topic within the Kafka cluster. We then instantiate a producer class and student data. This is succeeded by iteratively adding the test bed of data. Upon serialization, the data is published in the "student_data" topic within the Kafka cluster.

The "SerializeToString()" method is provided by the Python-based Protobuf library. It is a direct way to serialize a Protobuf into a binary string. It does not integrate well with the schema registry.

Here's how to install the Protobuf library:

```
pip3 install protobuf
```

CHAPTER 9 ENGINEERING REAL-TIME DATA PIPELINES USING APACHE KAFKA

Now let us look at the consumer or subscriber portion of this Kafka project:

```python
from confluent_kafka import Consumer
import model_pb2
from config import config, sr_config

def set_consumer_configs():
    config['group.id'] = 'student_group'
    config['auto.offset.reset'] = 'earliest'

set_consumer_configs()

consumer = Consumer(config)
consumer.subscribe(['student_data'])

while True:
    try:
        event = consumer.poll(1.0)

        if event is None:
            continue

        if event.error():
            print(f"Error occurred at {event.error()}")
            continue

        student_data = model_pb2.student_data()
        student_data.ParseFromString(event.value())

        for record in student_data.student_records:
            print(f"{record.student_id} named {record.student_name} has
            enrolled in {record.course}")

    except KeyboardInterrupt:
        break

    finally:
        consumer.close()
```

In this example, we have imported necessary classes and methods, set additional required configurations for the consumer class, and started listening on the given topic. We then write an infinite loop that would print the messages that are read from the topic within the Kafka cluster. The "ParseFromString()" method is from Protobuf Python library, which enables you to deserialize the binary object to structured data. I hope these illustrations help strengthen the foundational thinking and most commonly used classes and methods in Kafka programming.

Stream Processing

So far we have seen how Kafka can publish and subscribe to events. We also saw that Kafka stores events in a Kafka cluster. In this section, we will be looking at how Kafka can process and analyze these events that are stored in the Kafka cluster. This is what we call stream processing in Kafka. Let us define an event. It is simply something that occurred at a given point in time. It could be anything from a student signing up for a course in an online system to a television viewer who is voting for their favorite performer in a reality TV show. Events are basically commit logs that are immutable, and only append operation is permitted. You cannot undo an event; in the case of the student example, you can withdraw from the course that you have signed up for, but remember that becomes another event.

Stream processing can be performed in two approaches: one is Kafka Streams and the other is kSQL. Kafka Streams is a java library that provides various methods to stream processing applications, whereas kSQL is a module that enables building stream processing pipelines in a SQL dialect. ksqlDB is a database that is deliberately built for storing, processing, and analyzing events from stream processing applications.

ksqlDB is not a relational database and it does not replace relational database at all. There are no concept of indexes in ksqlDB. The tables in ksqlDB are different from tables in relational database management systems. The ksqlDB table can be seen more as a materialized view and not a table. Streams can also be defined as sets of events that you can derive a table from, for instance, the price of a company's stock listed/traded on a stock exchange in a given time interval. The sequence of stock prices over the time period is a stream. The same stream can be seen as a table to describe the most recent stock price of that company.

Stateful vs. Stateless Processing

There are two types of stream processing that can be performed within ksqlDB, namely, stateful and stateless processing. Stateless processing in Kafka is where an event is processed without having to rely on information relating to previous messages. As stateless processing does not require to have any knowledge about the messages that occurred in the past or will happen in the future, they are simpler and faster to process. Here are some examples:

- Transformation of a value that does not involve any knowledge from past or future messages, for instance, converting a Celsius to Fahrenheit

- Filtering a value based on a condition that does not require knowledge regarding other messages, for instance, a filter where the temperature is greater than 35 degrees Celsius

- Creating branches where the input streams are branched into multiple output streams based on criteria applied to the message

Stateful processing, on the other hand, requires information regarding previous messages. Typically, it involves performing operations that require one or more states. Common applications would include preparing aggregations over a given window (either time-based window or count-based window). Here are some examples:

- Performing aggregations over a window of time, for instance, obtaining the total count, sum, average, and median within a given time interval

- Performing joins based on the relationship between messages, for instance, joining from more than one stream

A stream processing application would typically involve ingesting data from one or more streams, applying transformations where applicable, and loading them into a storage. An average stream processing application can contain many moving parts, right from the data ingestion and performing transformations (which might mean that data is written in a staging area, and so that is another storage) to writing to a storage. ksqlDB simplifies the entire stream processing application and narrows down to only

CHAPTER 9 ENGINEERING REAL-TIME DATA PIPELINES USING APACHE KAFKA

two services: the ksqlDB that performs computation and Kafka that constitutes storage. Within ksqlDB, you can utilize Kafka to both stream data out of other systems into Kafka and from Kafka to other downstream systems.

It may be good to reiterate that ksqlDB is a separate package from Apache Kafka. And so, we will continue to use our existing clusters and may create new streams.

Let us try to set up a ksqlDB cluster for us to write streaming data pipelines. To get started, log in to Confluent Cloud, select the appropriate environment, choose the right cluster, and click ksqlDB.

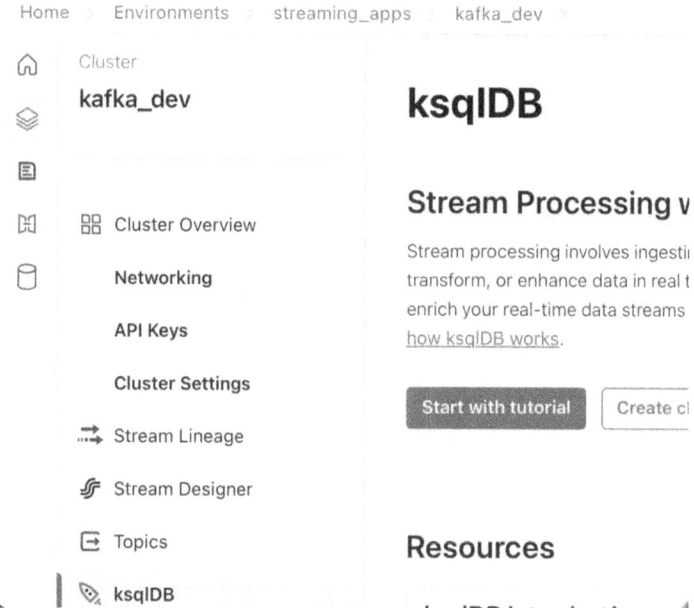

Figure 9-7. *ksqlDB within Confluent Kafka*

You may choose to create a cluster on your own or start with the tutorial. The goal is to set up a development instance to get started with the stream data processing pipelines. You may choose to enable global access for your ksqlDB cluster and ensure appropriate privileges to the schema registry.

CHAPTER 9   ENGINEERING REAL-TIME DATA PIPELINES USING APACHE KAFKA

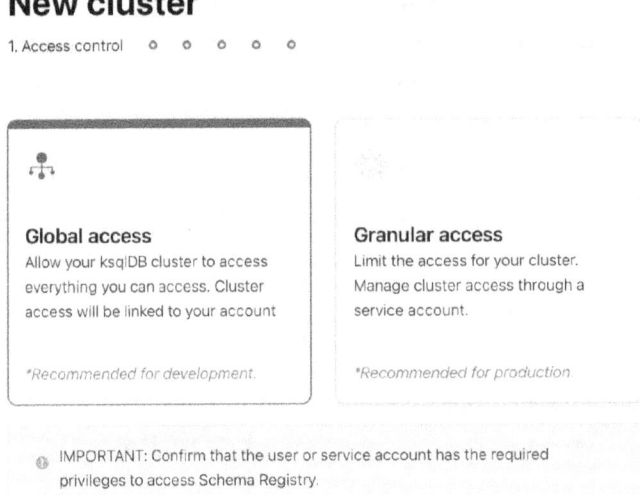

Figure 9-8. Creating a new ksqlDB cluster

You may choose the lowest Confluent streaming units and choose to launch clusters. Here is how that looks:

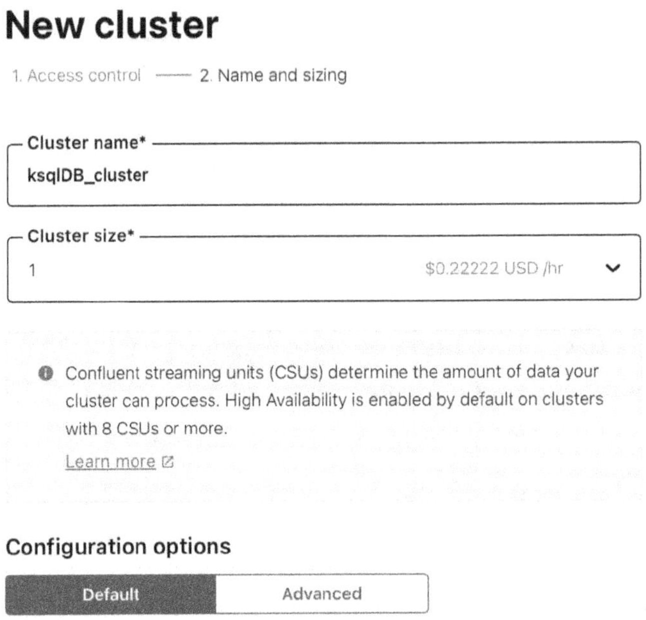

Figure 9-9. Naming and sizing the ksqlDB cluster

Do wait a few minutes for your cluster to be provisioned. You'll know the cluster is ready when the status changes to "Up" from Provisioning. Once we have the ksqlDB up and running, we can try to set up the command line interface for both Confluent and ksqlDB.

The easiest and straightforward method is to spin up a Docker container containing the Confluent platform. You can visit the appropriate page, copy the Docker compose file, and run those commands to get up and running. However, it would be nice to install these packages in the terminal, since we will also get hands on installing a package without an executable. Let us start looking at installing Confluent CLI in a Debian operating system.

The first step is to install the prerequisites:

```
sudo apt update
sudo apt install -y software-properties-common curl gnupg
```

Let's check if you have Java installed; observe the output for the following command:

```
java -version
```

If java is not installed, then kindly proceed with

```
sudo apt install -y default-jdk
```

Let us import the public key of the ksqlDB package. This is to verify the authenticity of the package, part of a security mechanism known as cryptographic signing and verification:

Note In the world of cryptographic signing and verification, a package is digitally signed with a private key. A public key that is available in public (which we are importing right now) is used by developers to verify this signature.

```
wget -qO - https://packages.confluent.io/confluent-cli/deb/archive.key | sudo apt-key add
```

Let us add the ksqlDB into our apt repository and run an update:

```
sudo add-apt-repository "deb https://packages.confluent.io/confluent-cli/deb stable main"
```

```
sudo apt update
```

Let us install the CLI package:

```
sudo apt install confluent-cli
```

In my case, I also installed the Confluent platform locally. It would mean downloading the source code, setting the environment variable and path, and checking to see if the installation worked.

Let us obtain the source code and extract the tarball at your preferred destination:

```
curl -O https://packages.confluent.io/archive/7.5/confluent-7.5.3.tar.gz
```

```
tar xzvf confluent-7.5.3.tar.gz
```

Copy the file path to set the environment variable and path:

```
export CONFLUENT_HOME=<your local installation path>
```

```
export PATH=$PATH:$CONFLUENT_HOME/bin
```

Enter the following help command on the terminal to check whether installation succeeded:

```
confluent –help
```

```
Manage your Confluent Cloud or Confluent Platform. Log in to see all available commands.

Usage:
  confluent [command]

Available Commands:
  cloud-signup    Sign up for Confluent Cloud.
  completion      Print shell completion code.
  configuration   Configure the Confluent CLI.
  context         Manage CLI configuration contexts.
  help            Help about any command
  kafka           Manage Apache Kafka.
  local           Manage a local Confluent Platform development environment.
  login           Log in to Confluent Cloud or Confluent Platform.
  logout          Log out of Confluent Cloud or Confluent Platform
  plugin          Manage Confluent plugins.
  prompt          Add Confluent CLI context to your terminal prompt.
  secret          Manage secrets for Confluent Platform.
  shell           Start an interactive shell.
  version         Show version of the Confluent CLI.

Flags:
      --version          Show version of the Confluent CLI.
  -h, --help             Show help for this command.
      --unsafe-trace     Equivalent to -vvvv, but also log HTTP requests and responses which might contain plaintext secrets.
  -v, --verbose count    Increase verbosity (-v for warn, -vv for info, -vvv for debug, -vvvv for trace).

Use "confluent [command] --help" for more information about a command.
```

Figure 9-10. Help screen for the Confluent command line interface

CHAPTER 9 ENGINEERING REAL-TIME DATA PIPELINES USING APACHE KAFKA

If you get something similar to this screen, then you succeeded in installing a Confluent platform.

Now it is time to set up the environment so you can interact with the ksqlDB cluster in the cloud. Within the dashboard, click ksqlDB and choose CLI instructions. You will see instructions on logging in to Confluent Cloud CLI and setting the context to your environment and cluster:

```
confluent login
confluent environment use env-0jvd89
confluent kafka cluster use lkc-555vmz
```

Create an API key and secret for your ksqlDB cluster:

```
confluent api-key create --resource lksqlc-8ggjyq
```

You will get something like the following:

```
It may take a couple of minutes for the API key to be ready.
Save the API key and secret. The secret is not retrievable later.
+------------+-----------------------------------------------------------------+
| API Key    | IA6NCZZHXVMEA3YT                                                |
| API Secret | huUVi2ySi8GH8ZnamFzAeanIFZz+CTihu+iRkTGyfEoo9BQ8y+VEnGDqLxhkOC0J |
+------------+-----------------------------------------------------------------+
```

Figure 9-11. *ksqlDB cluster API key and secret*

Now, enter the following so you can access the ksqlDB command line interface; don't forget to enter the appropriate key and secret:

```
$CONFLUENT_HOME/bin/ksql -u <KSQL_API_KEY> -p <KSQL_API_SECRET> https://pksqldb-pgkn0m.us-east-2.aws.confluent.cloud:443
```

Here is how that looks:

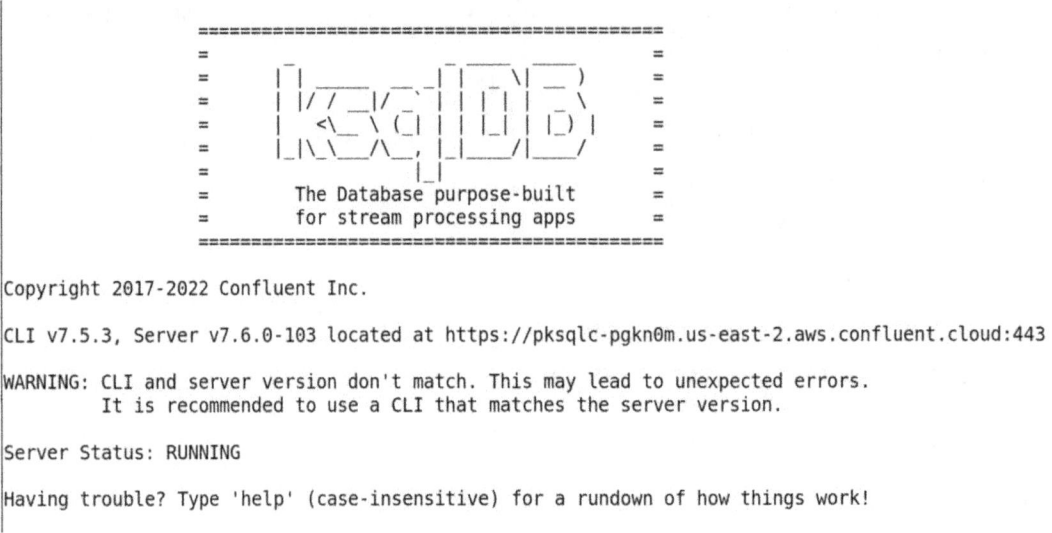

Figure 9-12. ksqlDB on the command line interface

Now let us start with some basic stream applications. A stream associates a schema with an underlying Kafka topic. Let us begin by creating a topic in the Kafka cluster. Visit the cluster's dashboard and click "Topics." Here is how that looks:

CHAPTER 9   ENGINEERING REAL-TIME DATA PIPELINES USING APACHE KAFKA

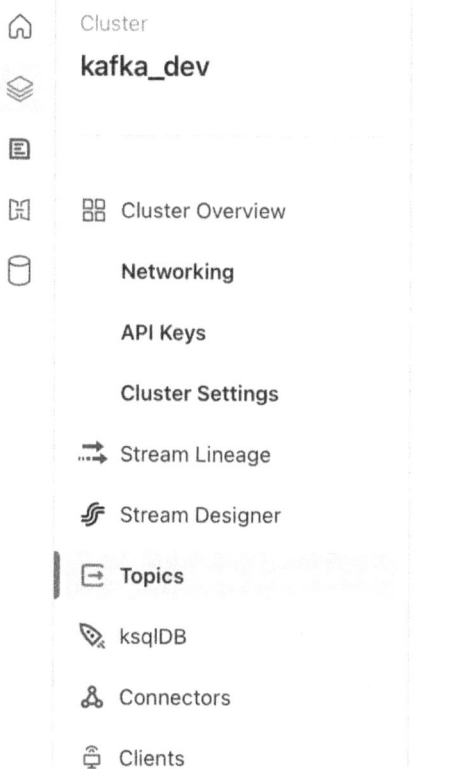

Figure 9-13. *Creating a new topic in the Kafka cluster*

Select "Add topic" and name the topic as "students." Leave other fields as is and select "Create with defaults." Here is how that looks:

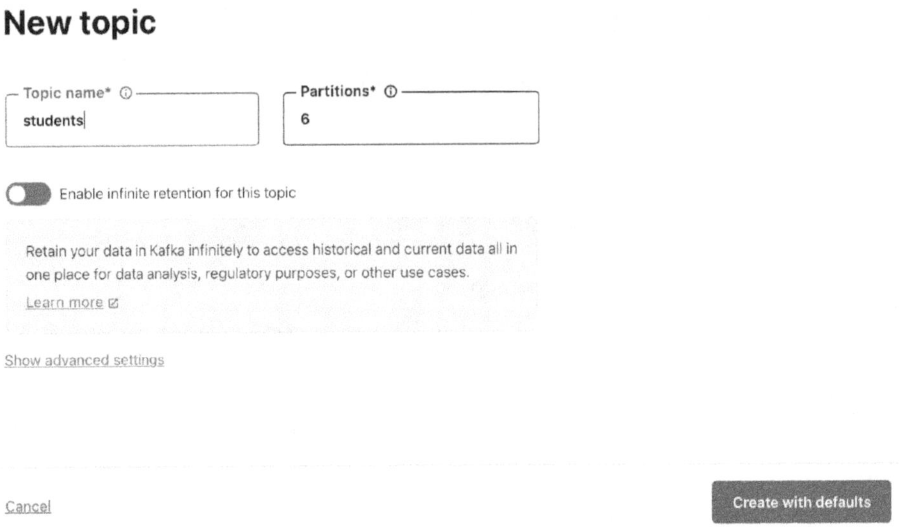

CHAPTER 9 ENGINEERING REAL-TIME DATA PIPELINES USING APACHE KAFKA

Figure 9-14. Naming and specifying partitions for the new topic

Select the option of creating a schema for message values. The option for creating a schema for message keys may help in the production environment when it comes to achieving consistency on structure of keys and routing to appropriate partitions (messages with the same key are sent to the same partition).

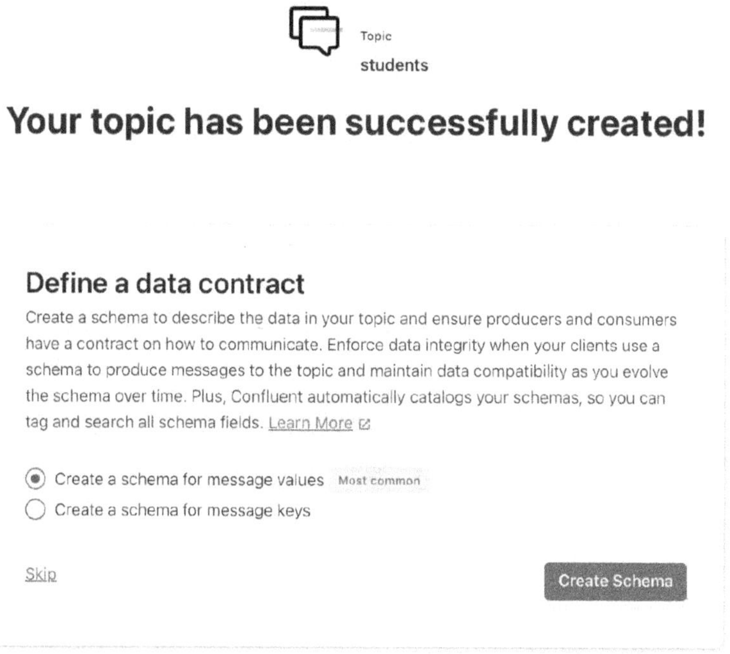

Your topic has been successfully created!

Define a data contract

Create a schema to describe the data in your topic and ensure producers and consumers have a contract on how to communicate. Enforce data integrity when your clients use a schema to produce messages to the topic and maintain data compatibility as you evolve the schema over time. Plus, Confluent automatically catalogs your schemas, so you can tag and search all schema fields. Learn More

- ● Create a schema for message values Most common
- ○ Create a schema for message keys

Skip   Create Schema

Figure 9-15. *Creating a schema for the topic*

Click "Create Schema" and paste the following:

```
syntax = "proto3";

message students {
    int32   id = 1;
    string  name = 2;
    string  course = 3;
}
```

CHAPTER 9 ENGINEERING REAL-TIME DATA PIPELINES USING APACHE KAFKA

Figure 9-16. Creating a new schema

There is also the option of creating a stream from the ksqlDB command line interface. Let us look at that approach in detail. Using CLI, we will create a new stream called students. Locate the Editor tab within the ksqlDB and enter the following code to create a new students stream. Ensure you set the "auto.offset.reset" property to earliest.

Here is how that looks:

```
set 'auto.offset.reset' = 'earliest';

CREATE STREAM students (
  id INTEGER KEY,
  name string,
  course string)
WITH (kafka_topic = 'students',
      PARTITIONS=1,
      value_format='JSON');
```

CHAPTER 9   ENGINEERING REAL-TIME DATA PIPELINES USING APACHE KAFKA

Now, insert data into the stream:

```
INSERT INTO students (id, name, course) VALUES (5, 'John', 'Physics');
INSERT INTO students (id, name, course) VALUES (6, 'Joe', 'Math');
INSERT INTO students (id, name, course) VALUES (7, 'Jane', 'Physics');
INSERT INTO students (id, name, course) VALUES (8, 'Jan', 'Physics');
INSERT INTO students (id, name, course) VALUES (10, 'Alice', 'Math');
INSERT INTO students (id, name, course) VALUES (12, 'Al', 'Chemistry');
INSERT INTO students (id, name, course) VALUES (4, 'Moe', 'Chemistry');
```

Now we can try to run a query on this stream. Here is a sample query for you to get started:

```
SELECT * FROM STUDENTS WHERE COURSE IN ('Math','Physics');
```

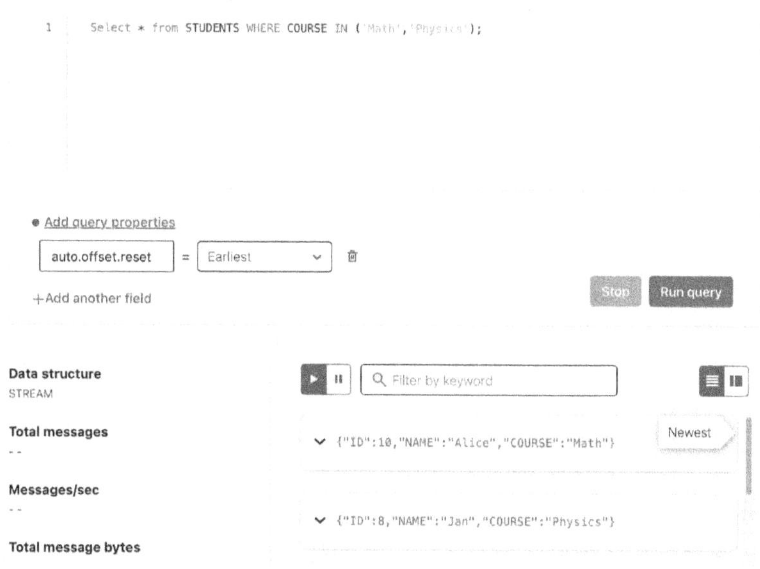

Figure 9-17. Executing a new query

Notice how you can view the results on the dashboard within your browser. Let us try to access the Kafka cluster through the command line interface. Log in to Confluent Cloud, which we installed earlier, and access ksqlDB using the same commands.

In the ksqlDB prompt, enter the following query:

```
ksql> select * from students where course = 'Physics' EMIT CHANGES;
```

You will see the following results:

CHAPTER 9   ENGINEERING REAL-TIME DATA PIPELINES USING APACHE KAFKA

```
ksql> select * from students where course = 'Physics' EMIT CHA
NGES;
+------------------+------------------+------------------+
|ID                |NAME              |COURSE            |
+------------------+------------------+------------------+
|5                 |John              |Physics           |
|7                 |Jane              |Physics           |
|8                 |Jan               |Physics           |
```

Figure 9-18. Results of the query

Let us navigate to the dashboard on browser to insert the following query:

insert into students (id, name, course) values (16, 'EnteredFromBrowser', 'Physics');

Click "Run query." Notice the terminal where you have the SQL results persistent. You will likely see the following:

```
ksql> select * from students where course = 'Physics' EMIT CHA
NGES;
+------------------+------------------+------------------+
|ID                |NAME              |COURSE            |
+------------------+------------------+------------------+
|5                 |John              |Physics           |
|7                 |Jane              |Physics           |
|8                 |Jan               |Physics           |
|16                |EnteredFromBrowser|Physics           |
```

Figure 9-19. Results of the query

We have just witnessed how these supposedly queries are not exactly relational database-like queries but rather streams. As soon as we insert data into the stream, that same data pops up in our persistent stream output.

Kafka Connect

We saw in our earlier discussion how Kafka Connect helps us scale and stream data reliably between Apache Kafka and other database systems. Kafka Connect offers a tool called connectors that would connect with other data systems to enable and coordinate data streaming. There are two types of connectors, namely, source connector and sink connector. Source connectors ingest the data systems to stream data to Kafka topics, whereas sink connectors deliver data to data systems from Kafka topics.

The Kafka Connect service can be installed on a standalone mode or be deployed in distributed mode for production purposes. Kafka Connect is an important aspect of streaming data/ETL (extract, transform, and load) pipelines, especially when you have Kafka in the architecture. You can connect to an external data system using the Kafka Connect service through ksqlDB. Please see the Oracle database example connector here; the fields with values enclosed in brackets need to be populated with respective database credentials:

```
CREATE SOURCE CONNECTOR IF NOT EXISTS CustomerDB WITH (
    'connector.class'           = 'OracleDatabaseSource',
    'kafka.api.key'             = '<your-kafka-api-key>',
    'kafka.api.secret'          = '<your-kafka-api-secret>',
    'connection.host'           = '<database-host>',
    'connection.port'           = '1521',
    'connection.user'           = '<database-username>',
    'connection.password'       = '<database-password>',
    'db.name'                   = '<db-name>',
    'table.whitelist'           = 'CUSTOMERS',
    'timestamp.column.name'     = 'created_at',
    'output.data.format'        = 'JSON',
    'db.timezone'               = 'UCT',
    'tasks.max'                 = '1'
);
```

Best Practices

As you can see, we have illustrated Kafka streaming data pipelines using a commercial distribution, Confluent. At the time of writing this chapter, the vendor provided hundreds of dollars on cloud credits that can be used to apply toward availing their services. This way, anyone who is looking to get their feet wet with Kafka can avail their services to run sample clusters. At the same time, it is also a good practice to be mindful of unused Kafka clusters running in the background. Here are some commands that can be used toward cleaning up unused Kafka resources:

```
DROP STREAM IF EXISTS customers;
```

You can also asynchronously delete the respective topic backing the table or stream:

```
DROP TABLE IF EXISTS t_customers DELETE TOPIC;
```

You can remove Kafka connectors as well:

```
DROP CONNECTOR IF EXISTS customerDB;
```

Conclusion

It is safe to say that Apache Kafka represents a paradigm shift on how we approach building a real-time data pipeline and stream analytics solution. The architecture is scalable, distributed, and fault tolerant, which provides a reliable foundation for building and managing complex data pipelines that can handle the modern data streams.
We discussed the Kafka architecture and various components and how they all work together. We looked at publisher-subscriber models and how topics are stored in a database. We then looked at streaming data pipelines using ksqlDB and how the data can be analyzed to gain business insights. I hope the skills and knowledge you gained from this chapter may serve you well in building real-time data pipelines.

CHAPTER 10

Engineering Machine Learning and Data REST APIs using FastAPI

Introduction

In the complex world of data engineering and machine learning modeling, where there are several moving parts (as you have seen so far), it is essential to deliver data services and deploy machine learning models for consumption. In this chapter, we will be discussing FastAPI. FastAPI is a Python library that primarily enables web application development and microservice development. We will look at FastAPI with the sole intention of using the library for delivering data services and machine learning models as services. Though Django has won many hearts in the world of Python-based application development, FastAPI is simple, strong, and very powerful. It is important to understand the concept of application programming interfaces for a ML engineer or data engineer.

By the end of this chapter, you will learn

- Fundamentals of web services and APIs, various types of APIs, and concepts
- FastAPI and its setup process
- Pydantic integration with FastAPI with robust data validation
- FastAPI dependency injection, concept, and implementation

- Database integration, including SQL, NoSQL, and object relational mapping (ORM) approaches
- Building a RESTful data API using FastAPI
- Machine learning API development and deployment using FastAPI

Introduction to Web Services and APIs

Businesses tend to offer value to their internal or external customers by providing valuable data products. The data products are built by the data engineers by identifying various data sources, performing various options, and combining data from these multiple sources, not necessarily in the same order. The data assets then are shared to various internal and external clients. One of the effective ways businesses and IT systems can share data and communicate with each other is the concept of APIs. APIs stands for application programming interface. APIs are built by various companies and organizations to allow various constituents to utilize their services. A good example of an organization providing data as a service (DaaS) is OpenWeather API.

OpenWeather API

OpenWeather API is a data service provided by OpenWeather, a company that provides weather data to individuals and organizations by delivering through REST API calls. The OpenWeather API enables engineers to access a wide range of weather data programmatically. Some of the data include current weather data, forecasts, and historical data for various locations across the world. They provide various API endpoints for current weather data, forecast weather data, daily aggregations, historical data, and so on. You can register for the OpenWeather API in the following website to get an access key. You can obtain various data through the use of the access key:

`https://openweathermap.org/api`

Types of APIs

Not all APIs are made the same way; neither do they serve the same constituents. The most common class of APIs is the Open API. These APIs are provided by various organizations and they are free to access and utilize. Engineers can access and utilize the services without any restrictions. The other class of APIs are internal APIs. These APIs once again deliver services; however, they are internal to an organization. They are not exposed to the public. Instead, they are accessed and utilized to fetch data and functionality by various teams, within a given organization only. There is another variation to internal APIs, which is called partner APIs. These APIs again pose some level of confidentiality; however, they may cater to the organization and some of their partner organizations as well.

In addition to various classes, and the purposes they serve, there are various methods of accessing APIs; some of them are as follows.

SOAP APIs

SOAP stands for simple object access protocol. SOAP APIs have long been in existence and used in enterprise-grade applications. SOAP is primarily meant for exchanging structured data between two applications. SOAP uses either HTTP or SMTP to transport the message. They use an XML template for message formatting.

REST APIs

REST stands for representational state transfer. REST is an architectural style for building lightweight and scalable APIs. They leverage HTTP methods for communication. Some of the HTTP methods are GET, PUT, POST, etc. REST APIs return data in JSON or XML format. These APIs are commonly used in web-based applications and mobile applications.

GraphQL APIs

GraphQL stands for graph query language. As the name stands, GraphQL uses query-based requisition for fetching, making sure they fetch just enough data they need. The emphasis of GraphQL is that it avoids overfetching. While that information may be a bit too far-fetched for you (depending upon your background), it is best to note that such an API exists.

Webhooks

Webhooks are another type of APIs that enable real-time communication between systems. Like REST APIs, webhooks also communicate using HTTP requests.

Typical Process of APIs

APIs serve as a middle person between software applications, enabling smoother interaction with each other. The communication can either be between two software applications or an engineer looking to fetch services from a host.

It all starts off when one of the parties, let's call them client, sends a request to the host through the API. The API carries the message and submits to the host. The host would process that message, prepare the necessary procedures, and provide the requested information.

Now, the host (let's call them server) responds to the client. The response contains the information that is intended for the appropriate request. This interaction or communication, whichever you may prefer to call it as, happens by exchange of data. This data is structured in a certain way for APIs and servers to understand. It is commonly referred to as data format. The most common data formats are JSON and XML. These data formats ensure that all the three entries—client, API, and server—can understand the interaction with each other.

Endpoints

As mentioned, APIs are intermediaries; they serve as a bridge between two parties. APIs provide a mechanism for clients initiating a conversation with a server. One such mechanism is called URL or URI, namely, uniform resource locator or uniform resource identifier. It contains the address of the server so the requests are made to the right address.

Let's take the following example:

> Jane Doe,
>
> Accounting Department,
>
> ABC Company,
>
> Any town, USA

Say, if one sends a physical mail to this address, then it would go to the respective town, to the concerned organization (hypothetical one, in this case), to the respective department, and to the person addressed in that said department.

A REST API endpoint is similar to the address of a person just mentioned.

The structure of a URL or URI is this:

`https://developer.MyFavoriteMovieTheatre.com/api/movie-theatre/showtimes?movie_id=5&date=2023-06-06`

In here

`https://developer.MyFavoriteMovieTheatre.com` is the base URL of the API.

`/api/movie-theatre/showtimes` represents the path parameters for retrieving showtimes.

`movie_id=5` and `date=2023-06-06` represent query parameters. They allow you to specify which movie you want to see and what date. The movie_id parameter corresponds to a movie shown on a certain day.

This whole URL is called an endpoint. Endpoints are addresses just like the physical address mentioned previously. By stating the correct base URL, correct path, and appropriate query parameters, the movie theatre server will be able to process your request and retrieve the response.

APIs usually require authentications. This is done so as to ensure only authorized users or applications access their functionality. There are several ways to authenticate with an API. Some of the common ways are the usage of API keys, tokens, and other authentication mechanisms.

APIs move data from source to sink and provide an easier way to ingest data.

API Development Process

To build an API, we begin by defining the API requirements like the data it may need to access, operations it will perform, and outcomes it will produce, to name a few. Once you have the requirements defined, you can move toward building data models, methods, and endpoints and defining response data formats for requests, authentication, authorization, etc. Using the requirements, implement the API, where you set up the required servers and their infrastructure, implement the code and dependencies (if any), and other relevant tasks. Performing adequate testing on features and functionalities may be a good step, either during the implementation or post-implementation phase. The final step would be to deploy the API in production servers; performance and usage monitoring may benefit addressing any improvements that may be needed.

CHAPTER 10   ENGINEERING MACHINE LEARNING AND DATA REST APIS USING FASTAPI

REST API

As discussed, REST expands to representational state transfer. It is one of the types of web service that complies with REST architectural standards. REST architectural standards leverage client–server communication over a given network. REST architecture is stateless, meaning that the server will not maintain session information for a request from a client. Instead, the request message itself would contain just enough information for the server to process the request.

RESTful services listen for HTTP methods and retrieve the response from the server. HTTP stands for HyperText Transfer Protocol. The HTTP methods are such that they tell the server which operation to perform in order to retrieve the response for a request. There are two categories of HTTP methods, safe HTTP methods and idempotent HTTP methods. Safe HTTP methods are such that they always return the same or similar response regardless of the number of times they have been called. Safe HTTP methods do not change the data on the server. Idempotent methods may be able to change the data on the server. Some of the more common HTTP methods (or verbs) include

- GET: Reads data from the server and does not alter or change resources in any way

    - Here is a simple GET request to request all movies at a multiplex:

    GET /multiplexapi/movies/

- HEAD: Similar to GET, retrieves only the header data without the content

    - Here is a HEAD request to see the last update time for movies:

    HEAD /multiplexapi/theatreslist

- POST: Triggers an action or can be used to create a new data; makes changes on the server

    - Here is an example of creating a new booking for a moviegoer:

    POST /multiplexapi/bookings

- PUT: Replaces current data with new data; can be seen as performing an update operation

    - Here is an example of updating showtime for a movie:

        PUT /multiplex/movies/SampleMovie { "showtime": "5:55PM" }

- PATCH: similar to PUT but partially replaces or updates the current data with new data

    - Here is an example of updating the number of seats left for "SampleMovie" that is shown on the third theater:

        PATCH /multiplex/threatre/3 { "seats_available" : 5 }

- OPTIONS: To obtain a possible set of permitted HTTP methods for a given URL to the server

    - Here is an example of examining various booking options:

        OPTIONS /multiplexapi/bookings

- CONNECT: Method used to establish a tunnel to a server

    - Here is an example of establishing TCP connection to multiplex:

        CONNECT app.multiplexapi.com:3690

HTTP Status Codes

When a server receives a request from a client, it processes and prepares a response. An HTTP status code is a code that is received along with the response. Here are some common status codes received by the client:

Table 10-1. *Some common HTTP status codes and their descriptions*

HTTP Status Code	Description
200 OK	Request successful, server returned the requested data.
201 Created Description	Request successful, new resource was created as a result.
204 No Content	Request successful, but nothing to send back (often for DELETE operations).
301 Permanent Redirect	Requested page has permanently moved to a new URL.
400 Bad Request	Server cannot process the request due to an error from client.
401 Unauthorized	Server requires user authentication (valid credentials) to process the request.
403 Forbidden	Server understood your request but refuses to authorize it, in spite of valid credentials (not enough privilege maybe).
404 Not Found	Whatever is requested, cannot be found on the server.
500 Internal Server Error	Due to some unexpected issue, server can't seem to find the requested.
503 Service Unavailable	Server unable to handle the request at the moment.

These web services expose their data or a service to the outside world through an API.

To build APIs in Python, there are quite a few packages available. There is Flask, a Python library known for and meant for building simple, easy-to-setup, and smaller APIs. The popular ones are Django and FastAPI. Django has been around for a long time, has more functionalities, and is a better option than Flask. In this chapter, we will take a deep dive into FastAPI.

FastAPI

FastAPI provides certain advantages over other Python libraries; first off, FastAPI supports Python's asynchronous programming through asyncio. FastAPI automatically generates documentation that is based on Open API standards, allowing the users of the

API to explore the documentation and test various endpoints. Most important of them all, as the name suggests, FastAPI is high performance and faster than the many other Python libraries.

Setup and Installation

Here's how to set up FastAPI in your machine:

```
pip install fastapi
```

FastAPI does not come with a web server; let us also install a web server:

```
pip install "uvicorn[standard]"
```

Assuming you are using a virtual development environment, I also encourage you to consider this command:

```
pip install "fastapi[all]"
```

This will install all the optional dependencies used by FastAPI, Pydantic, and Starlette Python libraries, which include the "uvicorn" web server and asynchronous client "http."

Now let us consider a simple hello world application using FastAPI. Here is how that looks:

```python
from fastapi import FastAPI, APIRouter

app = FastAPI()

api_router = APIRouter()

@api_router.get("/")
def root() -> dict:
    return {"msg": "Hello, World!"}

app.include_router(api_router)
```

Core Concepts

FastAPI is based on AGSI standards and Starlette. FastAPI utilizes AGSI servers to deliver its asynchronous capabilities. AGSI stands for Asynchronous Server Gateway Interface. AGSI is a relatively new web standard for servers to asynchronously communicate with clients. FastAPI supports Hypercorn and Uvicorn web servers as they are compatible with AGSI standards.

Starlette, on the other hand, is an AGSI standard framework that helps with core functionality of FastAPI. Starlette provides the core routing infrastructure, handling requests and responses, database integration, and many other features that are leveraged by FastAPI. On top of Starlette and AGSI standards, FastAPI natively provides data validation, documentation generation, etc. so engineers can focus on the business logic of things.

Path Parameters and Query Parameters

As you may know, path parameters are components of an endpoint, whereas query parameters are optional filter-like parameters. Let's look at an example; here is a URL:

https://www.amazon.com/Sweetcrispy-Ergonomic-Computer-Comfortable-Armrests/dp/B0C61DLQNH/ref=mp_s_a_1_5?crid=1CE83JNY3DEDI&keywords=chair&qid=1703648878&refinements=p_36%3A1253524011&rnid=386465011&s=home-garden&sprefix=cha%2Caps%2C478&sr=1-5

The base URL is
https://www.amazon.com/
The path parameters are

/Sweetcrispy-Ergonomic-Computer-Comfortable-Armrests
/dp
/B0C61DLQNH
/ref=mp_s_a_1_5

The query parameters are

?
crid=1CE83JNY3DEDI &
keywords=chair &

```
qid=1703648878 &
refinements=p_36%3A1253524011 &
rnid=386465011 &
s=home-garden &
sprefix=cha%2Caps%2C478 &
sr=1-5
```

When you are writing a function, you can pass variables to the function that can get called. Here is how that looks:

```
from fastapi import FastAPI, APIRouter
```

Let us create a list of sample students and course data:

```
Students = [
    {
        "Student_id": 1,
        "Student_name": "StudentPerson1",
        "Course" : "BIO 101",
        "Department" : "Biology"
    },
    {
        "Student_id": 2,
        "Student_name": "StudentPersonTwo",
        "Course" : "CHEM 201",
        "Department" : "Chemistry"
    },
    {
        "Student_id": 3,
        "Student_name": "StudentPerson3",
        "Course" : "MATH 102",
        "Department" : "Mathematics"
    },
]
```

Let us instantiate FastAPI (with a suitable title) and APIRouter objects:

```
app = FastAPI(title="Student/Course Data")

api_router = APIRouter()
```

The main function defined earlier has not changed much in its structure. Let us add a new function called "get_student" to retrieve student records. Notice the use of the decorator that includes variables. The value of that variable or parameter will be passed during runtime:

```
@api_router.get("/")
def hello() -> dict:
    return {"msg": "Welcome to the root page"}

@api_router.get("/student/{Student_id}")
def get_student(Student_id: int) -> dict:
    for s in Students:
        if s["Student_id"] == Student_id:
            result = s
    return result

app.include_router(api_router)
```

Let's run this code real-time. Here is how we are running this code:

```
uvicorn main3:app –reload
```

would return the following stack trace:

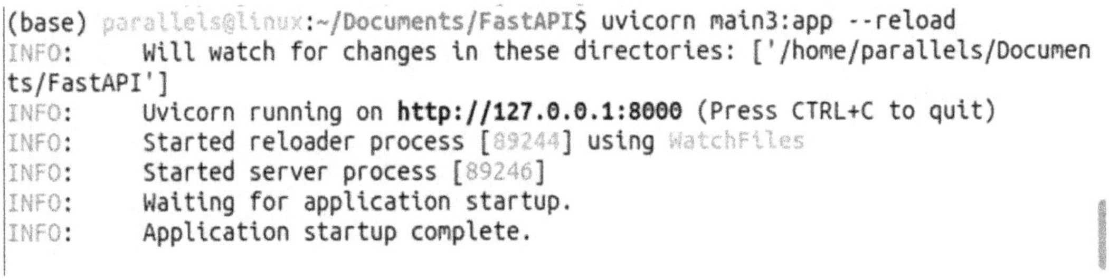

```
(base) parallels@linux:~/Documents/FastAPI$ uvicorn main3:app --reload
INFO:     Will watch for changes in these directories: ['/home/parallels/Documen
ts/FastAPI']
INFO:     Uvicorn running on http://127.0.0.1:8000 (Press CTRL+C to quit)
INFO:     Started reloader process [89244] using WatchFiles
INFO:     Started server process [89246]
INFO:     Waiting for application startup.
INFO:     Application startup complete.
```

Figure 10-1. *Executing a simple FastAPI app*

Now, let us open a browser and enter the address http://127.0.0.1:8000/student/3:

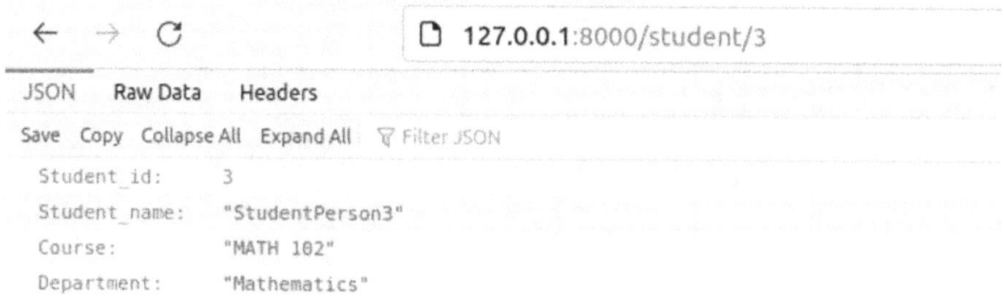

CHAPTER 10   ENGINEERING MACHINE LEARNING AND DATA REST APIS USING FASTAPI

```
← → C                    🗋 127.0.0.1:8000/student/3
JSON    Raw Data    Headers
Save  Copy  Collapse All  Expand All  ▼ Filter JSON
    Student_id:      3
    Student_name:    "StudentPerson3"
    Course:          "MATH 102"
    Department:      "Mathematics"
```

Figure 10-2. *Viewing the FastAPI app in a browser*

Note   You can even mention the data type of the path parameter. In the earlier example, where the get_student() function has a path parameter mentioned, the data type of the path parameter is also defined.

If you enter a data type that is not defined in the code, FastAPI would throw an HTTP error, like the following one:

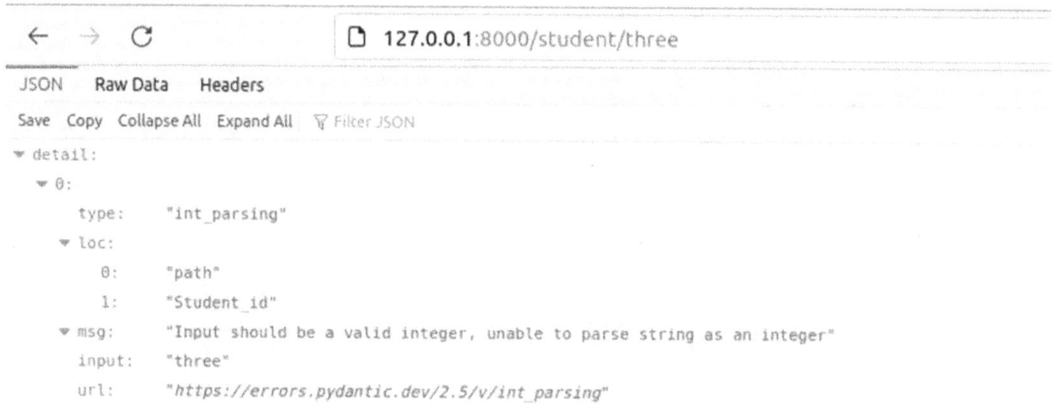

```
← → C                    🗋 127.0.0.1:8000/student/three
JSON    Raw Data    Headers
Save  Copy  Collapse All  Expand All  ▼ Filter JSON
▼ detail:
  ▼ 0:
      type:    "int_parsing"
    ▼ loc:
        0:     "path"
        1:     "Student_id"
    ▼ msg:     "Input should be a valid integer, unable to parse string as an integer"
      input:   "three"
      url:     "https://errors.pydantic.dev/2.5/v/int_parsing"
```

Figure 10-3. *FastAPI throwing an error*

FastAPI is expecting a student_id with 3 (integer) and not three (a string).

In addition to defining path parameters, you can also define query parameters within the function. Recall our preceding example of an Amazon URL. Query parameters come in at the end of a URL and are used for filtering or providing additional information to the endpoint. Query parameters are specified after the question mark symbol (?). Multiple query parameters are separated by the ampersand symbol (&).

335

Here is an example:

```python
from fastapi import FastAPI

app = FastAPI()

Students = [
    {
        "Student_id": 1,
        "Student_name": "StudentPerson1",
        "Course" : "BIO 101",
        "Department" : "Biology"
    },
    {
        "Student_id": 2,
        "Student_name": "StudentPersonTwo",
        "Course" : "CHEM 201",
        "Department" : "Chemistry"
    },
    {
        "Student_id": 3,
        "Student_name": "StudentPerson3",
        "Course" : "MATH 102",
        "Department" : "Mathematics"
    },
]

@app.get("/students/")
def read_item(skip: int = 0, limit: int = 10):
    return Students[skip : skip + limit]
```

This example would perform two filters, namely, skipping a number of records and limiting the number of results.

Here is how the URL looks like:

http://127.0.0.1:8000/items/?skip=1&limit=2

The query parameters are skip and limit; we intend to skip the first record and limit the total results to two records only. So the output would look something like this:

Figure 10-4. Defining the query parameters in the URL address

Pydantic Integration

As we saw in a previous chapter, Pydantic is a simple and powerful data validation tool. FastAPI integrates well with Pydantic.

Let us look at an example of Pydantic integration:

```
from fastapi import FastAPI, APIRouter
from pydantic import BaseModel
from typing import Sequence

app = FastAPI()
api_router = APIRouter()

Students = [
    {
        "Student_id": 1,
        "Student_name": "StudentPerson1",
        "Course" : "BIO 101",
        "Department" : "Biology"
    },
```

```
    {
        "Student_id": 2,
        "Student_name": "StudentPersonTwo",
        "Course" : "CHEM 201",
        "Department" : "Chemistry"
    },
    {
        "Student_id": 3,
        "Student_name": "StudentPerson3",
        "Course" : "MATH 102",
        "Department" : "Mathematics"
    },
]

class Student(BaseModel):
    Student_id: int
    Student_name: str
    Course: str
    Department: str

class retrieveStudent(BaseModel):
    Students: Sequence[Student]

class newStudent(BaseModel):
    Student_id: int
    Student_name: str
    Course: str
    Department: str

@app.get("/")
def hello():
    return {"welcome to the root"}

@api_router.get("/student/{Student_id}", response_model=Student)
def get_student(Student_id: int) -> dict:
    for s in Students:
```

```
        if s["Student_id"] == Student_id:
            result = s
    return result

@api_router.post("/student", response_model=Student)
def create_student(incomingStudent: newStudent) -> dict:
    StudentEntry = incomingStudent(
        Student_id = incomingStudent.Student_id,
        Student_name = incomingStudent.Student_name,
        Course = incomingStudent.Course,
        Department = incomingStudent.Department,
    )
    Students.append(StudentEntry.dict())
    return StudentEntry

app.include_router(api_router)
```

The code utilizes similar examples as previous instances. The get method retrieves students by their student_id. The post method would create a new student and the required parameters. Please note that this example is for illustrative purposes only. The data (that is posted) does not persist in the disk.

Response Model

In the get and post functions we have defined, one can see the use of the "response_model" attribute. The response_model would convert the given data into Pydantic and still return the type that the function is designated to do. The response_model does not actually change the return type of the function. The response_model parameter tells FastAPI to convert the returned data into a Pydantic object and perform necessary validations (wherever required). You can use the "response_model" parameter of the FastAPI decorator methods. Some of these methods are

```
app.get()
app.post()
app.delete()
```

   … and so on.

CHAPTER 10   ENGINEERING MACHINE LEARNING AND DATA REST APIS USING FASTAPI

There is also an option to set the response_model to None. This will enable FastAPI to skip performing the validation or filtering step that response_model intends to do. By specifying the response model to None, you can still keep the return type annotation and continue to receive type checking support from your IDE or type checkers.

Note   Furthermore, FastAPI provides Open API standards–based documentation. At any point, should you wish to access it in your development, suffix "docs" with your endpoint. For instance, `http://127.0.0.1:8000/docs` would retrieve the following documentation page (Figure 10-5).

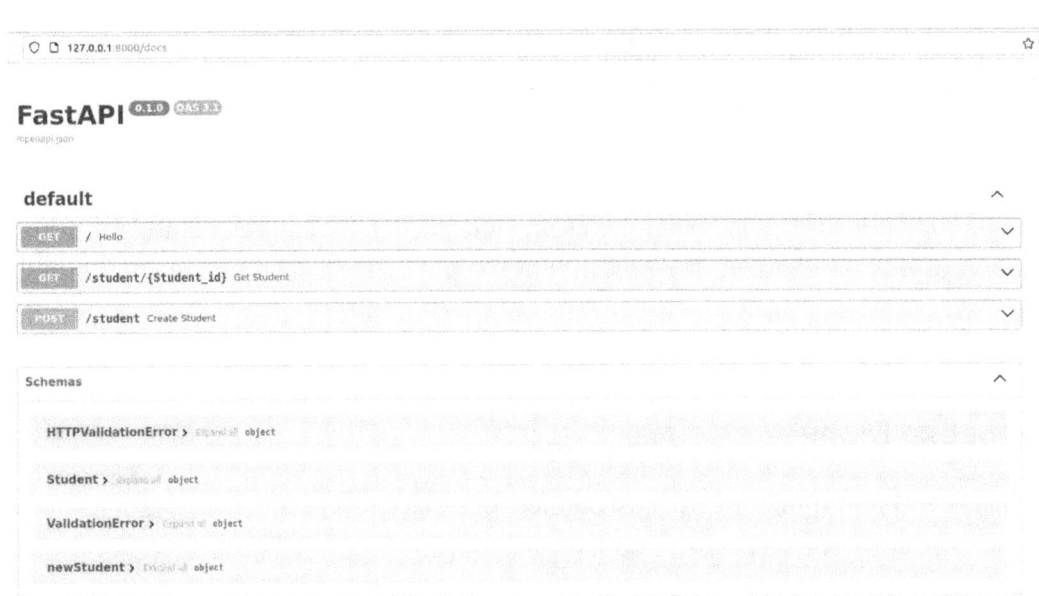

Figure 10-5. FastAPI documentation based on Open API standards

Within that window, do expand any of the HTTP methods to try out their functionality. Here is how that looks for "Get Student":

CHAPTER 10 ENGINEERING MACHINE LEARNING AND DATA REST APIS USING FASTAPI

Figure 10-6. GET request documentation for this app

When you click "Try it out," the text box would be live, and so you can enter a given request for the server to process and retrieve responses. The server would also generate a Curl command for the request, the endpoint for the appropriate request, status response message, and the response body. Here is how that looks:

CHAPTER 10   ENGINEERING MACHINE LEARNING AND DATA REST APIS USING FASTAPI

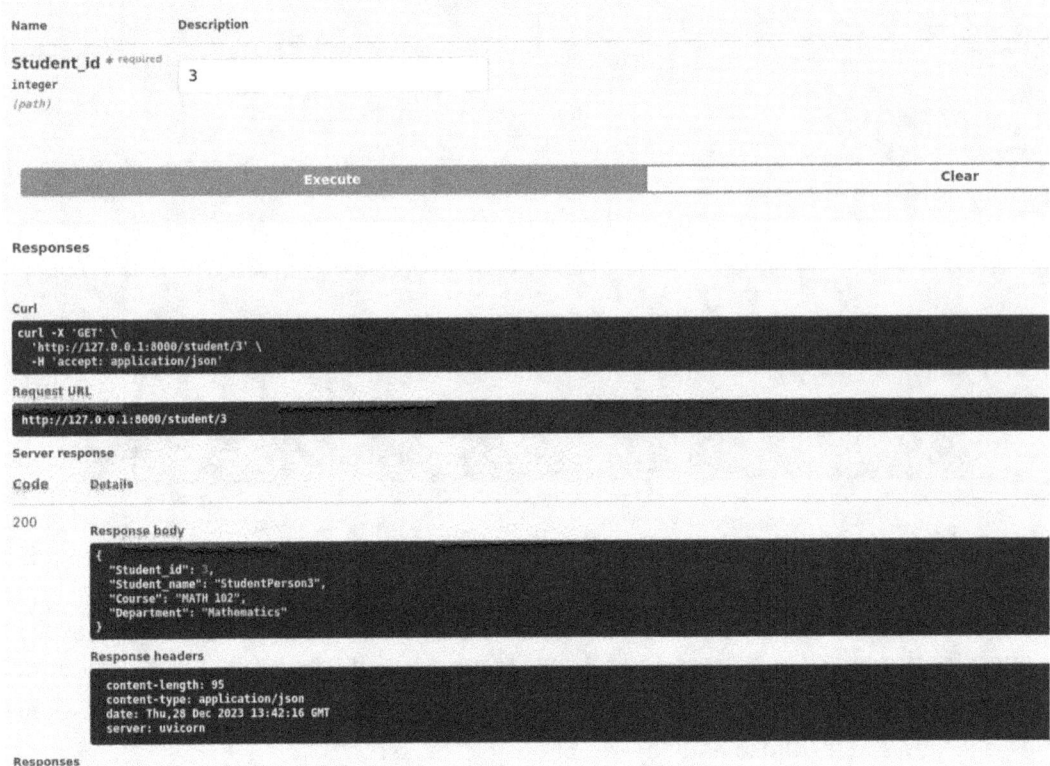

Figure 10-7. GET request illustration on FastAPI

Dependency Injection in FastAPI

When you are writing a data processing pipeline, you might come across a functionality or a set of functionality that is basically the same code but applies to multiple data pipelines. The easiest and quickest example is connecting to a specific database. It is possible that there may be "n" number of data pipelines that require connection to a specific database. Obviously, the said data pipeline is dependent on connecting with this specific database. The process of providing your code with this said database connectivity dependency is called dependency injection.

Note Injection is basically supplying the functionality here. Instead of writing code within the data processing pipeline for the functionality, the code is written outside and gets injected into the data processing pipeline.

FastAPI has a simple and intuitive dependency injection system. The method "Depends" provides the dependency injection functionality for FastAPI. Let's look at the following example:

```
from fastapi import FastAPI, Depends
from databases import Database
import sqlalchemy

app = FastAPI()

database = Database("Your database credentials here")

def dbConnect():
    return database

@app.get("/users")
def get_users(db=Depends(dbConnect)):
    conn = db
    users = conn.fetch_all("SELECT * from users")
    return {"users": users}
```

In the preceding example, we have a simple API design where we have designed a GET request, which would retrieve "users" when requested.

In order for the GET method to retrieve and respond with that piece of information, it requires a connection to a database. As you may notice the database connection is defined as a separate function.

And so, that database connection gets passed as a parameter within the get_users function with the FastAPI method "Depends". This way, the "get_users" function calls the function "dbConnect()" function, which establishes a connection to a database.

Database Integration with FastAPI

FastAPI integrates with any and all relational databases. There are different ways of interacting with databases. Let us begin by looking at a simple and straightforward way of interacting with a relational database. For our illustration, we will be looking at working with an open source relational database called MySQL on a Debian workstation. Let us execute the following commands:

```
sudo apt update
sudo apt install mysql-server
```

Check to see if the MySQL has been installed correctly:

```
mysqld –version
```

Let us log in as root:

```
mysql -u root -p
```

Enter the appropriate password and you will get the mysql prompt:

```
mysql>
```

Let us create a new database user:

```
create user 'newscott'@'localhost' identified with caching_sha2_password by 'Tiger@#@!!!9999';
```

To list all the databases, use

```
show databases;
```

To select a database, use

```
use your-database-name;
```

To create a database, use

```
create database your-database-name;
```

Let us create a database, create a table within that database, and insert data into that table:

```
create database stu;
Use stu;

CREATE TABLE students (
    id INT(11) NOT NULL AUTO_INCREMENT,
    name VARCHAR(100) NOT NULL,
    course VARCHAR(100) NOT NULL,
    PRIMARY KEY (id)
    );
```

The following command is also helpful:

```
show tables;
```

would yield

```
+---------------+
| Tables_in_stu |
+---------------+
| students      |
+---------------+
```

```
insert into students (id, name, course) values (1, "test", "tescourse");
insert into students (id, name, course) values (2, "JohnDoe", "Physics101");
insert into students (id, name, course) values (3, "Joe", "Chemistry101");
```

Let us look at the table now:

```
select * from stu.students;
```

```
+----+---------+---------------+
| id | name    | course        |
+----+---------+---------------+
| 1  | test    | tescourse     |
| 2  | JohnDoe | Physics101    |
| 3  | Joe     | Chemistry101  |
+----+---------+---------------+
```

Let us also grant privileges to the new database user we have created:

```
grant all privileges stu.* to 'newscott'@'localhost';
flush privileges;
```

Note These commands are meant for simple database setup for development purposes. You would have to use mysql_secure_installation and follow DBA best practices for setting up production instances of databases.

Now create a new Python script called "db1.py" and enter the following code:

CHAPTER 10   ENGINEERING MACHINE LEARNING AND DATA REST APIS USING FASTAPI

```python
from fastapi import FastAPI
from mysql import connector

dbapp = FastAPI()

database_creds = connector.connect(
  host="localhost",
  user="newscott",
  password="Tiger@#@!!!9999",
  database="stu",
  auth_plugin='mysql_native_password')

# GET request to list all students
@dbapp.get("/AllStudents")
def get_students() -> dict:
    cursor = database_creds.cursor()
    cursor.execute("Select * from students")
    result = cursor.fetchall()
    return { "students" : result }

# GET request to obtain a name from student id
@dbapp.get("/student/{id}")
def get_student(id: int) -> dict:
    cursor = database_creds.cursor()
    cursor.execute(f"Select name from student where id = {id}")
    result = cursor.fetchone()
    return { "student" : result }

# GET request to obtain a course from student name
@dbapp.get("/stucourse/{name}")
def stucourse(name: str) -> dict:
    cursor = database_creds.cursor()
    cursor.execute(f"Select course from student where name = {name}")
    result = cursor.fetchall()
    return { "student" : result }

# POST request to insert a new student
```

```
@dbapp.post("/newstudent")
def newstudent(name: str, course: str) -> dict:
    cursor = database_creds.cursor()
    sql = "INSERT INTO student (name, course) VALUES (%s, %s)"
    val = (name, course)
    cursor.execute(sql, val)
    database_creds.commit()
    return {"message": "student added successfully"}
```

What we have here is a very straightforward illustration of FastAPI interacting with a MySQL database instance. We have created functions with specific SQL queries, decorated the function with appropriate FastAPI methods, and passed parameters where required. Let us look at another method of FastAPI interacting with a given database instance.

Object Relational Mapping

Object relational mapping is a programming concept that is used to create a bridge between object-oriented programming code and relational databases. As the name suggests, it "maps" the objects in your code to relational database objects. The most common case is where you create a class in your code that represents a table in your relational database. The table's field names and their respective types are represented by the class's attributes. And finally, each instance of these objects would represent a row in the table within the relational database. ORM libraries simplify the interaction between the Python script and relational database. The advantage of using ORM is that it is designed to eliminate any possibility of SQL injection attacks.

Some of the ORM libraries that are utilized more often are Peewee, SQLAlchemy, and Django ORM. Django, being a web framework, has its own ORM library built in. Currently, the most adopted or widely used ORM library is SQLAlchemy.

SQLAlchemy

SQLAlchemy consists of two major modules, core and ORM. The core module provides a SQL abstraction that enables smooth interaction with a variety of database application programming interfaces (also known as DBAPI). The core module does not require ORM to be adopted to connect and interact with a database. The ORM module builds on top of the core module to provide object relational mapping functionality to SQLAlchemy. Let us look at some key concepts in SQLAlchemy.

Engine

To connect with any database using SQLAlchemy, you need to get acquainted with both the concepts of engine and session. In order to connect with a database, you need to create an engine. The engine is the starting point of a SQLAlchemy application that attempts to connect and communicate with a database. The create_engine object provides a method to create new database connections.

Here is an illustration to connect with a PostgreSQL database:

```
from sqlalchemy import URL

databaseURL = URL.create(
    "mysql+mysqlconnect",
    username="your_username_here",
    password="your_password_here",
    host="localhost",
    database="testdatabase",
)
engine = create_engine(databaseURL)
```

Session

The other concept is session. As the name suggests, the session establishes conversations with the database, after a connection is established. The session creates a workplace for all your queries and transactions (commits, rollbacks, etc.). If you are building a bigger application, it is recommended to utilize the sessionmaker class to create a session. Here is an illustration of how to utilize "sessionmaker" to create a new session:

```
from sqlalchemy import URL
from sqlalchemy.orm import sessionmaker

databaseURL = URL.create(
    "mysql+mysqlconnect",
    username="your_username_here",
    password="your_password_here",
```

```
        host="localhost",
        database="testdatabase",
)
engine = create_engine(databaseURL)
sampleSession = sessionmaker(autoflush=False, bind=engine)
# queries, transactions here
sampleSession.close()
```

The argument "bind" is a way to associate an engine with the session. The argument "autoflush" flushes or commits the database transactions automatically when a query is invoked to send SQL to the database or when you are using certain methods of session class.

Query API

SQLAlchemy also provides a query API that enables performing various operations to aid with data retrieval and manipulation. Some of these methods include but are not limited to sorting, filtering, joining, fetching data from tables, etc. Let's look at some SQL to ORM examples.

Let us say we have a table called "students" in a relational database. Here is how the SQL query translates:

```
Select * from Students
```

would translate to

```
Students.select()
```

Let us look at a slightly detailed query where we choose to retrieve students who have taken Chemistry 101 during Fall 2024:

```
Select
      *
from
      students
where
```

CHAPTER 10   ENGINEERING MACHINE LEARNING AND DATA REST APIS USING FASTAPI

```
    term = "Fall2024" and
    course = "Chemistry101" and
    plan = "something"
fetch first 25 rows only
```

would translate to

```
students \
 .select() \
 .filter( \
    students.has( \
        term="Fall2024", \
        course="Chemistry101", \
        plan="something" \
        )
    ).fetchmany(25)
```

Alembic

Furthermore, it is also important to know about database migrations and the Alembic library. Database migration does not necessarily mean moving large amounts of data between databases. Instead, database migration is about creating and maintaining both the current and evolving schemas with the application. Alembic is a Python library that helps manage and maintain these evolving schemas within the SQLAlchemy application.

Building a REST Data API

Let us leverage FastAPI and SQLAlchemy to build a RESTful data API that delivers data as a service. Data as a service (DaaS) is a business concept and a model that describes delivering high-quality data on demand no matter how geographically the provider and consumer are separated. In this example we will also incorporate Pydantic, the data validation library, to illustrate how data can be validated and sent downstream in building a RESTful data API. We are leveraging RESTful architecture to ensure statelessness, scalability, and simplicity of delivering data over the Web.

To begin, let us look at initiating the database connection with SQLAlchemy:

```
from sqlalchemy import create_engine
from sqlalchemy.ext.declarative import declarative_base
from sqlalchemy.orm import sessionmaker

SQLAlchemy_DB_URL = "mysql+mysqlconnector://newscott:Tiger$$$!!!9999@
localhost/stu"
engine = create_engine(SQLAlchemy_DB_URL)

SessionLocal = sessionmaker(autocommit=False,
                            autoflush=False,
                            bind=engine)

Base = declarative_base()
```

Next, we will define the class that corresponds to the respective table in the database:

```
from sqlalchemy import Column, Integer, String
from database import Base

class Student(Base):
    __tablename__ = "students"

    id = Column(Integer, primary_key=True)
    name = Column(String, unique=True, index=False)
    course = Column(String, unique=False, index=True)
```

This will be followed by creating a data validation class using Pydantic library. Here is how that looks:

```
import pydantic as pyd

class NewStudent(pyd.BaseModel):
    id: int
    name: str
    course: str
```

Let us now develop various functions that would become the functionality of our API. In this case, we can obtain a list of all students, obtain details of a student by name or id, or create a new student and commit the transaction as well:

CHAPTER 10   ENGINEERING MACHINE LEARNING AND DATA REST APIS USING FASTAPI

```
from sqlalchemy.orm import Session
import model, pydanticSchema

def get_AllStudents(db: Session, offset: int = 0, limit: int = 10):
    return db.query(model.Student).offset(offset).limit(limit).all()

def get_student_by_name(db: Session, name: str):
    return db.query(model.Student).filter(model.Student.name == name).first()

def get_student_by_id(db: Session, id: int):
    return db.query(model.Student).filter(model.Student.id == id).first()

def new_student(db: Session, student: pydanticSchema.NewStudent):
    #id = rand number gen
    db_user = model.Student(id = id, name = model.Student.name,
    email=model.Student.email)
    db.add(db_user)
    db.commit()
    db.refresh(db_user)
```

Once we have all the functionalities and moving parts, let us now create the FastAPI application:

```
from fastapi import Depends, FastAPI, HTTPException
from sqlalchemy.orm import Session

from database import SessionLocal, engine

import model
from crud import get_AllStudents, get_student_by_name, get_student_by_id,
new_student
from pydanticSchema import NewStudent

model.Base.metadata.create_all(bind=engine)

app = FastAPI()

# Dependency
def get_db():
    db = SessionLocal()
```

```python
    try:
        yield db
    finally:
        db.close()

@app.get("/student/{name}")
def get_stud_by_name(student: NewStudent,
                     db: Session = Depends(get_db)):
    try:
        byname = get_student_by_name(db, name=student.name)
        return byname
    except Exception:
        raise HTTPException(status_code=400)

@app.get("/students/")
def get_allstudents(db: Session = Depends(get_db)):
    try:
        allstudents = get_AllStudents(db)
        return allstudents
    except Exception:
        raise HTTPException(status_code=400)

@app.get("/student/{id}")
def get_stud_by_id(student: NewStudent,
                   db: Session = Depends(get_db)):
    try:
        by_id = get_student_by_id(db, id=student.id)
        return by_id
    except Exception:
        raise HTTPException(status_code=400)

@app.post("/students/")
def new_student(student: NewStudent,
                db: Session = Depends(get_db)):
    student_id = get_student_by_id(db, id=student.id)
    if student_id:
        raise HTTPException(status_code=400)
```

CHAPTER 10   ENGINEERING MACHINE LEARNING AND DATA REST APIS USING FASTAPI

```
return new_student(db, student=student)
```

Once we have the functionalities and code ready, let us test the RESTful data API application. Simply run

```
uvicorn main:app –reload
```

You can open a browser tab and enter http://127.0.0.1:8000/students/ to GET all the students.

Here is how that may look like:

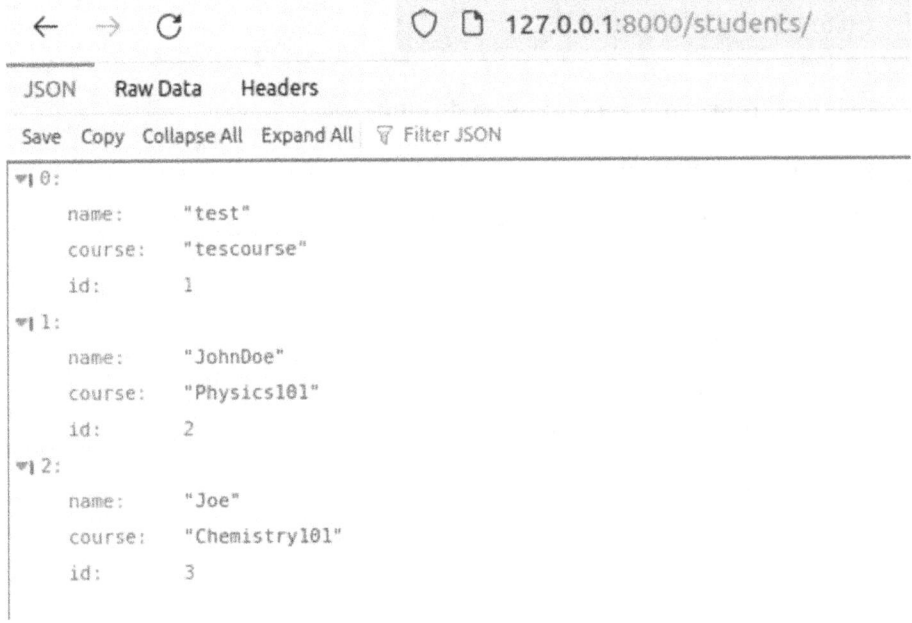

Figure 10-8. GET request illustration for this FastAPI app

Middleware in FastAPI

A middleware is a small function or set of programs that is placed between provider and consumer. It works with both request and response processes. With every request that is received prior to being sent to the server for processing, the middleware can be configured to process a specific operation before sending it to the server. Likewise, it can take a response, perform certain operations, and send that response to the consumer. Common examples of middleware applications are performing authentications and writing a metadata log entry on what has been requested and responded. Middleware also helps limit the number of requests from a specific consumer.

Let us look at a simple example of a middleware implementation. Imagine we have machine learning deployed as a service that is open to the public. We want to implement a rate-limiting middleware that limits the number of requests originating from a specific consumer to 25 per hour. On a side note, this is a good example of async/await implementation.

ML API Endpoint Using FastAPI

Let us consider a simple example of predicting (classifying) a flower's species. We shall obtain the Iris dataset. The Iris dataset consists of three species of the Iris flower with 50 samples for each species widely used for testing machine learning methods. (source: https://archive.ics.uci.edu/dataset/53/iris).

Here is a simple model that is employed on top of the dataset. We begin by constructing a simple machine learning model. We utilize the Iris dataset that was used by R. A. Fisher's research paper "The Use of Multiple Measurements in Taxonomic Problems," published in the year 1936. The dataset is Creative Commons licensed and can be found in the machine learning repository maintained by UC Irvine. The dataset consists of measurements of sepal and petal (parts of a flower) of three flowers in the iris plant along with the name of the respective species.

We have extracted the plant measurements as dependent variables and the name of the plant as independent variable to identify the type of iris plant given a set of measurements. The model is stored as a pickle file so we can import the same in our FastAPI application:

> **Note** Please note that the focus is more on the FastAPI implementation for our machine learning app illustration.

```
import pandas as pd
from sklearn.model_selection import train_test_split
from sklearn.linear_model import LogisticRegression
from sklearn.metrics import accuracy_score, confusion_matrix
import pickle
import numpy as np

df = pd.read_csv("/<your-path-to>/iris.csv")
```

CHAPTER 10 ENGINEERING MACHINE LEARNING AND DATA REST APIS USING FASTAPI

```
x = df.iloc[:,:4]
y = df.iloc[:,4]

x_train, x_test, y_train, y_test = train_test_split(x,
                                                    y,
                                                    random_state=0)

model=LogisticRegression()
model.fit(x_train,y_train)

y_pred = model.predict(x_test)

#print(confusion_matrix(y_test,y_pred))
#print(accuracy_score(y_test, y_pred)*100)

export_model = pickle.dump(model, open('irismodel1.pkl', 'wb'))

input_data_reshaped = np.asarray(input_data).reshape(1,-1)
prediction = model1.predict(input_data_reshaped)
```

Once we have the model pickle file, we can proceed to build our application in FastAPI with a custom middleware that limits access to the API after a few number of requests. We start off with importing the necessary classes and methods, followed by instantiating the FastAPI, loading the model, and creating a dictionary to store the IP addresses and number of hits. Then we define the middleware function, where we obtain the IP address from the request class and increment the hits for each IP address. The middleware is coded to ensure that if the number of hits goes beyond 5 in a given hour, then it will throw an exception:

Note It is important to understand that the middleware code is meant for illustration purposes. I also suggest looking into a Python library called "SlowAPI" for implementing a rate limiter. Furthermore, it is also suggested to use an in-memory store like Redis.

```
from pydantic import BaseModel
import pickle
import json
from fastapi import FastAPI, Request
```

```python
from datetime import datetime, timedelta

app = FastAPI()

model = pickle.load(open('irismodel1.pkl','rb'))
# creating a new register for the ipaddresses
registry: dict[str, (int, datetime)] = {}

@app.middleware("http")
async def rateThreshold(request: Request, call_next):
    ip_address = request.client.host
    rightnow = datetime.now()
    if ip_address in registry:
        hits, timestamp = registry[ip_address]
        if rightnow - timestamp > timedelta(hours=1):
            hits = 0
    else:
        hits = 0
    hits += 1
    if hits > 5:
        raise Exception("too many requests")
    registry[ip_address] = (hits, rightnow)
    response = await call_next(request)
    return response

class ModelInput(BaseModel):
    SLength: float
    SWidth: float
    PLength: float
    PWidth: float

@app.post("/predict/")
def predict(modelparam: ModelInput):
    modelparameters = json.loads(modelparam.json())
    sl = modelparameters['SLength']
    sw = modelparameters['SWidth']
    pl = modelparameters['PLength']
    pw = modelparameters['PWidth']
```

```
predictionclass = model.predict([[sl,sw,pl,pw]])
return predictionclass[0]
```

Once we have the application, we can launch by

```
uvicorn MLPipeline:app –reload
```

This will start up the application from the terminal.

To access or test the API, it is suggested to open a browser tab and enter the FastAPI documentation link:

http://127.0.0.1:8000/docs#/default/predict_predict__post

or simply

http://127.0.0.1:8000/docs

would yield

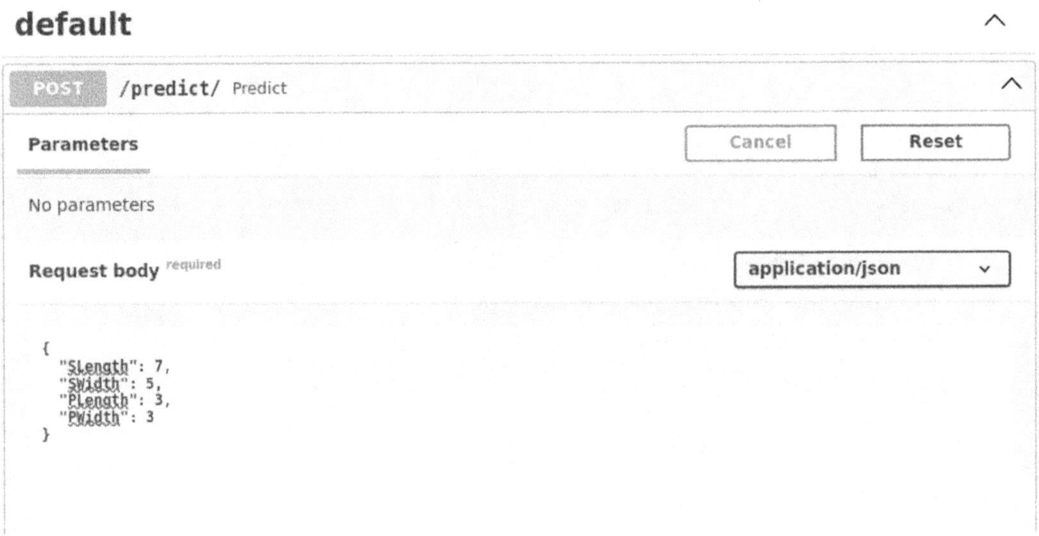

default

POST /predict/ Predict

Parameters

No parameters

Request body *required*   application/json

```
{
  "SLength": 7,
  "SWidth": 5,
  "PLength": 3,
  "PWidth": 3
}
```

Figure 10-9. A POST request for the machine learning model API

You can start inputting the values and let the system classify (predict) the iris plant. Here is how that looks:

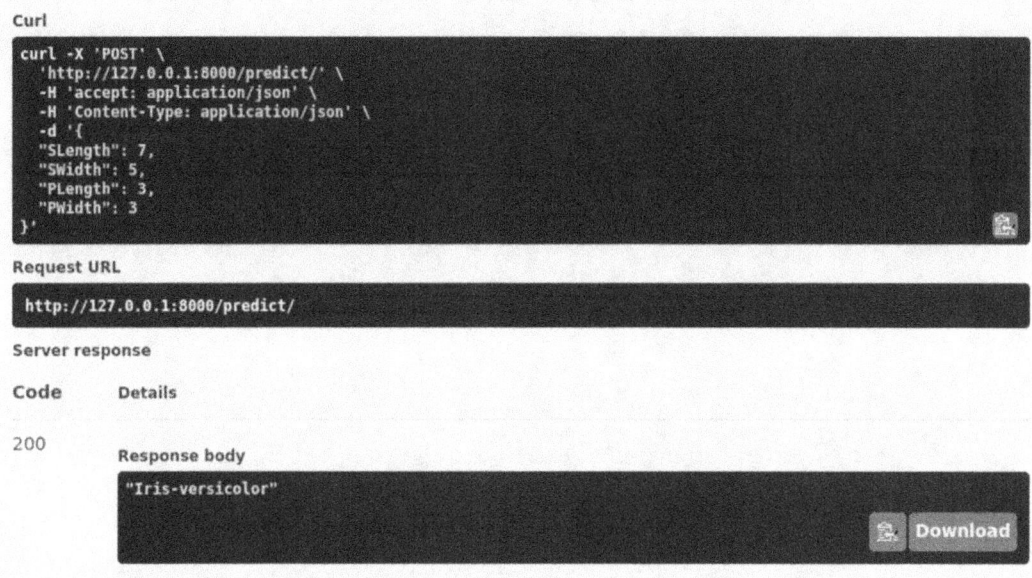

***Figure 10-10.** FastAPI computing the prediction for the input parameters and returning the response body*

Conclusion

APIs, application programming interfaces, are one of the growing ways of consuming data analytics and machine learning services. API-based development and deployment is an ocean of its own; it is good for a ML engineer or a data engineer to set up and build an API. In this chapter, we looked at the concepts and architecture of APIs. We went into greater detail on FastAPI architecture and looked at a variety of illustrations. We moved toward database integration, building a data service REST API. We also looked at deploying a machine learning model as a REST API. Not only have you mastered the core concepts of FastAPI, but also you have acquired a skillset that bridges the gap between complex engineering and service delivery. Overall, I hope this served you greatly in understanding how to engineer and deploy data pipelines using FastAPI.

CHAPTER 11

Getting Started with Workflow Management and Orchestration

Introduction

Organizations are increasingly leveraging data, creating more data teams, and performing various data projects. In addition to building and deploying products and services, companies are increasingly creating more data teams and performing various data projects from data discovery to gaining value to ensuring best security and privacy data practices. Data projects can be based on IT products, IT services, shared services, and other tasks and activities in organizations. Depending upon the nature and type of data projects, the data pipelines can get complex and sophisticated. It is important to efficiently manage these complex tasks, making sure to orchestrate and manage the workflow accurately to yield desired results.

Every data project requires transparency and observability as to what the project does, how it is working out, and what systems it is sourcing from and feeding to. Every data job or data pipeline is essential enough that in case of a job failure, one is expected to restart the job and troubleshoot whatever the systems that caused failure. With the big data revolution, organizations are increasingly looking to leverage various kinds of machine-generated data. As the number of data pipelines increases, it presents another complexity of managing them.

CHAPTER 11   GETTING STARTED WITH WORKFLOW MANAGEMENT AND ORCHESTRATION

By the end of this chapter, you will learn

- Fundamentals of workflow management and its importance to engineering data pipelines

- Concepts of workflow orchestration and how it optimizes the data pipeline operations

- The cron scheduler and its role in task automation

- Creating, monitoring, and managing cron jobs and cron logging

- Cron logging and practical applications of scheduling and orchestration of cron jobs

Introduction to Workflow Orchestration

Workflow

Workflow orchestration has long existed even before the IT and data revolution. A workflow, itself, could mean a set of tasks or activities that are performed in a specific order to produce a product. Each task has a set of specific input, process, and output. The output of a given task serves as the input of subsequent tasks. Conversely, the input of a given task contains the output of a preceding task. This way, the tasks must follow a specific set of steps in a certain order.

Workflow, in the context of any given data engineering or a machine learning job, is basically a set of data pipelines that consists of tasks and activities that are to be performed in a specific order. The input of a given pipeline is dependent on output of a preceding pipeline.

To illustrate this further, let's take an example of Acme bread company, a fictitious bread manufacturing shop. Yes, it is a reference from *The Road Runner Show*. Let us say that the basic process of making bread involves the following steps that need to be followed in the following specific order:

1. Ingredient mixing, where the dough, yeast, water, and salt are mixed together in specific, measured proportions in giant mixers.

2. Proofing, where the kneaded mixture is divided into smaller units and placed in room temperature, so that the dough gets more airy and rises (becomes bigger).

3. Baking, where the dough is baked in ovens, at controlled temperature and pressure parameters, and the humidity is controlled at appropriate levels to obtain desired consistency of bread.

4. Quality control, to make sure the bread produced is even and consistent and meets the quality standards.

5. Slicing and packaging, where the bread that is made is being sliced, packaged with a paper bag, and stamped when it was baked and how long until it is safe for consumption.

6. Distribution, where the bread packed in paper bags is once again packed into boxes and loaded into various trucks. These trucks then transport the bread boxes to various destinations that include grocery stores and restaurants.

Here is how this may look:

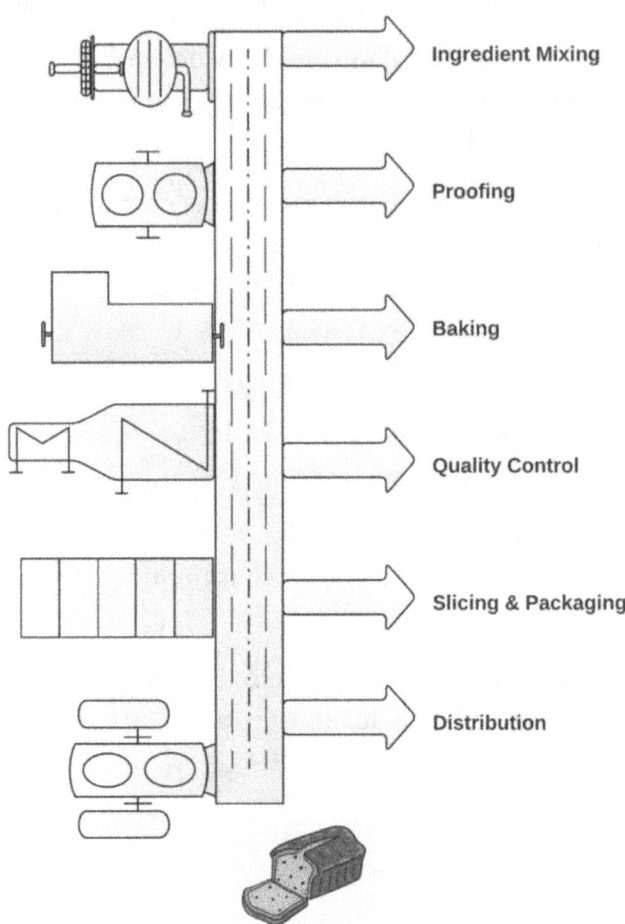

Figure 11-1. An example of an assembly line workflow

This is the most basic example of a workflow. In the steps listed earlier, one cannot interchange given steps. These steps have to be followed in exactly the same order. Furthermore, the output of a given step constitutes the input of the next step. And so, the output of the next step becomes the input of the next subsequent step, and so on.

ETL and ELT Data Pipeline Workflow

The process of baking a bread loaf involves a set of steps that need to be followed in a specific order. Similarly, in the context of data engineering or machine learning projects, there exists a set of functions/jobs/tasks that are to be executed in a specific order. There are many ways of writing these data pipelines; one of the most common and well-known types of workflow is called the ETL workflow. ETL stands for extract, transform, and load. In this type of workflow, the data is extracted and placed on a staging area, necessary transformations are applied, and the final work product is transported to another system for downstream processing and consumption purposes.

A simple example is a payroll processing data pipeline. Let's say the payroll data is extracted from an ERP system like PeopleSoft and other expenses data from other source systems; this data then gets cleaned, wrangled, and integrated into a unified dataset in a staging area. The next phase is the transformation phase where key metrics like net pay, taxes, and deductions are calculated. The resultant dataset then gets loaded into the payroll system for payment processing and reporting.

Here is how this may look:

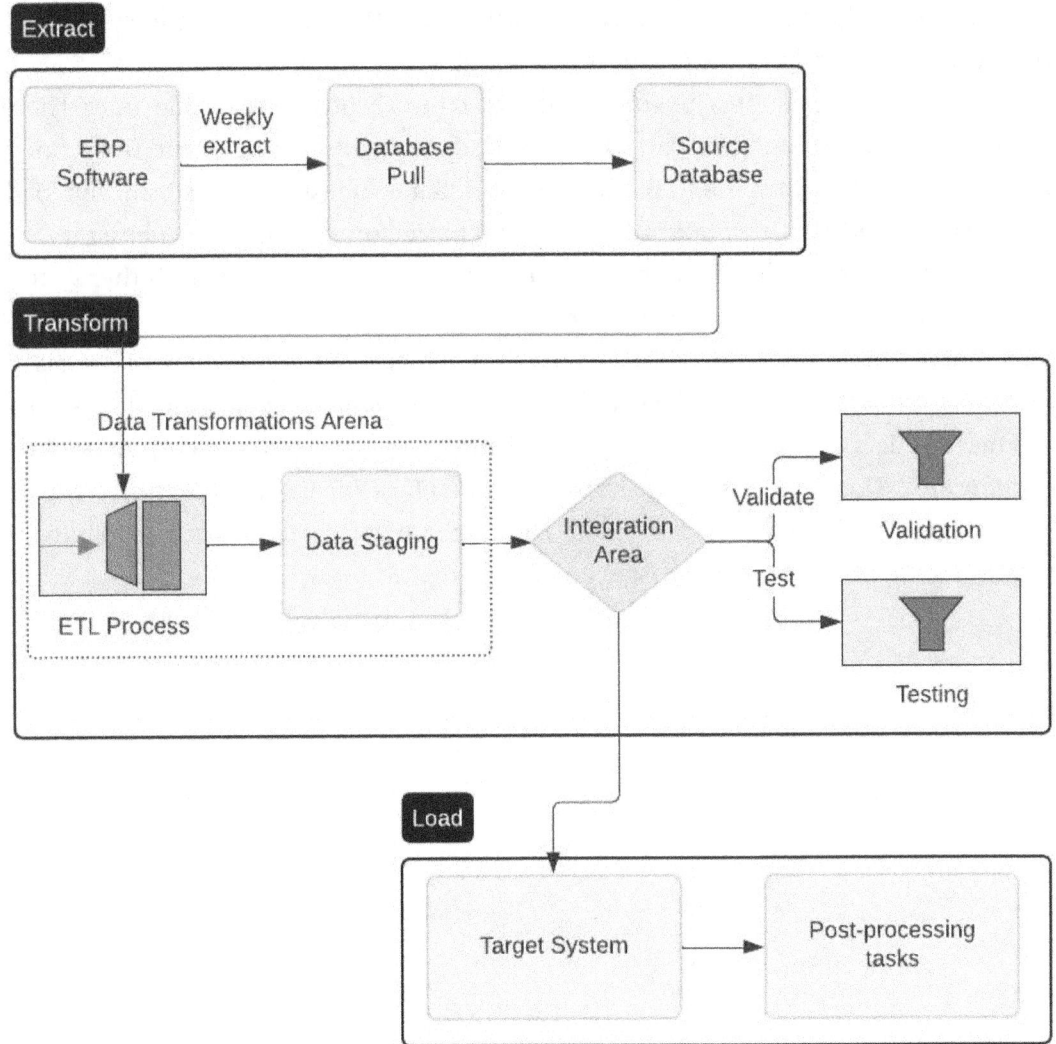

Figure 11-2. A sample ETL architecture for a payroll processing pipeline

A recent variant of ETL is referred to as ELT. ELT stands for extract, load, and transform. In this case, data from various sources are extracted and sourced into the target system. From there on, various transformation and business rules are applied directly onto the data. There may or may not be a staging layer in the ELT workflow. ELT workflows are more common in the case of cloud data stores. ELT is more suited when you are dealing with diverse data types. Most cloud providers offer serverless data lake/warehouse solutions that serve ELT pipelines, making it cost effective and easy to maintain.

Let us take an example of what an ELT data pipeline may look like. Let's say relational database data is being sourced from an inventory system on various products, along with JSON data of log sensor readings of the warehouse, along with few other data. Obtaining this data at frequent intervals would constitute the extract stage. For the load stage, imagine if we have a Google BigQuery where you dump all the data. The transform stage is where the data gets cleaned, wrangled, and either integrated into a unified dataset or loaded into dimensional tables for further analysis.

Here is how this may look:

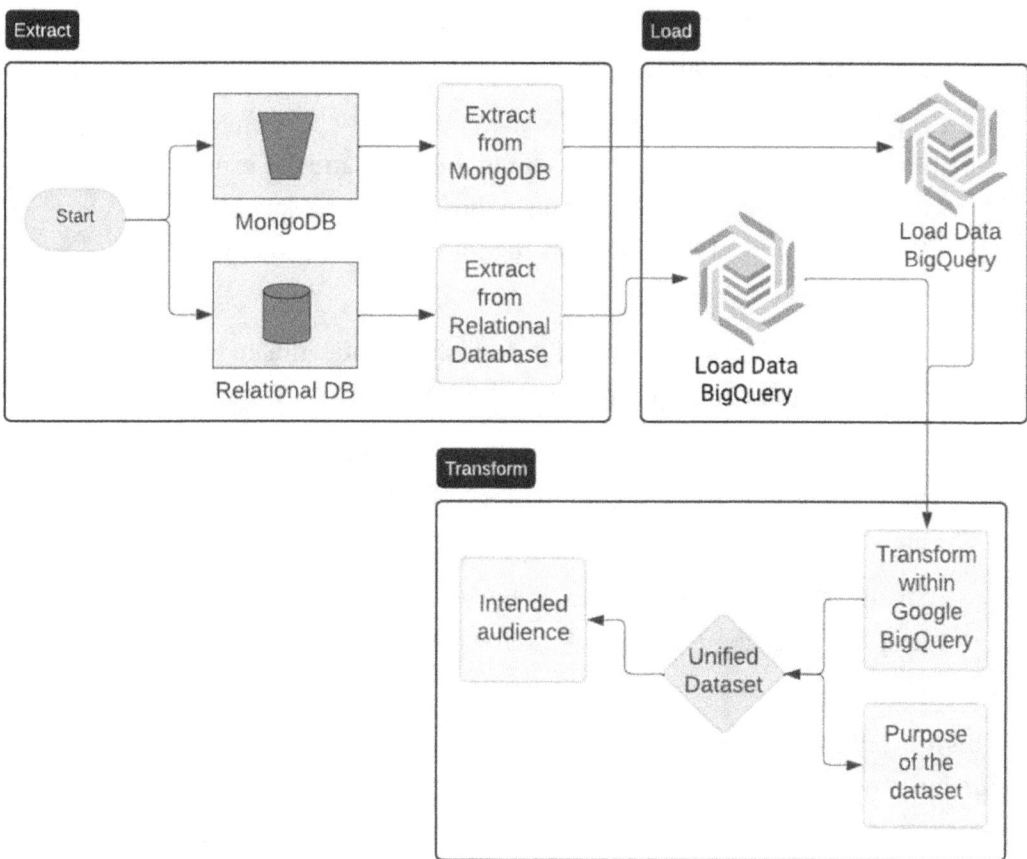

Figure 11-3. *A sample ELT architecture for inventory data processing pipelines*

CHAPTER 11　GETTING STARTED WITH WORKFLOW MANAGEMENT AND ORCHESTRATION

Workflow Configuration

Workflow configuration is another key concept to understand. Workflow configuration is basically a set of configuration parameters that would drive the workflow processes. Workflow configuration is essential and crucial to the success of delivering a product or solution, in this case, bread.

In this example, workflow configuration might involve the following aspects:

1. Ingredient mixing: The exact quantity of each ingredient, and the measurement is on the same scale.

2. Proofing: The controlled temperature of the proofing room, the exact amount of oil that is applied on the proofing sheet, and an automatic timer that will alarm once the number of hours of proofing has been completed.

3. Baking: The calibration of the temperature sensor, the total amount of time needed to be spent in the oven, and the exact temperature that the oven will be in during the time spent in it (first 10 minutes, 170 °C; second 10 minutes, 200 °C; cooling 10 minutes, 150 °C).

4. Quality control: The number of defects should be reduced to 3 loaves for every 1,000 produced and so on.

5. Slicing and packaging: The slicers should be configured to specific measurement so the bread is cut evenly; packing paper should be clean and without defects.

6. Distribution: The various types of boxes and how many loaves they can fit in; temperature within the truck; delivery time should be within x hours.

This preceding workflow configuration helps the organization in streamlining the production process, improving and establishing evenness and consistency, planning procurement and manpower needs, and optimal utilization of resources. This level of detailed configuration helps one to understand the steps needed to follow to produce quality bread.

Similarly, in the case of a data engineering or machine learning pipeline, workflow configuration parameters exist. Some of them could be as follows:

1. If a job is scheduled to run periodically, then the job may require the most recent or fresh data; so the time and date when the most recent source system refresh happened is very essential.

2. Organizations are starting to adopt data governance best practices. One such practice is to create a service account or specialized user account (depending upon the size, nature, and complexity of the data shop). The appropriate credentials must be configured in order to access the right data, which is crucial for building better data products.

3. A data pipeline may consist of several transformation and business rules. These transformations are basically functions that are applied to the data that was sourced from the system. Such transformation logic and business rules are workflow configurations as well.

4. While attempting to load the final data product into an appropriate system, there may be constraints that would enforce loading the data into a schema. A schema is simply a table structure where all the fields and their data types are defined. The schema can also be another workflow configuration parameter.

Workflow Orchestration

Workflow orchestration refers to the centralized administration and management of the entire process. Workflow orchestration involves planning, scheduling, and controlling of each step of the production process and their respective workflow configuration parameters, in order to ensure that all tasks are executed within the controls defined and in the appropriate order. Workflow orchestration optimizes the utilization of resources, helps plan procurement of raw materials, and establishes high quality and consistency among each of the products developed, eventually leading to seamless functioning of the organization. In this context, workflow orchestration would refer to

1. Defining the entire workflow: Where each step of the processes is defined with configuration parameters baked in.

2. Scheduling: Where the defined workflow will be scheduled, wherein a process will start upon completion of a process in the assembly chain.

3. A central control room that oversees the entire shop floor from ingredient mixing to packing bread in boxes where a forklift operator would place these boxes in trucks.

4. Automations: Where a computer specifies and controls configuration parameters for each of the industrial appliances like ovens, proofing rooms, etc.

5. Error handling and routing: In case a machine in the assembly chain fails to process its task, the workflow orchestration system would implement retries (automatic restart of the machine) for specific appliances and trigger certain events to avoid bottlenecks.

6. Integration with external systems: So the workflow orchestration keeps track of the number of boxes shipped, the number of loaves produced, which cluster of market received how many boxes, the number of units of raw materials consumed, the number of hours the machines were switched on, and other such data and feeds them to external systems.

Workflow orchestration enables a factory shop to run streamlined, with minimized errors, efficiency, and predictability.

Similarly, in the context of data pipelines, some of the workflow orchestration can be seen as the following:

1. Defining the entire workflow, where each step of the data engineering pipeline is configured with appropriate parameters to successfully execute.

2. In case a task fails to connect to a source or target system, a mechanism is available to be able to automatically re-execute a function after a few seconds with the hopes of connecting to that system.

3. In-built or well-integrated secrets management, so the development or production data pipeline is leveraging the appropriate credentials to execute the job.

4. Automatically logging the events of the job execution that could later be utilized for data analysis, especially identification of any patterns in certain types of events.

5. Centralized monitoring and observability, so one can view, manage, monitor, and control all the data pipelines under one roof.

Note A data pipeline is quite similar to a task or a process in a manufacturing shop. The data pipeline would have an input, where raw data gets sourced, a transformation layer that would perform various operations and incorporate business logic, and an output, where the final product is being produced and being sent to the next process in line. A data pipeline is called a data pipeline because the input of a process is basically an output of a different process and the output of that given job becomes the input of another data process. Generally, data pipelines are transient in nature.

Introduction to Cron Job Scheduler

One of the earliest workflow orchestrators is cron job scheduler in the Unix operating system. Cron stands for command run online and is a daemon process that runs in the operating system background. Originating in the 1970s, cron schedulers are still used today by many in their organizations. Cron job scheduler enables task automation and scheduling an execution of a task at a specific time interval in the future.

Concepts

At the heart of the cron scheduler, we have something called the cron daemon process. This process runs on system startup and in the background. The cron daemon executes commands that are mentioned in the file called crontab. There are two files that specify the users and their access to the cron scheduler environment.

These may be located at

```
/etc/cron.allow
/etc/cron.deny
/etc/cron.d/cron.allow
/etc/cron.d/cron.deny
```

CHAPTER 11   GETTING STARTED WITH WORKFLOW MANAGEMENT AND ORCHESTRATION

> **Note** In the case of Ubuntu and Debian operating systems, cron.allow and cron.deny files do not exist, meaning all users of Ubuntu and Debian operating systems are provided access to the cron scheduler by default.

The most basic unit of a cron scheduler environment is a cron job. A cron job represents a task or job that is scheduled to be executed. A cron job is one line in a crontab, representing one execution of a shell command.

Crontab File

Crontab is short for cron table. Crontab is a utility that maintains crontab files for users. A crontab file is automatically generated for a given Linux user. The cron daemon process runs this crontab file regardless of whether a task or a job is scheduled to run or not. A crontab file can be seen as a registry that contains a list of jobs scheduled. A crontab file belongs to root or a superuser. You can become a superuser to create a crontab file or edit an existing crontab file. The default location for crontab files is

/var/spool/cron/crontabs/

To create a new crontab file, use "crontab -e" from the command line. Here is how that looks:

Figure 11-4. *Creating a new crontab file*

CHAPTER 11 GETTING STARTED WITH WORKFLOW MANAGEMENT AND ORCHESTRATION

You will be presented with the following template to write your own cron jobs:

```
GNU nano 6.2                    /tmp/crontab.6AnzcG/crontab
# Edit this file to introduce tasks to be run by cron.
#
# Each task to run has to be defined through a single line
# indicating with different fields when the task will be run
# and what command to run for the task
#
# To define the time you can provide concrete values for
# minute (m), hour (h), day of month (dom), month (mon),
# and day of week (dow) or use '*' in these fields (for 'any').
#
# Notice that tasks will be started based on the cron's system
# daemon's notion of time and timezones.
#
# Output of the crontab jobs (including errors) is sent through
# email to the user the crontab file belongs to (unless redirected).
#
# For example, you can run a backup of all your user accounts
# at 5 a.m every week with:
# 0 5 * * 1 tar -zcf /var/backups/home.tgz /home/
#
# For more information see the manual pages of crontab(5) and cron(8)
#
# m h  dom mon dow   command


^G Help          ^O Write Out     ^W Where Is      ^K Cut           ^T Execute
^X Exit          ^R Read File     ^\ Replace       ^U Paste         ^J Justify
```

Figure 11-5. *Writing tasks to the crontab*

To list all the current crontab files, you can try the following:

`ls -l /var/spool/cron/crontabs/`

Let us look at the syntax of a crontab file.

Each line in a crontab file represents an entry for a given cron job. A crontab file entry contains six fields, separated by spaces.

Here is how the format looks:

`minute hour day-of-month month weekday command`

The minute column represents the specific minute when the job should run. It can contain numeric values only. The values can range any number between 0 and 59.

The hour column represents the hour at which the job should run and takes only numerical values. The values can range anything between 0 and 23.

The day-of-month column represents the day in a given calendar month. As you know, months contain 28, 29, 30, and 31 days depending upon the month and whether it's leap year or not. You can mention a numerical value between 1 and 31.

The month column represents a calendar month and can take numerical values between 1 and 12. These numerical values represent calendar months from January to December, respectively.

The weekday column represents a given day of the week starting from Sunday through Saturday. It can take any value between 0 and 6, where 0 represents Sunday and 6 represents Saturday.

The shell command is where the command that requires it to be run is issued.

Note In the crontab file, we have an asterisk operator. The "*" operator can be applied in the first five columns of the crontab entry. When a "*" is mentioned for the appropriate column, it means that it can run on all the intervals of the given allowable range. Here is an example:

15 6 * * * /user/parallels/bin/helloworld

The abovementioned crontab entry would run the hello world script every day at 15 minutes past 6 AM in the morning. Specifically, the script would run every day of the month, every month, and every weekday as well.

Cron Logging

The cron daemon process writes a log entry every time the system starts up, it executes a new job, and it has trouble interpreting the command or faces an error during execution of a job. Unfortunately, the log entry does not contain the source of the error. If you are working on a Debian-based operating system (Ubuntu and others), then you can find the logs here:

/var/log/syslog

And here's for Fedora-based distributions:

/var/log/cron

You can obtain the most recent cron log entries by the following command:

```
grep CRON /var/log/syslog | tail -100
```

Logging in cron stores a lot of key information. The log entry starts off with a date timestamp when the job was started, the host of the workstation, the user account of that workstation, and the command that is executed. Here is an example of a log entry:

```
Jan 29 15:55:07 linux CRON[1257]: (parallels) CMD (/home/parallels/helloworld.py)
```

In the preceding example, the log entry begins with the date timestamp, followed by the hostname (in this case, linux), followed by the user on the host (in this case, parallels) and finally the command that is executed (in this case, executing a Python script named helloworld.py).

Cron Job Usage

In this section, we will look at some simple examples of cron job writing. Let us begin by writing a simple shell script and schedule the same using simple cron job scripts.

We will create a shell script called cronexample.sh. Enter the following in the terminal:

```
nano cronexample.sh
```

And enter the following code in the nano editor:

```
#!/bin/bash

echo "Hello from the world of Cron scheduling"
echo "Execution at: $(date)"
```

CHAPTER 11   GETTING STARTED WITH WORKFLOW MANAGEMENT AND ORCHESTRATION

It should look like the following:

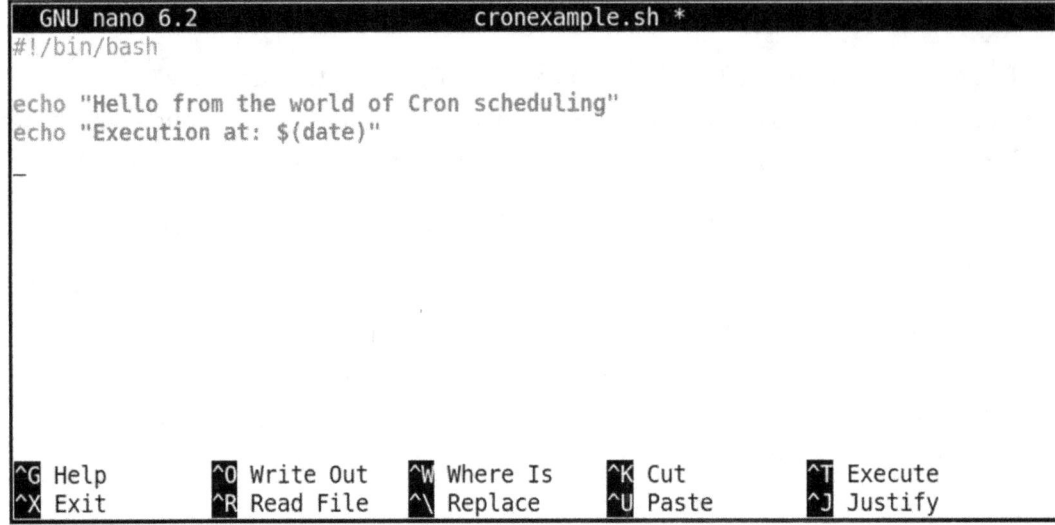

Figure 11-6. Writing a simple shell script

Save the file and exit the editor. Let us provide execute permissions to this file:

chmod +x cronexample.sh

Now enter the crontab to schedule the execution of this shell script:

crontab -u parallels -e

CHAPTER 11   GETTING STARTED WITH WORKFLOW MANAGEMENT AND ORCHESTRATION

Here is how all of these would look like:

```
  GNU nano 6.2                    /tmp/crontab.oBDdrb/crontab
# Edit this file to introduce tasks to be run by cron.
#
# Each task to run has to be defined through a single line
# indicating with different fields when the task will be run
# and what command to run for the task
#
# To define the time you can provide concrete values for
# minute (m), hour (h), day of month (dom), month (mon),
# and day of week (dow) or use '*' in these fields (for 'any').
#
# Notice that tasks will be started based on the cron's system
# daemon's notion of time and timezones.
#
# Output of the crontab jobs (including errors) is sent through
# email to the user the crontab file belongs to (unless redirected).
#
# For example, you can run a backup of all your user accounts
# at 5 a.m every week with:
# 0 5 * * 1 tar -zcf /var/backups/home.tgz /home/
#
# For more information see the manual pages of crontab(5) and cron(8)
#
# m h  dom mon dow   command
* * * * * /home/parallels/Documents/cronexample.sh
                         [ Read 26 lines ]
^G Help      ^O Write Out  ^W Where Is  ^K Cut      ^T Execute   ^C Location
^X Exit      ^R Read File  ^\ Replace   ^U Paste    ^J Justify   ^/ Go To Line
```

Figure 11-7. *Scheduling the execution of that previously written shell script*

What we have is

* * * * * /home/parallels/Documents/cronexample.sh

This script line would execute the command at every minute of every day of every week of every month.

Now let us introduce another operator through another example:

*/6 * * * * /home/parallels/Documents/cronexample.sh

We have a "/" operator, which is called the step value. In this case we have incorporated the step value for the minute field. This example would mean to run the following script at every sixth minute regardless of hour, day, month, date, etc.

The "/" step operator can be extended to other fields as well, rendering much more complex applications.

Here is an example:

```
*/24 */4 * * * /home/parallels/Documents/cronexample.sh
```

This would enable the cron scheduler to run this script at the 24th minute of every 4th hour. Here is how a sample schedule looks like:

```
first execution at 2024-02-09 12:00:00
followed by execution at 2024-02-09 12:24:00
followed by execution at 2024-02-09 12:48:00
followed by execution at 2024-02-09 16:00:00
followed by execution at 2024-02-09 16:24:00
```

Let us look at providing a list for a given field. Here is an example:

```
*/30 */18 * * */1,4,7 /home/parallels/Documents/cronexample.sh
```

In the preceding example, we are running the "cronexample.sh" at every half hour past every 18th hour on Monday, Thursday, and Sunday. Here is how the schedule may look like:

```
first execution at 2024-02-09 18:00:00
followed by execution at 2024-02-09 18:30:00
followed by execution at 2024-02-10 00:00:00
followed by execution at 2024-02-10 00:30:00
followed by execution at 2024-02-10 18:00:00
```

The first execution starts at 18:00 hours followed by a 30-minute gap in the same hour. The subsequent execution happens at noon of the next day providing the 18-hour time window.

Note The clock starts at 12:00:00 for any given schedule.

We can also create a range in crontab.

Here is an example:

```
0 22 * */6 1-5 /home/parallels/Documents/cronexample.sh
```

CHAPTER 11   GETTING STARTED WITH WORKFLOW MANAGEMENT AND ORCHESTRATION

In this example, we have scheduled this script to run at 10 PM, every day from Monday through Friday, every six months.

Here is how the schedule may look like:

```
first execution at 2024-07-01 22:00:00
followed by execution at 2024-07-02 22:00:00
followed by execution at 2024-07-03 22:00:00
followed by execution at 2024-07-04 22:00:00
followed by execution at 2024-07-05 22:00:00
```

Now, let us try to combine the range operator and the step value operator:

```
36 9-17/2 * * * /home/parallels/Documents/cronexample.sh
```

In this example, we have scheduled the script to run at the 36th minute of the given hour, every 2 hours between 9 AM and 5 PM:

Here is how that would look like:

```
first execution at 2024-02-09 13:36:00
followed by execution at 2024-02-09 15:36:00
followed by execution at 2024-02-09 17:36:00
followed by execution at 2024-02-10 09:36:00
followed by execution at 2024-02-10 11:36:00
```

Let us look at an example that uses word strings instead of numbers. Here is an example:

```
0 9-17/3 */12 JAN,MAR,MAY,JUL,OCT,NOV,DEC MON /home/parallels/Documents/cronexample.sh
```

This crontab script would execute the shell script every 12th day of the month if that falls on Monday in January, March, May, July, October, November, and December, and on the day of executing the script, it would execute every third hour between 9 AM and 5 PM.

Here is an example schedule:

```
first execution at 2024-03-25 09:00:00
followed by execution at 2024-03-25 12:00:00
followed by execution at 2024-03-25 15:00:00
followed by execution at 2024-05-13 09:00:00
followed by execution at 2024-05-13 12:00:00
```

379

Let us look at a few more operators:

```
*/15 3 L 6-9 FRI#4 /home/parallels/Documents/cronexample.sh
```

Notice the use of the character "L" and the hash symbol "#". The "L" character indicates the last of the values for a given field. The "#" hash operator indicates the nth day of the month. In this case, the script is scheduled to be executed every 15 minutes, between 3 AM and 3:59 AM, on the last day of the month if that last day happens to be the fourth Friday of the month, in the months of June, August, and September.

Cron Scheduler Applications

In the world of standing up infrastructure and engineering data systems, there are many applications that involve utilizing cron schedulers. Let us look at some of these applications in detail.

Database Backup

Database backup is a process of creating a copy of an organization's master files and transactional tables. This is a critical operation, as data is the most valuable asset and data is required to run everyday business and operations. There are various types of database backup methods that exist depending upon the type and frequency of backups. There are differential backups, incremental backups, and complete backups.

Let us look at a simple example where a shell script executing database backup operation is scheduled:

```
0 2 * * * /home/parallels/mysql_backup.sh
```

This script would run every day at 2 AM carrying out instructions mentioned in the shell script file.

Data Processing

Many of the organizations are working with a variety of systems that generate data. The data team within the organization would usually perform report generation every day or perform a file transfer of certain data. These data processing tasks that arise within certain source systems may be automated using cron scheduler utility.

Let us look at scheduling a batch job that performs data processing or a report generation every day:

```
45 3 * * * /home/parallels/Documents/data_processing.sh
```

The abovementioned cron scheduler would process the batch operation every day at 3:45 AM, thereby preparing the report or performing file transfer to another location.

Email Notification

Email notification is a process of notifying one or more persons with a certain message. In the world of data systems and engineering, one can set up to receive email notifications about the status of the task that is processed (backing up a database, preparing a dataset, data mining, etc.). Several libraries exist to set up email server configuration and output messages to the appropriate email.

Let us look at a simple example:

```
45 7 * * MON-FRI /home/parallels/email_notification.sh
```

The abovementioned would send email notifications on various tasks every day at 7:45 AM from Monday to Friday.

Cron Alternatives

There are quite a few alternatives to cron job scheduler; some of these alternatives are still based on earlier distribution of cron scheduler utility. But there is one alternative that is utilized and well respected in the admin community, and that is called "systemd timer." Systemd is a software suite that provides a variety of components to manage system processes in Linux operating systems. Systemd timer is one of the components of systemd that provides a job scheduling mechanism. Jobs can be scheduled to run on a specific time interval or based off of a certain event.

Conclusion

Workflow management and orchestration are essential components in data engineering. In this chapter, we looked at crontab and cron scheduling and how we can orchestrate complex jobs and automate routine tasks using the same. What appears to be a simple

CHAPTER 11   GETTING STARTED WITH WORKFLOW MANAGEMENT AND ORCHESTRATION

concept is a powerful scheduling tool for automating recurring tasks. The foundations and key concepts you have learned in this chapter may serve you well when looking at modern data orchestration platforms, which we will see in the next couple of chapters. Efficient workflow management, scheduling, and orchestration are key to unlocking the complete potential of your data infrastructure.

CHAPTER 12

Orchestrating Data Engineering Pipelines using Apache Airflow

Introduction

Apache Airflow is an open source workflow orchestration platform. It enables engineers to define, schedule, and orchestrate workflows as directed acyclic graphs (DAGs). Airflow supports a wide range of integrations with popular data storage and processing systems and is suitable for a variety of use cases, from ETL (extract, transform, load) processes to sophisticated machine learning pipelines. With its scalable and extensible architecture, great graphical user interface, and collection of plugins and extensions, Airflow is widely used to improve productivity and ensure data reliability and quality. We will have a deep dive into the architecture of Apache Airflow, various components, and key concepts that make Apache Airflow a powerful workflow orchestrator.

By the end of this chapter, you will learn

- Fundamentals of Apache Airflow and its architecture
- Setup, installation, and configuration of Airflow in a development environment
- Key concepts such as DAGs, operators, and macros
- Inter-task communication using Xcom and flexible workflow configuration using params and variables
- Controlling complex workflows using triggers and task flows

CHAPTER 12 ORCHESTRATING DATA ENGINEERING PIPELINES USING APACHE AIRFLOW

Introduction to Apache Airflow

Apache Airflow is an open source, workflow management and orchestration platform. Airflow originated at Airbnb almost a decade ago in 2014. Airflow was created to programmatically schedule and orchestrate workflows.

Setup and Installation

Here is how we can get set up with Apache Airflow. The version I am working with is Apache Airflow 2.8.1. Let us begin by creating a new virtual environment:

```
python -m venv airflow-dev
```

Here's how to activate the virtual environment in Linux:

```
source airflow-dev/bin/activate
```

If you are using MS Windows, then please use

```
airflow-dev/Scripts/activate
```

Create a new folder somewhere in your development workspace:

```
mkdir airflow_sandbox
```

Now, copy the folder path and set the AIRFLOW_HOME environment variable. Ensure that you are performing this prior to installing Apache Airflow in your system:

```
export AIRFLOW_HOME=/home/parallels/development/airflow_sandbox
```

The following instructions are extracted from the Apache Airflow website. We shall obtain the Python version using the following command:

```
PYTHON_VERSION="$(python --version | cut -d " " -f 2 | cut -d "." -f 1-2)"
```

As mentioned earlier, we are looking to install Apache Airflow version 2.8.1. Here is the command to perform this step:

```
pip install "apache-airflow==2.8.1" \
--constraint "https://raw.githubusercontent.com/apache/airflow/constraints-2.8.1/constraints-${PYTHON_VERSION}.txt"
```

CHAPTER 12 ORCHESTRATING DATA ENGINEERING PIPELINES USING APACHE AIRFLOW

This command will kick-start installing Apache Airflow and its dependencies. Here is how that may look:

```
Collecting apache-airflow==2.8.1
  Using cached apache_airflow-2.8.1-py3-none-any.whl.metadata (36 kB)
Requirement already satisfied: alembic<2.0,>=1.6.3 in ./airflow-dev/lib/python3.11
/site-packages (from apache-airflow==2.8.1) (1.13.1)
Collecting argcomplete>=1.10 (from apache-airflow==2.8.1)
  Downloading argcomplete-3.2.1-py3-none-any.whl.metadata (16 kB)
Requirement already satisfied: asgiref in ./airflow-dev/lib/python3.11/site-packag
es (from apache-airflow==2.8.1) (3.7.2)
Requirement already satisfied: attrs>=22.1.0 in ./airflow-dev/lib/python3.11/site-
packages (from apache-airflow==2.8.1) (23.2.0)
Requirement already satisfied: blinker in ./airflow-dev/lib/python3.11/site-packag
es (from apache-airflow==2.8.1) (1.7.0)
Requirement already satisfied: colorlog<5.0,>=4.0.2 in ./airflow-dev/lib/python3.1
1/site-packages (from apache-airflow==2.8.1) (4.8.0)
Requirement already satisfied: configupdater>=3.1.1 in ./airflow-dev/lib/python3.1
1/site-packages (from apache-airflow==2.8.1) (3.2)
Requirement already satisfied: connexion<3.0,>=2.10.0 in ./airflow-dev/lib/python3
.11/site-packages (from connexion[flask]<3.0,>=2.10.0->apache-airflow==2.8.1) (2.1
4.2)
Collecting cron-descriptor>=1.2.24 (from apache-airflow==2.8.1)
  Downloading cron_descriptor-1.4.0.tar.gz (29 kB)
  Installing build dependencies ... done
  Getting requirements to build wheel ... done
  Preparing metadata (pyproject.toml) ... done
```

***Figure 12-1.** Pip installation of Apache Airflow*

Now, let us initialize the backend database, create an admin user, and start all components by entering the following:

```
airflow standalone
```

CHAPTER 12   ORCHESTRATING DATA ENGINEERING PIPELINES USING APACHE AIRFLOW

```
(airflow-dev) pk@mac airflow % airflow standalone
standalone | Starting Airflow Standalone
standalone | Checking database is initialized
INFO  [alembic.runtime.migration] Context impl SQLiteImpl.
INFO  [alembic.runtime.migration] Will assume non-transactional DDL.
INFO  [alembic.runtime.migration] Running stamp_revision  -> 405de8318b3a
WARNI [airflow.models.crypto] empty cryptography key - values will not be stored e
ncrypted.
standalone | Database ready
/Library/Frameworks/Python.framework/Versions/3.10/lib/python3.10/site-packages/fl
ask_limiter/extension.py:335 UserWarning: Using the in-memory storage for tracking
 rate limits as no storage was explicitly specified. This is not recommended for p
roduction use. See: https://flask-limiter.readthedocs.io#configuring-a-storage-bac
kend for documentation about configuring the storage backend.
WARNI [airflow.www.fab_security.manager] No user yet created, use flask fab comman
d to do it.
standalone | Creating admin user
standalone | Created admin user
 webserver | [2024-02-09T17:26:00.593+0530] {configuration.py:2065} INFO - Creatin
g new FAB webserver config file in: /Users/pk/Documents/de/airflow/webserver_confi
g.py
```

Figure 12-2. *Initializing the backend configuration of Apache Airflow*

Once this step is completed, let us initialize the Apache Airflow web server by the following command:

`airflow webserver -D`

This command will start the Airflow web server as a background process.

CHAPTER 12 ORCHESTRATING DATA ENGINEERING PIPELINES USING APACHE AIRFLOW

Here is how that looks:

Figure 12-3. Apache Airflow initialization

Once you have the web server running, let us try to run the scheduler. The following command will kick-start the Airflow scheduler:

`airflow scheduler -D`

You may open your browser and enter the following URL to access the Apache Airflow graphical user interface.

Here is the URL:

`http://localhost:8080`

Here is how that may look for you:

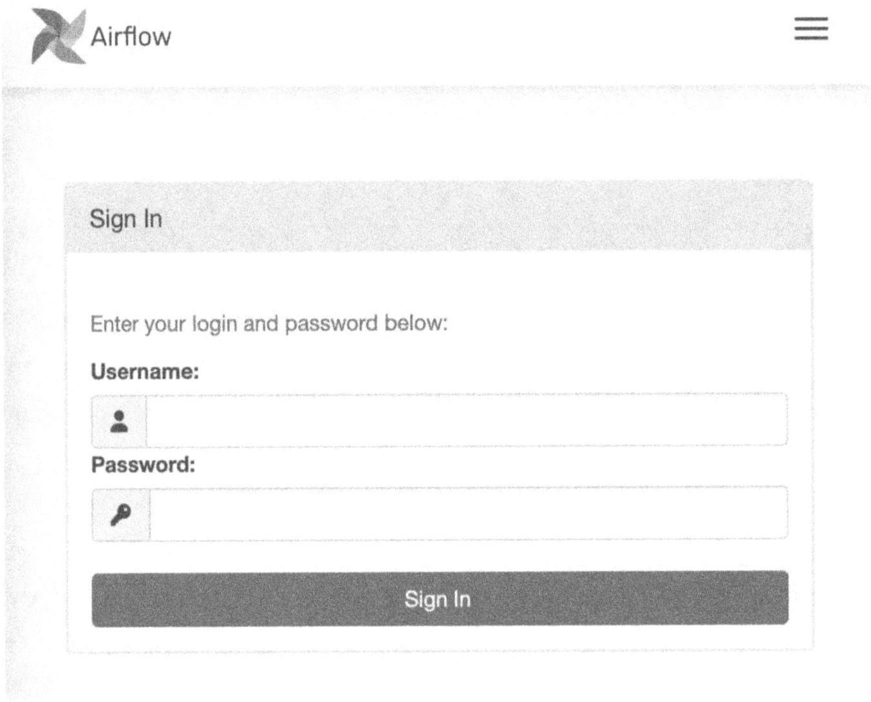

Figure 12-4. Apache Airflow GUI

When you ran the "airflow standalone" command, it must have created an admin account with the password stored in a text file within the location of AIRFLOW_HOME. Obtain the password from the text file "standalone_admin_password.txt". Here is how that may look:

```
[(airflow-dev)          airflow % ls
airflow-dev                     airflow.db
airflow-webserver-monitor.pid   docker-compose.yaml
airflow-webserver.err           logs
airflow-webserver.log           standalone_admin_password.txt
airflow-webserver.out           test
airflow-webserver.pid           webserver_config.py
airflow.cfg
[(airflow-dev)          airflow % cat standalone_admin_password.txt
```

Figure 12-5. Obtaining the user credentials from the text file

Once your login is successful, you may see the following screen with prebuilt jobs:

CHAPTER 12   ORCHESTRATING DATA ENGINEERING PIPELINES USING APACHE AIRFLOW

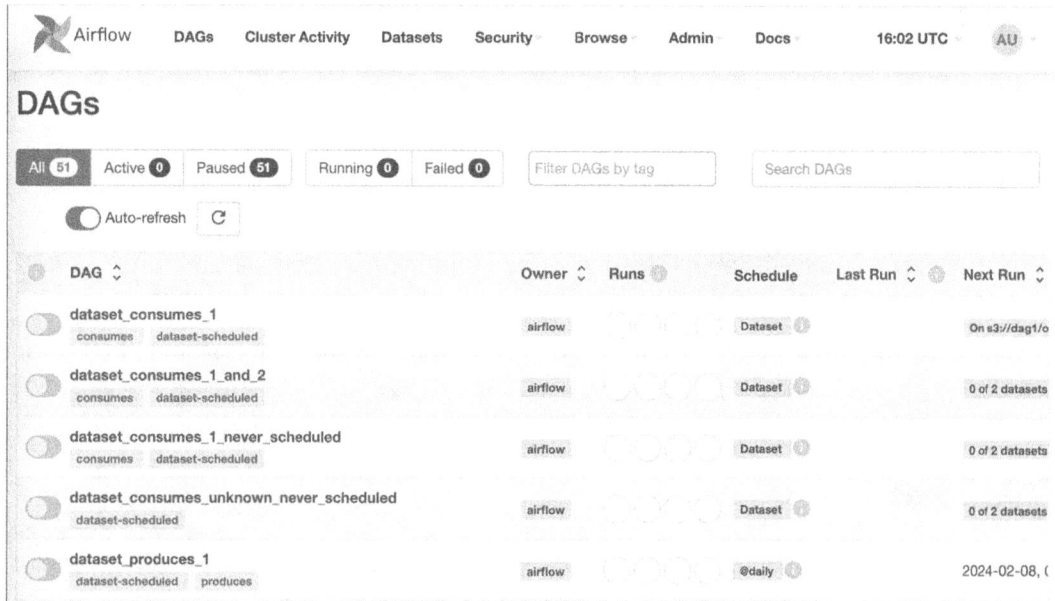

Figure 12-6. *Home screen of Apache Airflow*

Airflow Architecture

Airflow provides you a platform to define, schedule, and orchestrate jobs and workflows. As discussed earlier, a workflow can be a series of tasks where the output of a given task serves as the input of the subsequent task. In Apache Airflow, a workflow is represented by a diagonal acyclic graph. A diagonal acyclic graph is a mathematical term representing a series of nodes connected by edges that does not form a closed loop nowhere in its structure. In the context of Apache Airflow, a diagonal acyclic graph represents a series of tasks connected and dependent on each other but does not in any way form a loop. They are called simply as DAGs.

A DAG workflow has tasks represented as nodes; data sources and dependencies are represented as edges. The DAG represents the order in which the tasks should be executed. The tasks themselves contain programmatic instructions that describe what to do. The concept of DAG is at the heart of Apache Airflow's functionalities. In addition to that, there are some core components that are essential to Apache Airflow functionalities. Let us look at these core features in detail, starting with the components we are already familiar with.

Web Server

The web server component provides the user interface functionality of the Apache Airflow software. As we have seen in the previous section, the web server runs as a daemon process. The installation is shown as a local machine/single machine standalone process. You can deploy Apache Airflow, in a distributed process where the web server can have its own instance with heightened security parameters. The web server provides the user interface for a developer to look at various jobs that are currently scheduled, the schedules they were configured to run, the last time they were run, and the various tasks that are contained in the workflow of a given job.

Database

The database component of Apache Airflow contains metadata of various jobs. Some of this metadata information includes the last run of a task or workflow, status of previous runs, etc. The web server, which has a modern user interface, retrieves information from this database. Apache Airflow uses SQLAlchemy as its default data store. For a production instance, it is recommended to set up a different database backend like Postgres or MySQL. The backend database setup involves creating a separate database for an Airflow instance, a separate database user, providing all privileges on the Airflow database to the Airflow user, and adding the newly created Airflow user to the database's access control list. Lastly, initialize the new backend database and utilize appropriate database drivers in your connection string within the Airflow configuration file (we will see more on this later).

Executor

The executor is how the tasks within a given workflow are run. The executor is actually part of the scheduler component. It comes with several pluggable executors. These executors provide various different mechanisms for running your workflows. There are two major types of pluggable executors, namely, the local executors and the remote executors. As their namesakes, local executors run tasks within your workstation (locally), whereas remote executors run these tasks in multiple worker node environments (distributed environment). For instance, the LocalExecutor runs on your local workstation and supports parallelism and hyper-threading. Another honorable mention is the KubernetesExecutor, which runs each task in a separate pod in a Kubernetes cluster. A pod is like a container, and Kubernetes is a container orchestration platform. You have to have a Kubernetes cluster set up to utilize the KubernetesExecutor.

Scheduler

This is the most important component of the Apache Airflow architecture. The scheduler component monitors all the workflows and all the tasks within each workflow. The scheduler triggers execution of a task once all its underlying dependencies are complete. The scheduler is a separate service that needs to be started manually, as we have seen in the "Setup and Installation" section. Apache Airflow supports more than one scheduler component running concurrently to improve performance. The scheduler component makes use of the executor to execute tasks that are ready to go.

Configuration Files

The Airflow configuration file is not part of the Apache Airflow architecture; however, it controls the workings and functionalities. Apache Airflow comes with an option to fine-tune various options through the "airflow.cfg" file. This file gets created when you install Apache Airflow in your workstation. Each property that is configurable has been provided with descriptions and default configuration options as well. You can open the Airflow configuration file simply using a notepad or text editor. You can also get the list of options from the command line as well by entering the following command:

```
airflow config list
```

Here is how the configuration be displayed over the command line:

```
(airflow-dev)  ███████ dags % airflow config list
[core]
dags_folder = /Users/████████████airflow/dags/
hostname_callable = airflow.utils.net.getfqdn
might_contain_dag_callable = airflow.utils.file.might_contain_dag_via_default_heuristic
default_timezone = utc
executor = SequentialExecutor
auth_manager = airflow.auth.managers.fab.fab_auth_manager.FabAuthManager
parallelism = 32
max_active_tasks_per_dag = 16
dags_are_paused_at_creation = True
max_active_runs_per_dag = 16
# mp_start_method =
load_examples = True
plugins_folder = /Users/████████████airflow/plugins
execute_tasks_new_python_interpreter = False
fernet_key =
donot_pickle = True
dagbag_import_timeout = 30.0
dagbag_import_error_tracebacks = True
```

Figure 12-7. *Apache Airflow configuration*

If you want to change the location where you keep your DAGs, then here is where you would specify the location of the folder:

[core]
dags_folder = /user/parallels/<location-to-your-DAGs-folder>/

If you want to specify or change the executor that Airflow intends to use, then here is where you would do so:

[core]
executor = LocalExecutor

If you wish to change the default sqlite backend to either MySQL or Postgres database instance, then you may change the database string from sqlite driver and credentials to either mysqldriver or psycopg2 driver along with the respective credentials. Here is where you would replace the database string:

[database]
sql_alchemy_conn = sqlite:////

If you would like to gain visibility over your development or sandbox instance of Apache Airflow and enable ability to view the configuration, then set the following parameter to True:

```
[webserver]
expose_config=False
```

Once you have made the necessary changes, it is essential to restart the Airflow server. A good production practice is to create a service within the systemd service in order to start, stop, and restart the Airflow server. You may find a systemd service configuration file within the GitHub repository of Apache Airflow. However, for development instances, it is easier to identify the process id (also known as PID), attempt to kill the process, and start the Apache Airflow server by using the following command:

```
airflow webserver -p 8080 -D
```

> **Note** The "-p" refers to specifying the port, whereas "-D" is to run the server as a daemon process (commonly known as background process).

In addition, you can use the following command from the terminal to obtain a list of all current DAGs loaded in the environment:

```
airflow dags list
```

A Simple Example

Let us look at a simple example of Apache Airflow and examine various components in the following sections. Here is a simple example to get started:

```
from datetime import datetime
from airflow import DAG
from airflow.operators.python import PythonOperator

def say_hello():
    print("Hello, World!")

def print_time():
    now = datetime.today()
    print(now)
    return now
```

```python
first_airflow_dag = DAG(
    'hello_world',
    description='A simple DAG that prints Hello, World!',
    schedule=None,
)

hello_task = PythonOperator(
    task_id='say_hello_task',
    python_callable=say_hello,
    dag=first_airflow_dag,
)

display_time = PythonOperator(
    task_id='print_time',
    python_callable=print_time,
    dag=first_airflow_dag
)

hello_task >> display_time
```

We import the required classes and methods, to begin with. We have two functions in this code: the first function would print a string, and the other function would display the current date and time. We then define the DAG where we specify various details like name, description, and schedule. These DAGs can also take in much more complex configurations. Then, we define the tasks for each of those functions and associate them with our DAG. Finally, we define the order of the tasks to be executed. While this may not print anything in the command line, this is helpful to understand the various parts of workflow definition using Apache Airflow. To get the DAG displayed on your user interface, make sure you place the Python script in the appropriate folder.

Note You have to place your Python script in the folder that is referenced in the airflow.cfg configuration file in order to view and monitor your DAG. Make sure your "dags_folder" property is pointing to the appropriate folder where you have your scripts that need to be run. Moreover, locate the configuration parameter named "`expose_config`" and set that parameter as "`True`". This will expose all the configuration parameters in the web server. In development or sandbox instances, these may be helpful to troubleshoot.

CHAPTER 12   ORCHESTRATING DATA ENGINEERING PIPELINES USING APACHE AIRFLOW

Airflow DAGs

The diagonal acyclic graph or simply the DAG is the heart of a workflow management and orchestration system and, in this case, Apache Airflow. A typical DAG would consist of various tasks that are connected with each other through the database connections and dependencies such that there won't be a loop formation anywhere in the graph.

Here is a simple example of a DAG:

Connect to the database ➤ extract the table ➤ perform the transformations ➤ load the dataset ➤ close the database ➤ email notify

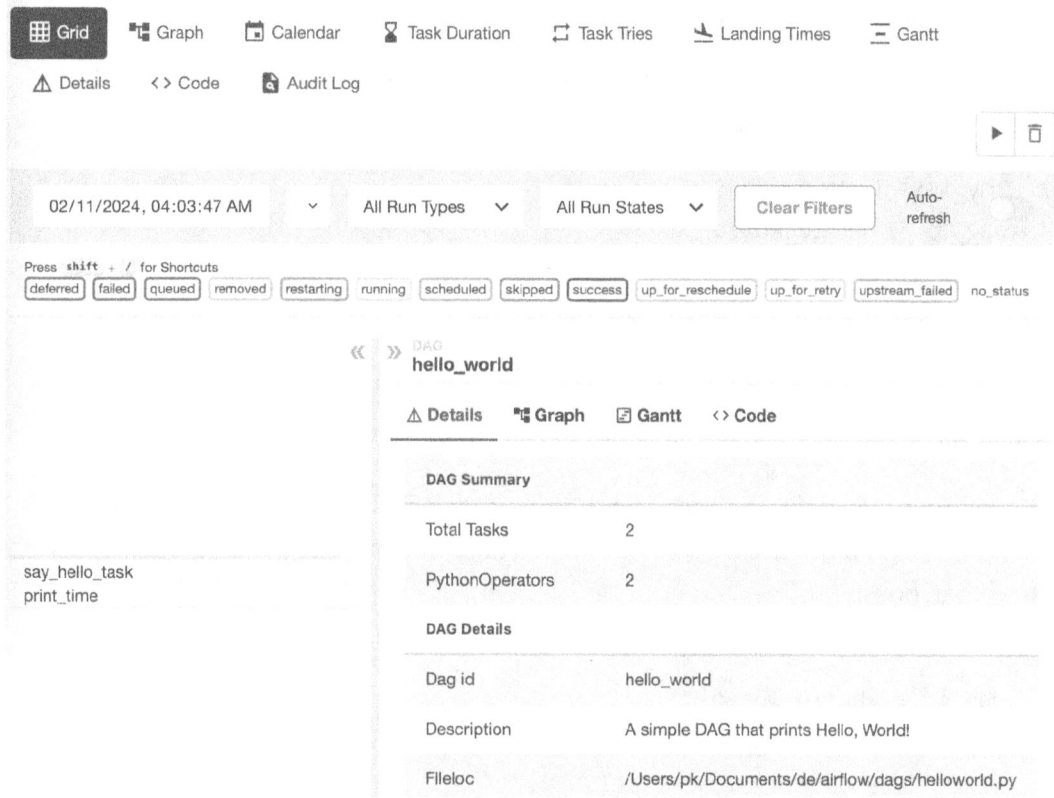

Figure 12-8.   Viewing the DAG in the Apache Airflow GUI

These DAGs will define and control how these tasks will be executed; however, they have no information about what these tasks actually do. In a way, these DAGs are task agnostic.

There are three ways of declaring a DAG within a data pipeline. The most basic and familiar way is using a constructor and passing the DAG into an operator that you plan to use. Here is how that looks:

```python
my_test_airflow_dag = DAG(
    dag_id = 'hello_world',
    description='A simple DAG that prints Hello, World!',
    start_date=datetime.datetime(2024,2,5)
    schedule="@weekly",
)
EmptyOperator(task_id="my_task", dag="my_test_airflow_dag")
```

You can also declare a DAG by using a Python context manager and adding the operators within the context:

```python
with my_test_airflow_dag(
    dag_id = 'hello_world',
    description='A simple DAG that prints Hello, World!',
    start_date=datetime.datetime(2024,2,5)
    schedule="@weekly",
):
PythonOperator(task_id="hello_world_task")
```

Lastly, you can define a function, where you specify all the operators within that function; then you can use a decorator to turn that function into a DAG generator. Here is how that looks:

```python
@dag(
    dag_id = 'hello_world',
    description='A simple DAG that prints Hello, World!',
    start_date=datetime.datetime(2024,2,5)
    schedule="@weekly",
)
def some_random_function():
    EmptyOperator(task_id="hello_world_task")
    PythonOperator(task_id="print_current_date")

some_random_function()
```

Whenever you run a DAG, Apache Airflow will create a new instance of a DAG and run that instance. These are called DAG runs. Similarly, new instances of the tasks defined within DAGs will also be created during runtime. These are called task instances.

Tasks

DAGs require tasks to run. Tasks are at the heart of any given DAG. They can be seen as a unit of work. During a run, you will be creating a new instance of a task, namely, task instance. These task instances have a begin date, an end date, and a status. Tasks have states, representing the stage of the lifecycle. Here is a sample lifecycle:

None ➤ Scheduled ➤ Queued ➤ Running ➤ Success

There are three kinds of tasks, namely, operators, sensors, and task flows. As discussed in previous sections, tasks are interconnected using relationships. In this context, the dependencies between tasks are represented as relationships. Here is an example of how relationships are represented in Apache Airflow. Let's assume we have three tasks, namely, extract, transform, and load. The output of an extract task serves as the input of a transform task, and the output of a transform task serves as the input of a load task. Here is how these tasks can be represented in relationships:

extract_data >> transform_data >> load_data

Task1 >> Task2 >> Task3

Operators

An operator in Apache Airflow represents a template for performing certain tasks. Apache Airflow has an extensive list of operators available that are already developed. Some of the operators cater to specific cloud environments, databases, etc. Let us look at a simple example of leveraging a bash operator to download a flat file and save it in a specific directory.

Here is the code for the same:

```
from airflow.decorator import dag
from airflow.operators.bash import BashOperator
from datetime import datetime
```

CHAPTER 12   ORCHESTRATING DATA ENGINEERING PIPELINES USING APACHE AIRFLOW

```
@dag(
        dag_id="extract_task",
        schedule=None,
        start_date=datetime(2024,2,2)
)
def extract_task():
    BashOperator(
        task_id="data_extract",
        bash_command="wget -c https://www.dol.gov/sites/dolgov/files/ETA/
        naws/pdfs/NAWS_F2Y197.csv -O /Users/<your-directory-here>/airflow/
        wash-electric-vehicles.csv",
    )

extract_task()
```

Every time you utilize Apache Airflow to schedule and orchestrate your data pipeline, you have to reinitialize the database. You can perform that by using the following command:

`airflow db init`

This command will pick up the DAG from the folder, and you can see the DAG listed in the user interface. Here is how that looks:

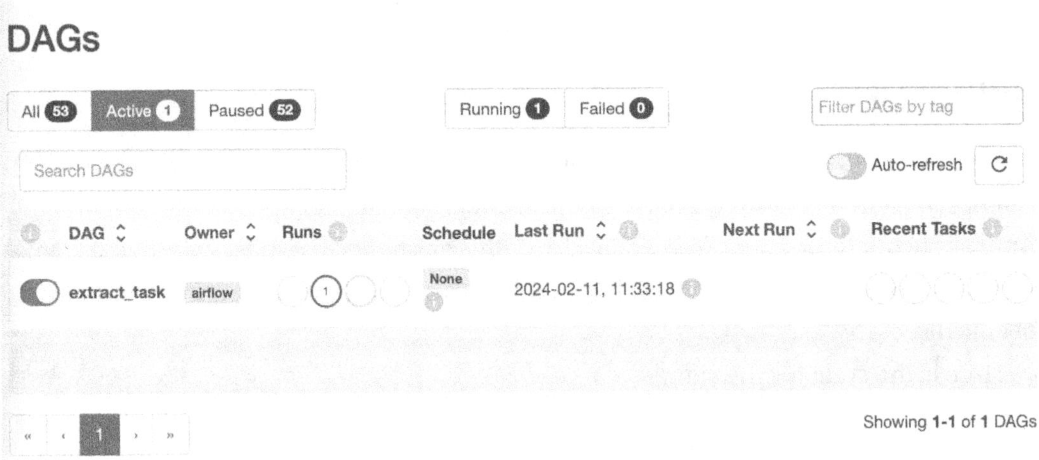

Figure 12-9. List of DAGs active in Apache Airflow

Sensors

A sensor is also a type of an operator, except it performs a very distinct function. The sensor can be used to create an event where it would wait for something to happen. Once the event happens, it would trigger a downstream job. If the event does not happen, it will continue to wait. Sensors are commonly used to wait for a file to arrive within a file system, monitor external APIs and databases, or simply wait for the completion of a different task. Here are some of the commonly used sensors and how they are utilized in data pipeline workflows:

> ExternalTaskSensor: The "ExternalTaskSensor" waits for an external task to complete in a different DAG.
>
> SQLSensor: The "SQLSensor" executes a SQL query and waits till the query returns the results.
>
> FileSensor: The "FileSensor" waits for a file or a document to be present at a given folder or file location.
>
> TimeSensor: Waiting for a certain point in time.

Let us look at a simple example of a sensor. We have a file sensor that continues to wait till a file arrives in a folder/file system for a specified amount of time. Whenever the file arrives into the respective folder location, the DAG would run the downstream task, which echoes that the file has arrived. In this context, it is assumed that the file will be placed in the respective folder by another process:

```python
from datetime import datetime
from airflow.decorators import dag
from airflow.sensors.filesystem import FileSensor
from airflow.operators.bash import BashOperator

@dag(
        dag_id = "file_sensor_example",
        start_date=datetime(2024,2,3)
)
def wait_for_file():
    FileSensor(
        task_id="wait_for_file",
        filepath="/Users/<your-location-here>/airflow/testfile.txt",
```

```
        mode="poke",
        timeout=300,
        poke_interval=25,
    )
@dag(
        dag_id = "file_sensor_example",
        start_date=datetime(2024,2,3)
)
def file_notification():
    BashOperator(
        task_id="file_notification_task",
        bash_command='echo "file is available here"'
)

wait_for_file >> file_notification
```

Task Flow

As you have seen in previous sections, you will have to start planning tasks and DAGs as you develop your data pipeline. If you have data pipelines that are mostly written in pure Python code without using Airflow operators, then task flow decorators will help create DAGs without having to write extra operator code:

```
from airflow.decorators import dag, task

@dag(schedule="@daily", start_date=(2024,2,2))
def taskflow_example():

    @task(task_id="extract", retries=5)
    def extract():
        """
        Extract data from the data source.
        If there is a connection issue, attempt to
        establish the connection again
        """

        return dataset

    @task(task_id="transform")
```

```python
    def transform(dataset):
        """
        Perform the necessary transformation for
        the extracted dataset
        """
        return processed_dataset

    @task(task_id="loaddata", retries=5)
    def load(processed_dataset):
        """
        Obtain the dataset and transport it to file system.
        Retry the task if there is an issue with connection.
        """
        pass

    extract_data = extract()
    transform_data = transform(dataset)
    load_data = load(processed_dataset)
taskflow()
```

In this example, we have three functions, each performing a separate activity. We have a task decorator that would make these functions a task, and the parent function has been decorated with a DAG.

Xcom

Tasks, by default, perform or execute the instructions they have been provided in isolation and do not communicate with other tasks or processes. Xcom is an Apache Airflow exclusive concept that enables cross-communications between tasks. With Xcom, you can obtain information or value from one task instance and utilize that in another task instance. While this can also be programmatically accomplished by exporting the value to a file or a database and then reading the file or database to obtain the value, Xcom functionality is native and can be considered as more secure.

From a development point of view, cross-communication can be achieved in two steps. First, you create an Xcom for a task and you pull the value from where you have created it. As far as creating an Xcom is concerned, you can simply return a value of a function, or you can utilize the Xcom push method. Here is how that may look:

```
@task
def some_function(value):
    result = some_other_function(*args)
    value.xcom_push(key="Result_Func1", value=result)
```

The Xcom pull method will attempt to obtain values using task_id and keys. Here is an example:

```
def another_function(value):
    value_from_some_function = value.xcom_pull(key="Result_Func1", task_id="some_function")
    pass
```

Hooks

Apache Airflow lets you connect with external data platforms and systems with ease, through the concept of Hooks. Hooks in Airflow, let you connect with external systems without having to write code to manually connect to these systems. Hooks also have the optional "retry" logic embedded in them. Airflow offers many built-in hooks and there is also a community of developers developing custom hooks to various applications. Let us look at a simple illustration.

Lets look at the following code that connects to a database:

```
import psycopg2

def querydb():
    conn_params = {
        "host": localhost,
        "database": testDB1,
        "port": 5432,
        "user": "username",
        "password": "password"
    }
    try:
        conn = psycopg2.connect(**conn_params)
        cur = conn.cursor()
```

```
    sql = "select * from table1"
    cur.execute(sql)

    rows = cur.fetchall()
    for row in rows:
        print(row)

except:
    print("error connecting to database")
```

Now let us use an Airflow Hook to connect to the same database. Here is how that code may look:

```
from airflow.hooks.postgres_hook import PostgresHook

def querydb():
    dbhook = PostgresHook(postgres_conn_id='Postgres_TestDB1')

    sql = "select * from table1"
    rows = dbhook.get_records(sql)

    for row in rows:
        print(row)
```

Notice the use of the Postgres database hook while connecting and retrieving rows; the database connection parameters are not required to be specified here.

Note Prior to using Hooks in Airflow, you must setup the credentials of the external data system in Airflow separately.

You can visit Airflow UI ➤ Admin ➤ Connections ➤ New Connection.

Once you have the external data system configured, you may then get a connection id that you can use it in the hook.

Variables

Apache Airflow variables are key-value pairs that can be defined and used within an Airflow DAG. These variables are scoped globally and can be accessed from any of the tasks. The tasks within a DAG can query these variables. You can also use these variables within your template. Here is how you can set a variable in Apache Airflow. A good application is storing some database or access credentials that remain static during the course of the project. When using Airflow variables to store API secrets, keys, passwords, and other sensitive information, Airflow will automatically mask the variables. In addition, Airflow also encrypts these variables. And so, it is suggested to use variables to store sensitive information over other options, which may lack masking and encryption features.

You can create Apache Airflow variables programmatically within the pipeline, through the command line interface, or even through the user interface (web server). Here is an example of setting up Airflow variables in the command line interface:

```
airflow variables set server_ip 192.0.0.1
airflow variables set -j aws_creds '{"AWS_SECRET": "your_secret_here"}'
```

You can also set up or bulk upload JSON from the Apache Airflow web server. To get started, navigate to the Admin menu in the web server and click Variables:

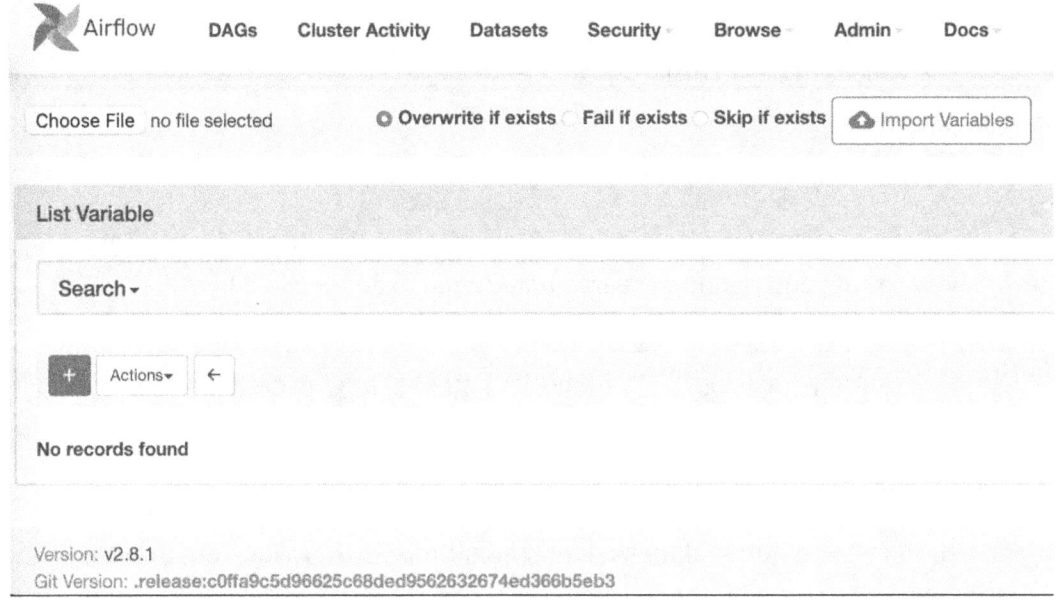

Figure 12-10. Creating variables in Apache Airflow

You can click the "+" button to create a new variable. It will render a new screen where you can enter the key, value, and description.

Here is how that looks:

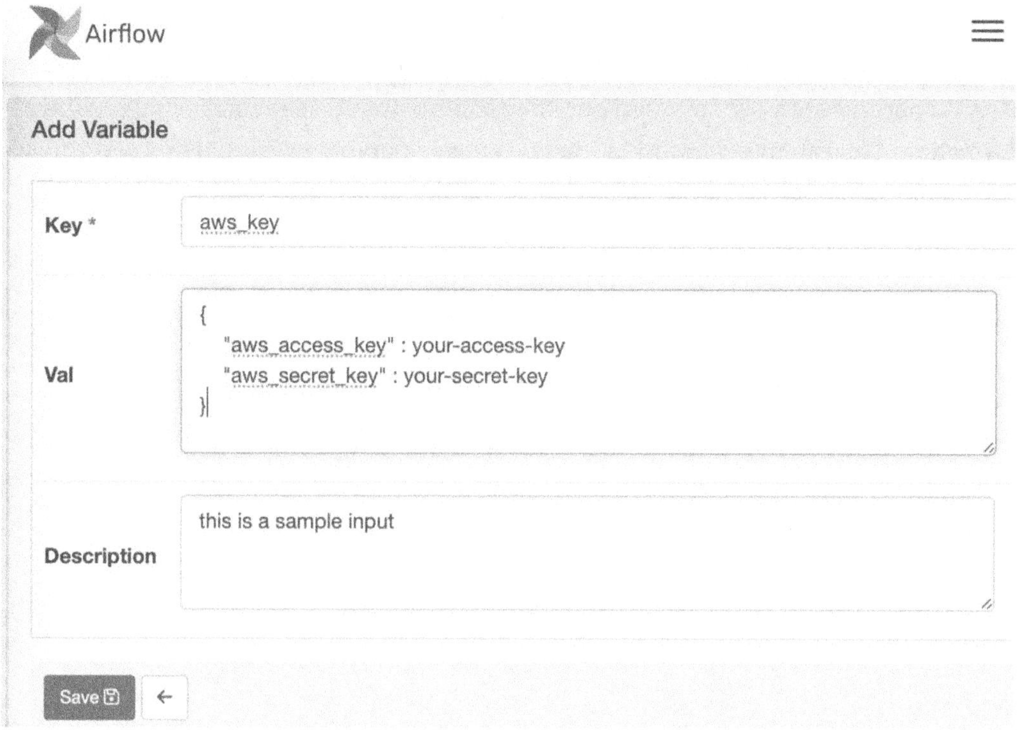

Figure 12-11. Populating the key, value, and description of the variable

You can retrieve these variables by using the following syntax:

var.value.<key-portion-of-variable>

Here is a sample illustration:

```
from airflow.models import Variable

def get_aws_parameters():
    aws_access_key = Variable.get("aws_key",default_var=None)["aws_
    access_key"]
    aws_secret_key = Variable.get("aws_key",default_var=None)["aws_
    secret_key"]

PythonOperator(
    task_id="get_aws_parameters",
    python_callable=get_aws_parameters,
)
```

Params

Params are short for parameters. Similar to variables, params are arguments you can pass to Airflow DAG runs or task instances during runtime. The difference is that values passed to params are not encrypted and they should not be used for storing sensitive information. You can pass params at the DAG level or at the individual task level. Task-level params take precedence over DAG-level params.

Let us look at passing params at the task level:

```
BashOperator(
    task_id="an_example_task",
    bash_command="echo {{ params.my_param }}",
    params={"my_param": "this is an example parameter"},
)
```

Here is another illustration of passing params at the DAG level:

```
from datetime import datetime
from airflow.decorators import dag, task

@dag(
    start_date=datetime(2024, 2, 2),
    schedule=None,
    catchup=False,
    params={
        "Person": "JohnDoe",
        "Value": 10
    },
)
def demo_dag():
    @task
    def demo_task(**context):
        print(context["params"]["Person"])
        print(context["params"]["Value"] + 10)

    print_all_params()

demo_dag()
```

Params are more localized when compared with variables in Apache Airflow. If you want to pass different values for a task at each run, params would be a good choice.

Templates

In Apache Airflow, the concept of templating utilizes a Python templating framework called "Jinja," which allows you to pass data into task instances during runtimes. These Jinja templates enable you to fill values during runtime. The syntax of this templating is as follows:

{{ your-value-here }}

The most commonly used application for templating is obtaining the execution information in Apache Airflow. These include the date of the task instance running, execution date, and start and end date interval of the current DAG. These templates are supported in most commonly used operators in Apache Airflow. You can also extend the functionality of these templates by adding filters or functions. Here is how you can use templates in Airflow:

```
BashOperator(
    task_id="begin and end date interval"
    bash_command='echo "the DAG started at {{ data_interval_start }} and
    finished at {{ data_interval_end }}"'
)
```

Macros

Macros in Apache Airflow enable you to inject predefined functions to be used within the templates. They live under the "macros" namespace in the templates. Currently, Airflow macros support methods from Python libraries, namely, time, datetime, uuid, random, and dateutil. Here is an example of utilizing macros:

```
from datetime import datetime

BashOperator(
    task_id="display_current_datetime",
    bash_command='echo "It is currently {{ macros.datetime.now() }}"'
)
```

In the preceding example, we have called a method from the Python library with the macros namespace. If you try to call it without the macros namespace, the command won't return the datetime.

Controlling the DAG Workflow

Often, production applications or data pipelines get more sophisticated, where the flow of tasks may be more complex than the illustrations we have seen. There may be a need to set up a decision loop at various stages of a production data pipeline. The typical scenario may look something like executing a subset of downstream tasks or executing only certain parts of the pipeline downstream, based on the result of a given upstream task. Other scenarios may look like skipping the entire downstream tasks based on a condition. And so, it becomes essential not only to schedule and orchestrate DAG data pipelines but also control the conditional execution logic of these DAGs. In Apache Airflow, this is called branching.

You can control the execution of a DAG in two different concepts, branching and triggering. In branching, you make use of operators to define the conditional execution. In this section we will look at two commonly used operators, BranchPythonOperator and ShortCircuitOperator.

The BranchPythonOperator is an operator in Apache Airflow that lets you evaluate a certain condition (or a decision loop) and decide which subset of downstream tasks you wish to execute. The way it works is that the operator would evaluate a condition from a task and return either a task id or list of task ids to execute downstream. Here is an example illustration of BranchPythonOperator:

```
def decision_loop(result):
    if value > 0.5:
        return ['third_task', 'fifth_task']
    return ['fourth_task']

branching = BranchPythonOperator(
    task_id='branch',
    python_callable=decision_loop,
)
```

This example utilizes the older syntax where you create an instance of the respective operator class, whereas the latest syntax makes use of a decorator that handles the instantiation and operation of the respective operator.

Let us look at an example with the new syntax:

```python
from airflow.decorators import dag, task
from datetime import datetime
from random import randint

@dag(
    start_date=datetime(2024, 2, 2),
    schedule=None,
    catchup=False
)
def branch():

    @task
    def cur_task():
        return randint(1,100)

    @task.branch
    def decision_loop(value_from_cur_task):
        if ((value_from_cur_task > 0) and (value_from_cur_task <= 40)):
            return ['i_am_next','next_to_that']
        elif ((value_from_cur_task > 40) and (value_from_cur_task <= 80)):
            return 'other_task'
        else:
            return 'hello_world'

    @task
    def i_am_next():
        pass

    @task
    def next_to_that():
        pass
```

```
@task
def other_task():
  pass

@task
def hello_world():
  pass

value_from_cur_task = cur_task()
decision_loop(value_from_cur_task)
```

This example uses task decorators. Depending upon the value a given task generates, a condition is checked, and based on the evaluation, a respective task id is returned, which would determine which downstream task to run. In the code, "`@task.branch`" is the decorator version of the "`BranchPythonOperator`".

The other operator that can be used in controlling the workflow is called the ShortCircuitOperator. The ShortCircuitOperator, if it satisfies a given condition, will skip all the downstream tasks. Let us look at an example:

```
from airflow.decorators import task, dag
from airflow.models.baseoperator import chain
from datetime import datetime
from random import getrandbits

@dag(
    start_date=datetime(2024, 2, 2),
    schedule='@daily',
    catchup=False,
)
def sc1():

    @task
    def begin():
        return bool(getrandbits(1))

    @task.short_circuit
    def decision_loop(value):
        return value
```

```
    @task
    def success():
        print('Not skipping the execution')

    chain(decision_loop(begin()), success())
sc1()
```

In this example, we have a simple function that returns a random true or false, which is provided to the decision loop function. If the random function returns true, then the downstream task success() gets executed; if the random function does not return true, then the downstream task is skipped. And so, every time this DAG is run, you would obtain different results.

Triggers

As we discussed in earlier sections, a given task will execute only after all its preceding tasks have been successfully completed. We have seen ways in which we can control execution based on conditions. Let us imagine a scenario where we have "n" number of tasks linearly scheduled with one beginning to execute after its preceding task successfully completed.

Let us say we have a requirement where if a task fails, all the other succeeding tasks are ignored and only the conclusion task must be run. You cannot short-circuit the remaining workflow as that would enable skipping all the tasks including the conclusion task. This is where the concept of triggers is beneficial.

You can configure triggers within the task itself, by adding the "trigger_rule" argument to the task. Let us look at an example:

```
from airflow.decorators import dag, task
from airflow.utils.trigger_rule import TriggerRule
from datetime import datetime
import random

@dag(
    start_date=datetime(2024,2,2),
    dag_id="trigger_illustration",
    schedule="@daily"
)
```

```python
def dag_trigger():

    @task(task_id="begin")
    def begin():
        print("Task execution begins")

    @task(task_id="random_task")
    def random_task():
        if random.choice([True,False]):
            raise ValueError("Task 2nd in queue unsuccessful")
        else:
            print("Task 2nd in queue executed")

    @task(
          task_id="task3",
          trigger_rule=TriggerRule.NONE_FAILED_MIN_ONE_SUCCESS
          )
    def task3():
        print("Task 3rd in queue executed")

    @task(
          task_id="task4",
          trigger_rule=TriggerRule.NONE_FAILED_MIN_ONE_SUCCESS
          )
    def task4():
        print("Task 4th in queue executed")

    @task(
          task_id="conclusion_task",
          trigger_rule=TriggerRule.ONE_FAILED
          )
    def conclusion_task():
        print("One of the tasks unsuccessful")

    #begin() >> random_task()
    #random_task() >> [task3(), task4()] >> conclusion_task()
    begin() >> random_task() >> [task3(), task4()] >> conclusion_task()

dag_trigger()
```

CHAPTER 12   ORCHESTRATING DATA ENGINEERING PIPELINES USING APACHE AIRFLOW

In the preceding example, we have five tasks, where one of the tasks has been configured to either execute or not execute, during which the conclusion task must be executed. By specifying various trigger rules, we are able to skip the downstream tasks and execute the conclusion task, in the event of the task not being executed.

Conclusion

As we explored throughout this chapter, Apache Airflow is a powerful workflow orchestration tool and a good solution for orchestrating data engineering and machine learning jobs. Apache Airflow has been present for close to ten years, enabling workflow orchestration and providing advanced features and customizations. All the cloud vendors do seem to offer Apache Airflow as a service, showing its wide adoption in the industry. We looked at setting up Airflow, the architecture of the Airflow system, components, and how they all work together. We then looked at DAGs, hooks, operators, variables, macros, params, triggers, and other concepts and how you can leverage them in orchestrating a data pipeline. In the next chapter, we will look at another workflow orchestration tool that is more robust and enables you to work without a DAG. The choice of the workflow orchestration tool depends on the project requirements, current infra, and team.

CHAPTER 13

Orchestrating Data Engineering Pipelines using Prefect

Introduction

Earlier we looked at Apache Airflow, the most widely adopted workflow orchestration solution. Apache Airflow is well adopted and supported by the community. In this chapter, we will look at another workflow orchestration solution called Prefect. We will look at its architecture and how you can create a workflow without DAGs. You can still define DAGs in Prefect though. The features, components, and typical workflow of Prefect appear different from what we have seen with Apache Airflow. As for which tool is a better choice, it depends on many factors like team, requirements, and infrastructure, among other things.

By the end of this chapter, you will learn

- Fundamentals of Prefect workflow orchestration and key concepts
- Setup, installation, and configuration of a Prefect development environment
- Core concepts of Prefect like flows, tasks, and results
- Artifacts, variables, and persisting results
- States, blocks, and state change hooks
- Task runners

CHAPTER 13   ORCHESTRATING DATA ENGINEERING PIPELINES USING PREFECT

Introduction to Prefect

Prefect is a second-generation, open source workflow orchestration platform, designed to help automate, orchestrate, execute, and manage various Python-based tasks and processes. Prefect can generate dynamic workflows at runtime. The learning curve for Prefect is shallow, meaning it takes relatively less time to use Prefect in the workflows. Using Prefect, one can transform Python functions into a unit of work that can be orchestrated, monitored, and observed. Some of the highlighting features that Prefect offers are

   Scheduling: Scheduling a job or a task is usually done through cron scheduling or using a dedicated orchestration service. With Prefect, one can create, modify, and view the schedules in a minimalistic web UI. Prefect offers cron scheduling, interval scheduling, and recurrence rule scheduling for more complex event scheduling.

   Retries: Data engineers often write code to retry a certain operation in case of a failure. Often, this is used in cases where one needs to connect to a database, as part of a data integration/engineering pipeline that runs periodically. With Prefect, one can simply use a Python decorator to specify the number of retries, eliminating the need for several lines of code and their associated maintainability.

   Logging: Logging is a process of registering or recording information when an event happens during an execution of a data pipeline. Logging can include high-level information, like registering the status of connecting to a database, and lower-level information, like capturing the value of a variable at a given point in time during execution. Prefect automatically logs events for various operations without having to perform additional configurations with Prefect.

   Caching: Caching is a method of storing copies of data in specific locations, often referred to as cache. The idea is to efficiently retrieve such items to reuse the same for other jobs. With Prefect, one can cache the results of a task and persist the same to a specified location.

   Async: Async is short for asynchronous and is a method of executing tasks, outside the main flow. With Prefect, one can run tasks asynchronously, meaning that Prefect can start another task while waiting for another task to finish, without stopping the execution of the program.

   Notifications: Notifications are messages that are sent by the data pipelines about an event or an update before, during, or after the execution of a data pipeline. Prefect Cloud enables notifications about the data pipeline by providing various alerts by various communication mechanisms like emails, text messages, and platforms like Slack, MS Teams, etc.

Observability: Prefect Cloud offers a dashboard-like user interface, where users can observe and monitor various tasks and flows that are scheduled, currently running, or finished.

In a traditional data stack, one has to write several lines of code, and in some cases, it would involve the utilization of another tool or software to enable such features. The fact that Prefect can offer these features out of the box is considered to be appealing when compared with similar orchestration platforms.

Prefect comes in two different editions, namely, Prefect Core and Prefect Cloud. Prefect Cloud is a commercial offering of Prefect Technologies, Inc., that comes with a dashboard-like user interface where one can observe and monitor the workflows along with few other features like role-based access control and integrations with other tools and technologies, to name a few.

Prefect Core is an open source framework that provides the core functionality. One can run Prefect Core within their own infrastructure. Prefect Core is a Python package that provides the core components of Prefect. One can obtain Prefect from the Python package installer.

Setup and Installation

To set up Prefect in your workstation, let us create a new virtual environment in your system:

```
sudo apt update / brew update
```

```
python3 -m venv prefect
```

```
source prefect/bin/activate
```

If you are using MS Windows operating system, then please use the following command:

```
prefect/Scripts/activate
```

This should be followed by installing Prefect in your system:

```
pip install -U prefect
```

Let us verify the installation process by running

```
prefect version
```

The terminal would print something like this:

```
(base) parallels@linux:~$ prefect version
Version:             2.14.21
API version:         0.8.4
Python version:      3.11.5
Git commit:          9a059bfe
Built:               Thu, Feb 8, 2024 5:35 PM
OS/Arch:             linux/aarch64
Profile:             default
Server type:         ephemeral
Server:
  Database:          sqlite
  SQLite version:    3.41.2
(base) parallels@linux:~$
```

Figure 13-1. Checking the installed version of Prefect

Prefect Server

Prefect Server is an open source backend that helps monitor and execute the Prefect flows. Prefect Server is part of the Prefect Core package. Prefect Server offers the following options in its user interface:

- Flow runs
- Flows
- Deployments
- Work pools
- Blocks
- Variables
- Notifications
- Task run concurrency
- Artifacts

Here is how you can set up a local server:

```
prefect server start
```

You would get the following screen if the web server successfully started:

```
Successfully installed Mako-1.3.2 alembic-1.13.1 apprise-1.7.2 asgi-lifespan-2.1
.0 async-timeout-4.0.3 asyncpg-0.29.0 cachetools-5.3.2 coolname-2.2.0 croniter-2
.0.1 dateparser-1.2.0 docker-6.1.3 google-auth-2.27.0 graphviz-0.20.1 griffe-0.4
0.1 h2-4.1.0 hpack-4.0.0 hyperframe-6.0.1 kubernetes-29.0.0 oauthlib-3.2.2 pendu
lum-2.1.2 prefect-2.14.21 pytzdata-2020.1 readchar-4.0.5 requests-oauthlib-1.3.1
 rich-13.7.0 rsa-4.9 sniffio-1.3.0 typer-0.9.0 ujson-5.9.0
(base) parallels@linux:~$ prefect server start

 ___ ___ ___ ___ ___ ___ _____
| _ \ _ \ __| __| __/ __|_   _|
|  _/   / _|| _|| _| (__  | |
|_| |_|_\___|_| |_____| |_|

Configure Prefect to communicate with the server with:

    prefect config set PREFECT_API_URL=http://127.0.0.1:4200/api

View the API reference documentation at http://127.0.0.1:4200/docs

Check out the dashboard at http://127.0.0.1:4200
```

Figure 13-2. *Starting the Prefect server*

Navigate to a browser and enter this URL to access the user interface: http://127.0.0.1:4200

CHAPTER 13 ORCHESTRATING DATA ENGINEERING PIPELINES USING PREFECT

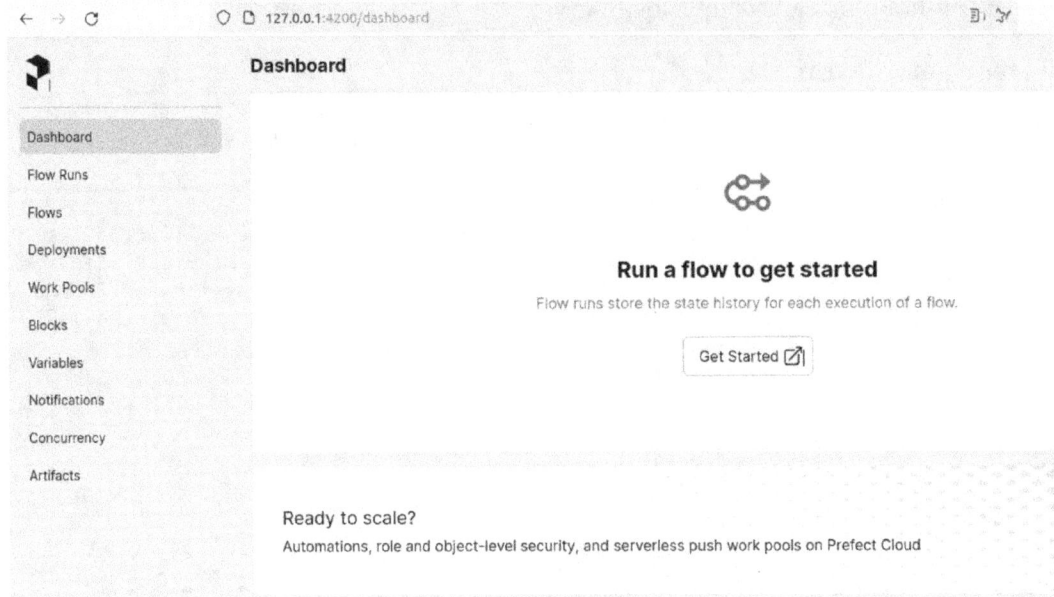

Figure 13-3. Prefect user interface in the web browser

From a software design point of view, there appear to be three major components:

- The Prefect core, which provides the core Prefect functionality and workflow orchestration engine

- A backend database for Prefect metadata, where sqlite is shipped by default and it is configurable to Postgres database

- A web server for observing and monitoring workflows, secrets management, and other orchestration-related functions

You can reset the backend Prefect database by issuing the following command:

`prefect server database reset -y`

Prefect is an open source workflow orchestration system that manages an organization's data pipelines and provides various out-of-the-box features that enable uniformity across all the data jobs and pipelines. Furthermore, Prefect provides a centralized view of the entire organization's data pipelines, including the status of the current run and history of runs. Prefect also provides a secrets management system where one can store secure login and other key credentials, integrates with the current toolset of an organization, and provides custom integrations as well.

Prefect Development

Flows

Flows are the most basic unit of Prefect development. A flow is a container for workflow logic and allows users to manage and understand the status of their workflows. It is represented in Python as a single function. To convert a Python function into a Prefect flow, one needs to add the following decorator to the function:

@flow

Once this decorator is added, the behavior of the function changes, rendering the following features:

- Changes in the state of flow are reported to the API, enabling real-time observation of flow execution.
- Input arguments can be validated, enabling data integrity and consistency.
- Retries can be performed, in the event of failure (needs to be mentioned as a parameter).
- Timeouts can be enforced to prevent unintentional, long-running workflows.

Flows can include calls to other flows or tasks, which Prefect refers to as "subflows."

Flow Runs

Flow runs represent a single instance of the flow execution. One can create a flow run, simply by running the Python script that contains the flow. A flow run can also be initiated by setting up deployment on Prefect Cloud or self-hosted Prefect Server. If a flow run consists of tasks or subflows, then Prefect will track the relationship of each child flow run to the parent flow run.

Here is an example of a basic Prefect flow:

```
from prefect import flow

@flow()
def hello_world():
    print("Hello World of Prefect!")

hello_world()
```

Let us look at an illustration of a Prefect flow that utilizes various parameters that can be passed to the decorator:

```
from prefect import flow

@flow(name = "MyFirstFlow",
    description = "This flow prints Hello World",
    retries = 3,
    retry_delay_seconds = 10,
    flow_run_name = "Hello World Flow",
    timeout_seconds = 300
    )
def hello_world():
    print("Hello World of Prefect!")

hello_world()
```

In addition to the name and description of the flow, we can also configure retries. The retries parameter is the number of times to retry execution upon a flow run failure. You can also specify the number of seconds to wait before retrying the flow after failure, by setting the retry_delay_seconds parameter. Furthermore, there is the timeout_ operator, which would mark the flow as failed if the flow exceeds the specified runtime.

To execute this Python program, simply execute the program from your terminal. Upon executing you will see a log trace similar to what is shown in the following:

CHAPTER 13   ORCHESTRATING DATA ENGINEERING PIPELINES USING PREFECT

```
(base) parallels@linux:~/Documents/prefect$ python prefect1.py
08:03:21.828 | INFO    | prefect.engine - Created flow run 'elated-turkey' for f
low 'MyFirstFlow'
Hello World of Prefect
08:03:21.876 | INFO    | Flow run 'elated-turkey' - Finished in state Completed(
)
(base) parallels@linux:~/Documents/prefect$
```

Figure 13-4. *Executing the Prefect flow*

As you can see, Prefect has executed an instance of the flow, and this flow run has been given a name. You can also see that the flow finished in the completed state. Let us look at the user interface:

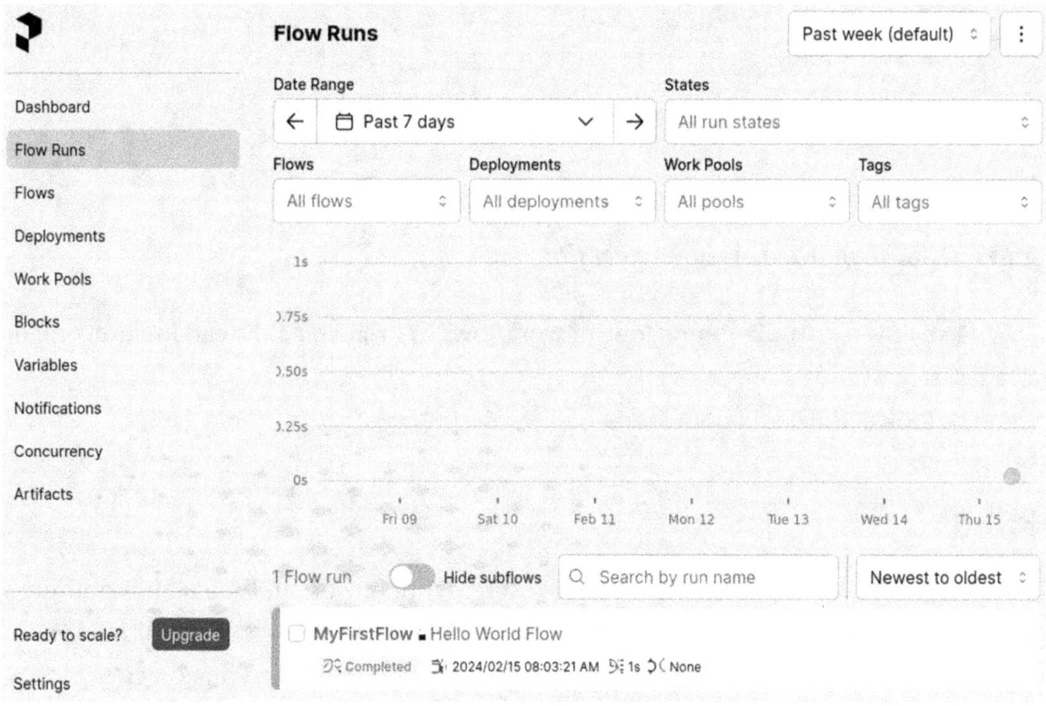

Figure 13-5. *Observing the Prefect flow in the user interface*

CHAPTER 13   ORCHESTRATING DATA ENGINEERING PIPELINES USING PREFECT

Interface

Recall that we had assigned a name for both the flow and the instance of the flow as "MyFirstFlow" and "Hello World Flow," respectively. You can obtain the same log trace and few other details when you click the flow run name. Here is how that looks:

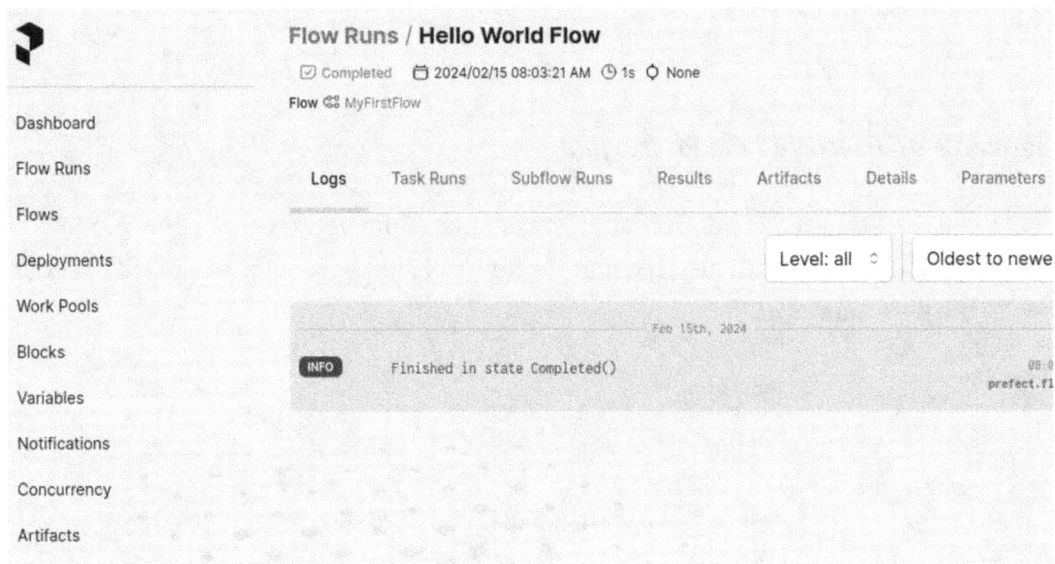

Figure 13-6. Log trace of the Prefect flow

While a flow is a fundamental idea of workflows in Prefect, a flow can include one or more tasks.

Let us look at the concept of tasks.

Tasks

A task is a function that represents a single discrete unit of work in Prefect. It is entirely possible to create a workflow, solely with flows, where the process is encapsulated as a Python function with a flow decorator. However, tasks enable breaking down the workflow logic in smaller packages that may be reused in various other flows and subflows. Tasks are similar to functions, where they take inputs, perform functions, and return an output. Tasks can be defined within the same Python file, where the flow is also defined. Tasks can also be defined within a module and can be imported to be used in flow definitions. It is essential that all tasks must be called from within a flow.

Tasks receive metadata from upstream dependencies, including the state of the dependencies prior to their execution, regardless of receiving data inputs from them. This enables the developers to make a task wait prior to the completion of another task. Usage of tasks is optional in Prefect. One can write all their code in a giant function and use a @flow decorator for that function. Prefect can execute that function. However, organizing the code into as many smaller tasks as possible may help in gaining visibility into the runtime state and also in debugging the code. In the absence of tasks, if a job runs in one flow and any line of code fails, then the entire flow will fail. If tasks are used to break the flow into several tasks, then it can be debugged relatively easily.

Each Prefect workflow must contain one primary, entry point @flow function. From that flow function, a number of tasks, subflows, and other Python functions can be called. Tasks, by default, cannot be called in other tasks; special syntax must be employed while calling tasks from other tasks:

```
from prefect import flow, task

@task
def task1():
    print("I am task1")

@task
def task2():
    task1.fn()
    print("I am task2")

@flow
def flow1():
    task1()
    task2()

flow1()
```

In the preceding example, we have seen two tasks that are incorporated within a flow. While Prefect does not allow triggering tasks from other tasks, there is an option to call the function of the task directly by using the "fn()" method. You can see the function "task1" has been called within the "task2()" function. It will only be a one-time execution of the function, and you cannot utilize retries among other Prefect functionalities when calling this function from a task.

Let us look at the output of this program.

When you run this code from the terminal, you may receive a log trace similar to following:

```
(base) parallels@linux:~/Documents/prefect$ python prefect2.py
08:21:08.572 | INFO    | prefect.engine - Created flow run 'charcoal-silkworm' for flow 'flow1'
08:21:08.607 | INFO    | Flow run 'charcoal-silkworm' - Created task run 'task1-0' for task 'task1'
08:21:08.607 | INFO    | Flow run 'charcoal-silkworm' - Executing 'task1-0' immediately...
I am task1
08:21:08.645 | INFO    | Task run 'task1-0' - Finished in state Completed()
08:21:08.655 | INFO    | Flow run 'charcoal-silkworm' - Created task run 'task2-0' for task 'task2'
08:21:08.655 | INFO    | Flow run 'charcoal-silkworm' - Executing 'task2-0' immediately...
I am task1
I am task2
08:21:08.688 | INFO    | Task run 'task2-0' - Finished in state Completed()
08:21:08.703 | INFO    | Flow run 'charcoal-silkworm' - Finished in state Completed('All states completed.')
(base) parallels@linux:~/Documents/prefect$ _
```

Figure 13-7. *Observing the output of a Prefect flow in the command line*

This Prefect flow has two tasks, namely, task1 and task2, where task2 calls the function of task1 directly. And so, you will see the task1() execution and the task2() execution, which includes the print function of task1.

Tasks provide parameters for further customization. Some of them are listed in the following. Here is an example of tasks with optional parameters included:

```
from prefect import flow, task

@task
def task1(
        name="Task1",
        description="This is the task1 returns a string",
        tags=[test],
        retries=2,
        retry_delay_seconds=10
        ):
    print("I am task1")
```

```
@task
def task2(
      name="Task2",
      description="This is the task2 returns a string and \
                  calls the task1 function directly",
      tags=[test],
      retries=2,
      retry_delay_seconds=10
      ):
   task1.fn()
   print("I am task2")

@flow
def flow1(
      name="MainFlow",
      description="Main flow with tasks included",
      retries=1
      ):
   task1()
   task2()

flow1()
```

In addition to providing the name and description of the task, you can also specify the number of times to retry the specific task upon the unsuccessful execution of the task and the duration to wait before retrying the task. There is also a Boolean parameter specifying whether to persist the task run results to storage.

Results

Results are quite straightforward. They represent data returned by a flow or a task. As mentioned previously, tasks or flows are basically Python functions that may return a value.

Here is a simple example of how results are returned:

```
from prefect import flow, task

@task
def my_task():
    return 1

@flow
def my_flow():
    task_result = my_task()
    return task_result + 1

result = my_flow()
assert result == 2
```

In the preceding example, you can see the output of a task (function with the task decorator) is saved to a variable. And the output of a flow is also saved to a variable. While this program is able to save the results of tasks and flows to variables, the program cannot persist them.

Persisting Results

Prefect does not store the results unless they help Prefect reduce the overhead of reading and writing to result storage. Results are persisted to a storage location, and Prefect stores a reference to the result.

The following features are dependent on results being persisted:

- Task cache keys
- Flow run retries
- Disabling in-memory caching

To persist the results in Prefect, you need to enable the result persistence parameter, specify the location of storage, and also have the option to serialize the output either to JSON or using pickle serializer. Depending upon the parameters used in the program, results can either be persisted to a given task or all tasks or to a flow. Consider the following example:

CHAPTER 13   ORCHESTRATING DATA ENGINEERING PIPELINES USING PREFECT

```
from prefect import flow, task

@task
def task1(a: int, b: int) -> int:
    c = a + b
    return c

@flow(retries=2)
def flow():
    ans = task1(2,3)
    print(ans)

flow()
```

In this example, the flow has the retries parameter enabled, which would require that results of all tasks to be persisted. However, flow retries do not require the flow to be persisted. Let us try to run this program:

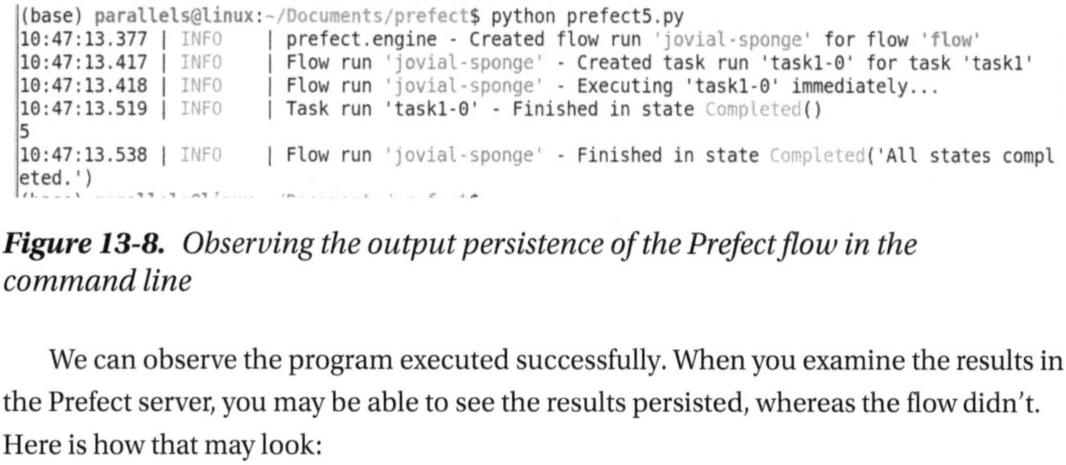

```
(base) parallels@linux:~/Documents/prefect$ python prefect5.py
10:47:13.377 | INFO    | prefect.engine - Created flow run 'jovial-sponge' for flow 'flow'
10:47:13.417 | INFO    | Flow run 'jovial-sponge' - Created task run 'task1-0' for task 'task1'
10:47:13.418 | INFO    | Flow run 'jovial-sponge' - Executing 'task1-0' immediately...
10:47:13.519 | INFO    | Task run 'task1-0' - Finished in state Completed()
5
10:47:13.538 | INFO    | Flow run 'jovial-sponge' - Finished in state Completed('All states compl
eted.')
```

Figure 13-8. *Observing the output persistence of the Prefect flow in the command line*

We can observe the program executed successfully. When you examine the results in the Prefect server, you may be able to see the results persisted, whereas the flow didn't. Here is how that may look:

CHAPTER 13   ORCHESTRATING DATA ENGINEERING PIPELINES USING PREFECT

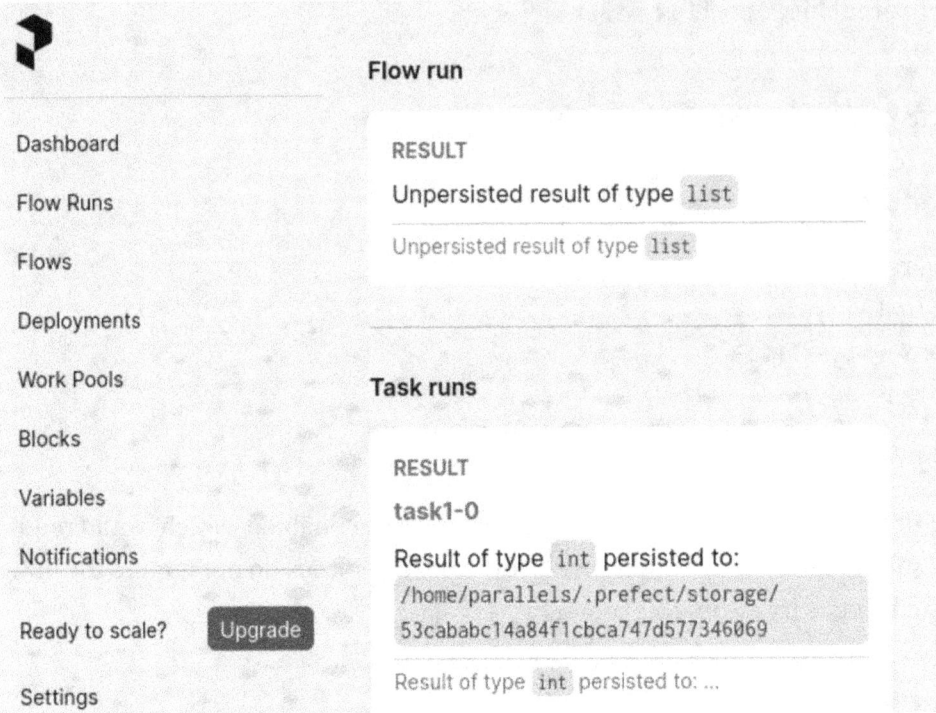

Figure 13-9. Results persist in storage

One can also manually toggle the result persistence. Toggling persistence will override any action that Prefect may arrive on its own. Here is an example of that:

```
from prefect import flow, task

@task(persist_result=False)
def my_task(a: int, b: int) -> int:
    c = a * b
    return c

@flow(persist_result=True)
def my_flow() -> int:
    result = my_task(5,6)
    return result

my_flow()
```

CHAPTER 13 ORCHESTRATING DATA ENGINEERING PIPELINES USING PREFECT

In the preceding example, the persist_result has been set to False for the task. This means the task will never persist. Any feature that depends on persistence may not work and result in error. However, for the flow, the persist_result has been set to True. The flow will persist the result even if it is not necessary for a feature.

Let us run this code:

```
(base) parallels@linux:~/Documents/prefect$ python prefect6.py
11:02:58.573 | INFO    | prefect.engine - Created flow run 'asparagus-panda' for flow 'my-flow'
11:02:58.672 | INFO    | Flow run 'asparagus-panda' - Created task run 'my_task-0' for task 'my_task'
11:02:58.672 | INFO    | Flow run 'asparagus-panda' - Executing 'my_task-0' immediately...
11:02:58.714 | INFO    | Task run 'my_task-0' - Finished in state Completed()
11:02:58.728 | INFO    | Flow run 'asparagus-panda' - Finished in state Completed()
(base) parallels@linux:~/Documents/prefect$
```

Figure 13-10. *Toggling the persistence*

The code ran without errors. Recall that we have asked the task not to persist the result and the flow to persist the result. Let us look at the flow run output in the Prefect server:

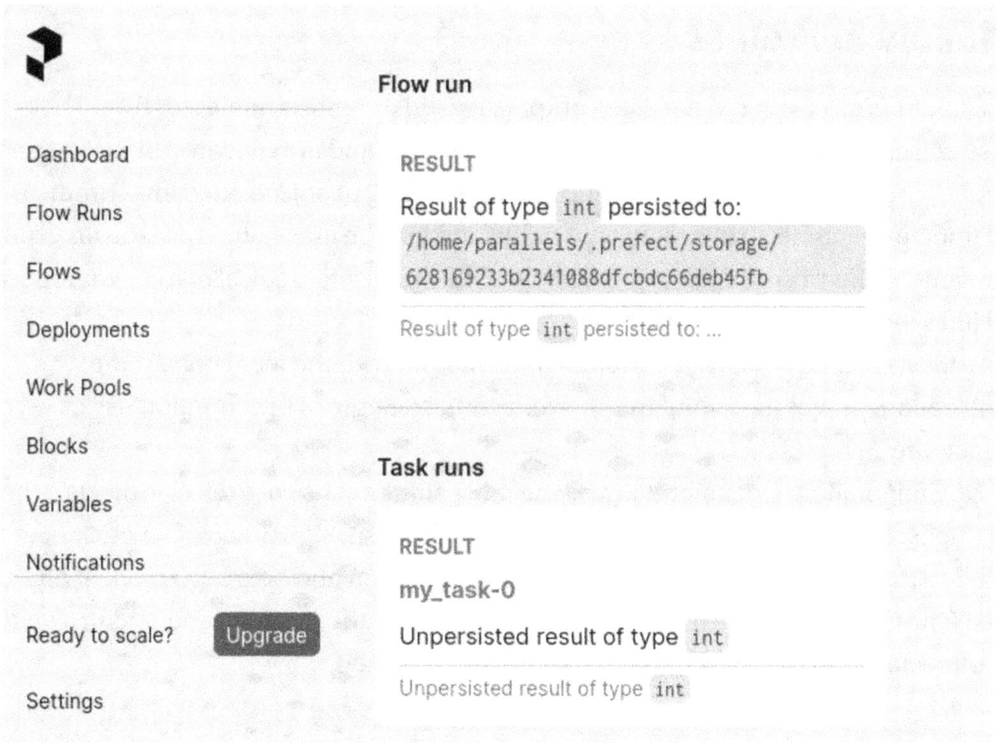

Figure 13-11. *Flow run output in the Prefect server*

Here, we see that while the task did not persist the result, the flow stored the result in a local storage.

Upon further examination, here is the serialized output of that file:

```
{
  "serializer": {
    "type": "pickle",
    "picklelib": "cloudpickle",
    "picklelib_version": "2.2.1"
  },
  "data": "gAVLHi4=\n",
  "prefect_version": "2.14.21"
}
```

You can choose to specify a specific location where you would like the results to be stored.

Artifacts in Prefect

Artifacts, in Prefect, refer to persisted outputs like links, reports, tables, or files. They are stored on Prefect Cloud or Prefect Server instances and are rendered in the user interface. Artifacts enable easy tracking and monitoring of objects that flows produce and update over time. Artifacts that are published may be associated with the task run, task name, or flow run. Artifacts provide the ability to display tables, markdown reports, and links to various external data.

Artifacts can help manage and share information with the team or with the organization, providing helpful insights and context. Some of the common use cases of artifacts are:

Documentation: Documentation is the most common use of Prefect artifacts. One can publish various reports and documentation about specific processes, which may include sample data to share information with your team and help keep track of one's work. One can also track artifacts over time, to help see the progress and updates on the data. To enable that, the artifacts must share the same key.

Data quality reports: Artifacts can help publish data quality checks during a task. In the case of training a large machine learning model, one may leverage using artifacts to publish performance graphs. By specifying the same key for multiple artifacts, one can also track something very specific, like irregularities on the data pipeline.

Debugging: Using artifacts, one can publish various information about when and where the results were written. One can also add external or internal links to storage locations.

Link Artifacts

Artifacts can be created by calling the create_link_artifact() method. Each create_link_artifact() method produces a distinct artifact and has the following parameters.

Here is a simple example of an artifact containing a link:

```
from prefect import flow
from prefect.artifacts import create_link_artifact

@flow
def my_flow():
    create_link_artifact(
        key="artifcatkey1",
        link="https://www.google.com/",
        link_text="Google",
        description="Here is a search engine"
    )
if __name__ == "__main__":
    my_flow()
```

When we run this code, it will generate an artifact for us. Let us try to locate the same. In your Prefect server, click Flow Runs and navigate to the Artifacts column:

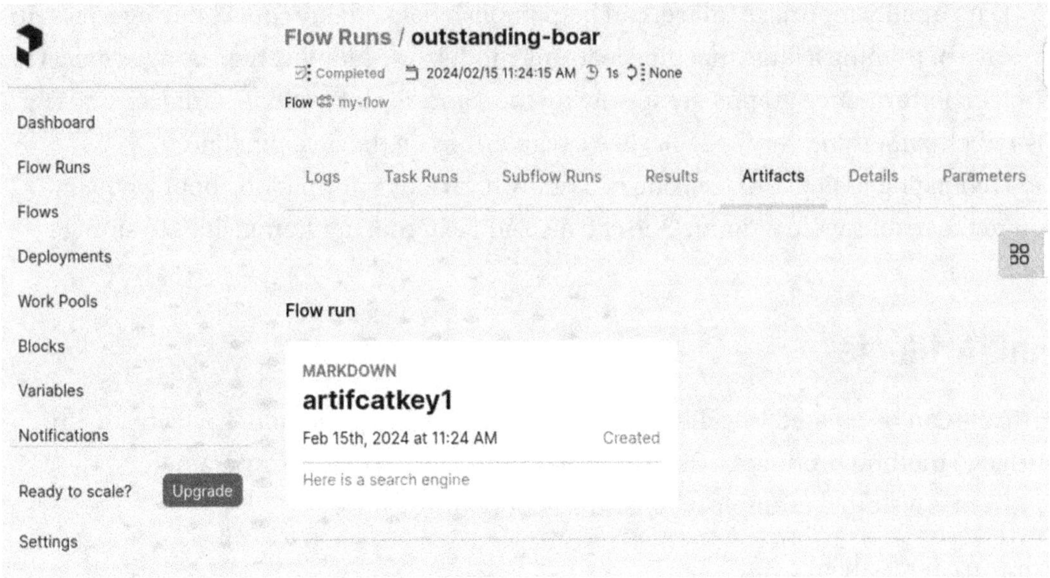

CHAPTER 13   ORCHESTRATING DATA ENGINEERING PIPELINES USING PREFECT

Figure 13-12. Artifact in the Prefect server

When you click the artifact, here is how that may render the link artifact:

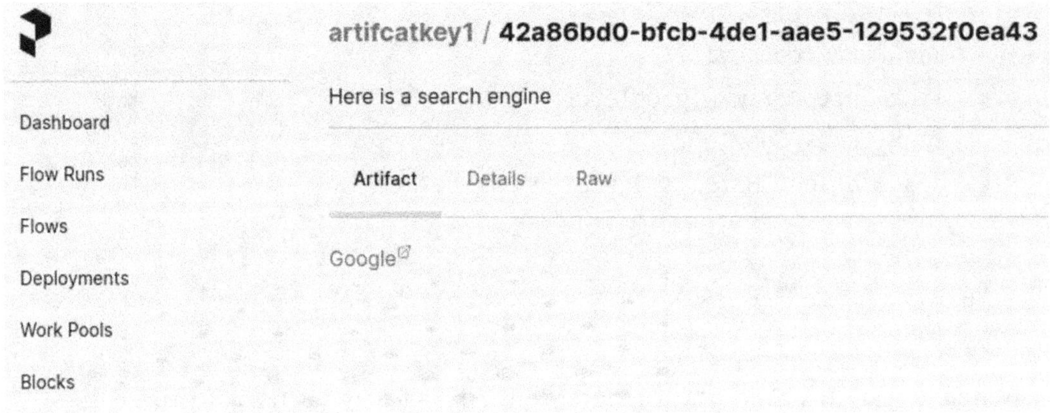

Figure 13-13. Rendering the link artifact

Markdown Artifacts

Markdown is a lightweight markup language that can be used to add formatting elements to plain text documents. The markdown syntax needs to be added to various texts to indicate which words or phrases look different and how. To create a markdown artifact, the create_markdown_artifact() function needs to be used:

```python
from prefect import flow, task
from prefect.artifacts import create_markdown_artifact

@task(name="Manufacturing Report")
def markdown_task():
    markdown_report = f"""

# Manufacturing Report

## Summary

At Pineapple Computers, we have manufactured the following items during this past quarter.

## Data pipelines by department

| Items          | Quantity |
|:---------------|-------:|
| Portable PC    | 60000  |
| Tablet         | 70000  |
| Phone          | 80000  |

## Conclusion

These numbers are very encouraging and exemplify our hardworking and resourceful manufacturing team.
"""
    create_markdown_artifact(
        key="manufacturing-report-1a",
        markdown=markdown_report,
        description="Quarterly Manufacturing Report",
    )

@flow()
def my_flow():
    markdown_task()

if __name__ == "__main__":
    my_flow()
```

Upon running the Python code, we may be able to access the markdown artifact in the same location. Here is how the artifact looks like:

Figure 13-14. Output of the markdown artifact

Table Artifacts

One can create a table artifact by using the function create_table_artifact(). Here is an example:

```
from prefect.artifacts import create_table_artifact
from prefect import flow

@flow
def my_fn():
    highest_churn_possibility = [
        {'customer_id':'12345', 'name': 'John Doe',
        'churn_probability': 0.85 },
        {'customer_id':'56789', 'name': 'Jane Doe',
        'churn_probability': 0.65 }
    ]
```

```
    create_table_artifact(
        key="table-artifact-001",
        table=highest_churn_possibility,
        description= "## please reach out to these customers today!"
    )
if __name__ == "__main__":
    my_fn()
```

Upon running this program, we may be able to locate the artifact at the similar location, within the Prefect server.

Here is how that may look:

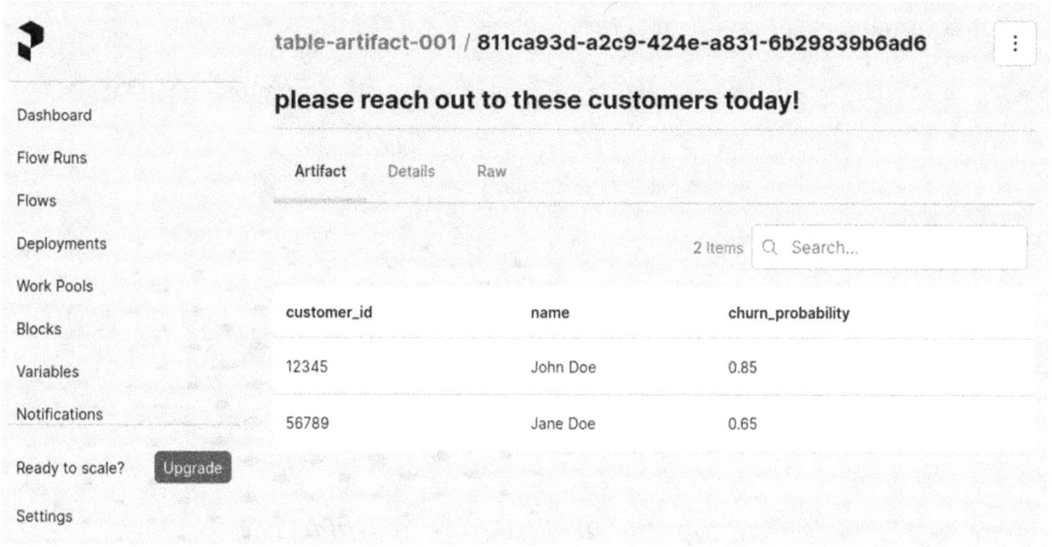

Figure 13-15. Artifacts in the Prefect server

Furthermore, one can list all the artifacts in the command line.

By entering "`prefect artifact ls`", we can list all artifacts like the following:

```
(base) parallels@linux:~/Documents/prefect$ prefect artifact ls
                                    Artifacts
```

ID	Key	Type	Updated
42a86bd0-bfcb-4de1-aae5-129532f0ea43	artifcatkey1	markdown	16 minutes ago
3efb0a6a-e215-41dc-b7f9-222360c378e9	manufacturing-report-1a	markdown	9 minutes ago
811ca93d-a2c9-424e-a831-6b29839b6ad6	table-artifact-001	table	3 minutes ago

```
              List Artifacts using `prefect artifact ls`
```

Figure 13-16. *Listing all artifacts in the terminal*

In addition, you can inspect the contents of these artifacts using the artifact key in the terminal.

The command is "`prefect artifact inspect <artifact-key>`".

```
(base) parallels@linux:~/Documents/prefect$ prefect artifact inspect table-artifact-001
[
    {
        'id': '811ca93d-a2c9-424e-a831-6b29839b6ad6',
        'created': '2024-02-15T06:06:57.411469+00:00',
        'updated': '2024-02-15T06:06:57.411469+00:00',
        'key': 'table-artifact-001',
        'type': 'table',
        'description': '## please reach out to these customers today!',
        'data': '[{"customer_id": "12345", "name": "John Doe", "churn_probability": 0.85}, {"customer_id": "56789", "name": "Jane Doe", "churn_probability": 0.65}]',
        'metadata_': None,
        'flow_run_id': 'fe328073-7300-4531-a9e7-8829278ea21e',
        'task_run_id': None
    }
]
```

Figure 13-17. *Inspecting a specific artifact using its artifact key*

States in Prefect

A state basically means the status or condition of an object at a given point in time in the system. In Prefect, states contain information about the status of a particular task run or a flow run. A reference to the state of a flow or a task means the state of a flow run or task run. For instance, if a flow is meant to be successful or completed or running, then it directly refers to the state of the flow run. Prefect states describe where the program currently is, during its lifecycle of execution. There are terminal and non-terminal states in Prefect.

Here are the states and their descriptions:

Table 13-1. *Table of various states and their descriptions*

State	Type	Description
SCHEDULED	Non-terminal	The run will begin at a particular time in the future.
PENDING	Non-terminal	The run has been submitted to run, but is waiting on necessary preconditions to be satisfied.
RUNNING	Non-terminal	The run code is currently executing.
PAUSED	Non-terminal	The run code has stopped executing until it receives manual approval to proceed.
CANCELING	Non-terminal	The infrastructure on which the code was running is being cleaned up.
COMPLETED	Terminal	The run completed successfully.
FAILED	Terminal	The run did not complete because of a code issue and had no remaining retry attempts.
CRASHED	Terminal	The run did not complete because of an infrastructure issue.

In Prefect, tasks of flow objects can return data objects, Prefect state objects, or Prefect futures, a combination of both data and state objects. Let us look at an illustration for data objects:

```
from prefect import flow, task

@task
def add_one(x):
    return x + 1

@flow
def my_flow():
    result = add_one(1) # return int
```

CHAPTER 13   ORCHESTRATING DATA ENGINEERING PIPELINES USING PREFECT

A state object is a Prefect object that indicates the status of a given flow run or task run. You can return a Prefect state by setting the "return_state" parameter to true. Here is how that may look:

```
from prefect import flow, task

@task
def square(x):
    return x * x

@flow
def my_flow():
    state = square(5, return_state=True)
    result = state.result()
    return result

my_flow()
```

A Prefect future is a Prefect object that contains both data and state. You can return a Prefect future by adding ".submit()" to the function:

```
from prefect import flow, task

@task
def square(x):
    return x * x

@flow
def my_flow():
    state - square.submit(5)
    data = state.result()
    pstate = state.wait()
    return data, pstate

print(my_flow())
```

The return value determines the final state of the flow.

State Change Hooks

State change hooks are a workflow-controlling mechanism concept. In the event that a flow or task changes its state or transitions state, state change hooks enable you to define response action based on the type of state change in a given task or flow. For a task, you can define the response action when the state is completed or failed. For a flow, you can define the response action when its state is completed, canceled, crashed, or failed.

Let us look at an example:

```
from prefect import flow, task

def success_hook(task, task_run, state):
    print("Task run succeeded")

def failure_hook(task, task_run, state):
    print("Task run failed")

def conclusive_hook(task, task_run, state):
    print("Irrespective of the task run succeeds or fails, this hook runs.")

def flow_cancellation_hook(task, task_run, state):
    print(f"flow {flow.run} cancelled")

def flow_crashed_hook(task, task_run, state):
    print("flow crashed")

def flow_completion_hook(flow, flow_run, state):
    print("flow has completed execution")

@task(
    on_completion=[success_hook, conclusive_hook],
    on_failure=[failure_hook, conclusive_hook]
)
def my_task():
    print("this task is successful")

@flow(
    on_completion=[flow_completion_hook],
    on_cancellation=[flow_cancellation_hook, conclusive_hook],
```

```
        on_crashed=[flow_crashed_hook]
    )
def my_flow():
    my_task()

my_flow()
```

You can see from the preceding code that, for a given state transition, we have more than a state change hook specified as a list. Also, you can specify the same state hook to multiple tasks and flows for the same state transition. This way, when you have several tasks and flows, you can easily control the workflow, with respect to state changes in tasks and flows.

Blocks

Blocks are one of the important aspects of Prefect. Blocks help store configuration and authentication information and provide an interface to interact with other systems.

With Prefect blocks, one can securely store authentication, pull values from environment variables, and configuration information for various source systems, downstream systems, communication channels, and other systems. Using blocks, one can query data from a database, upload or download data from a cloud data store like AWS S3, send a message to Microsoft Teams channels, and access source code management tools like GitHub. Blocks are available both in Prefect Server and Prefect Cloud.

It is important to note that blocks are used for configurations that do not change or modify during runtime. What is defined in blocks may not be altered during the runtime of flows. Blocks can be created and existing blocks can be modified using Prefect UI.

Here is how it appears on the Prefect UI:

CHAPTER 13 ORCHESTRATING DATA ENGINEERING PIPELINES USING PREFECT

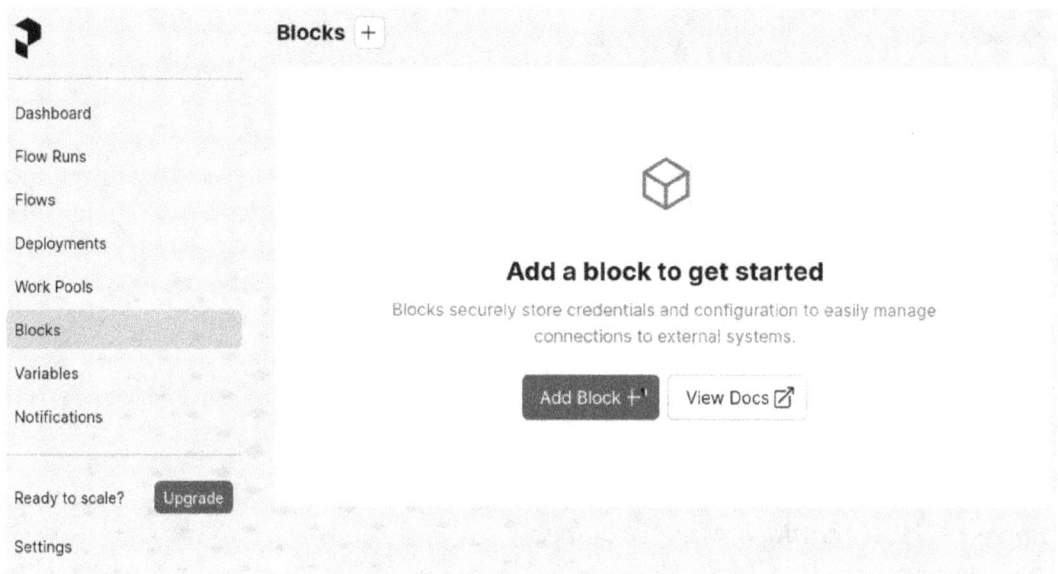

Figure 13-18. *Blocks page in the Prefect server*

Prefect provides blocks for so many cloud data stores and other external systems. It is also possible to create a custom block in Prefect. Prefect's custom blocks are built off of BaseModel class from Pydantic library. While all blocks are encrypted before being stored, one can also obfuscate fields using Pydantic's SecretStr field type.

Let us create a new block. In here, click Add block; you will see a list of various blocks. Here is an example of creating a custom AWS S3 block. S3 is an object storage cloud service, offered by Amazon Web Services. To create a new S3 block, simply type in "s3" in the search bar and you will get the S3 block. Enter the required details in the block and save it. Here is how that may look:

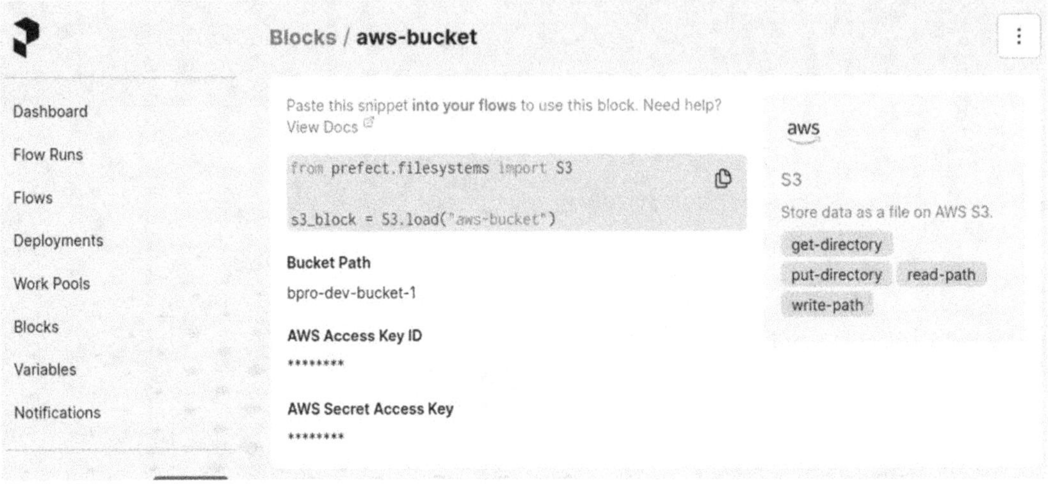

CHAPTER 13   ORCHESTRATING DATA ENGINEERING PIPELINES USING PREFECT

Figure 13-19. Creating a new AWS block

One can also register the custom blocks, so they can be used in the Prefect UI, using the command

```
prefect block register --file my_minio_block.py
```

Prefect Variables

In Prefect, variables are named and mutable string values. Variables can be created or modified at any time and they may be cached for faster retrieval. Variables can be used to store information of nonsensitive nature, such as configuration information. It is important to note that Prefect variables are not encrypted. To store sensitive information like credentials, one would use blocks, as opposed to variables.

One can create, read, write, and delete variables via the Prefect user interface, Prefect API, and command line interface. To access Prefect variables using the command line interface, here's what you need to do:

To list all the Prefect variables

```
prefect variable ls
```

To inspect a Prefect variable named prefect_var_1

```
prefect variable inspect prefect_var_1
```

To delete a Prefect variable named prefect_var_1

CHAPTER 13   ORCHESTRATING DATA ENGINEERING PIPELINES USING PREFECT

```
prefect variable delete prefect_var_1
```

To create Prefect variables via the user interface, go to the Prefect server, choose Variables, and click New variable.

Here is how that may look for you:

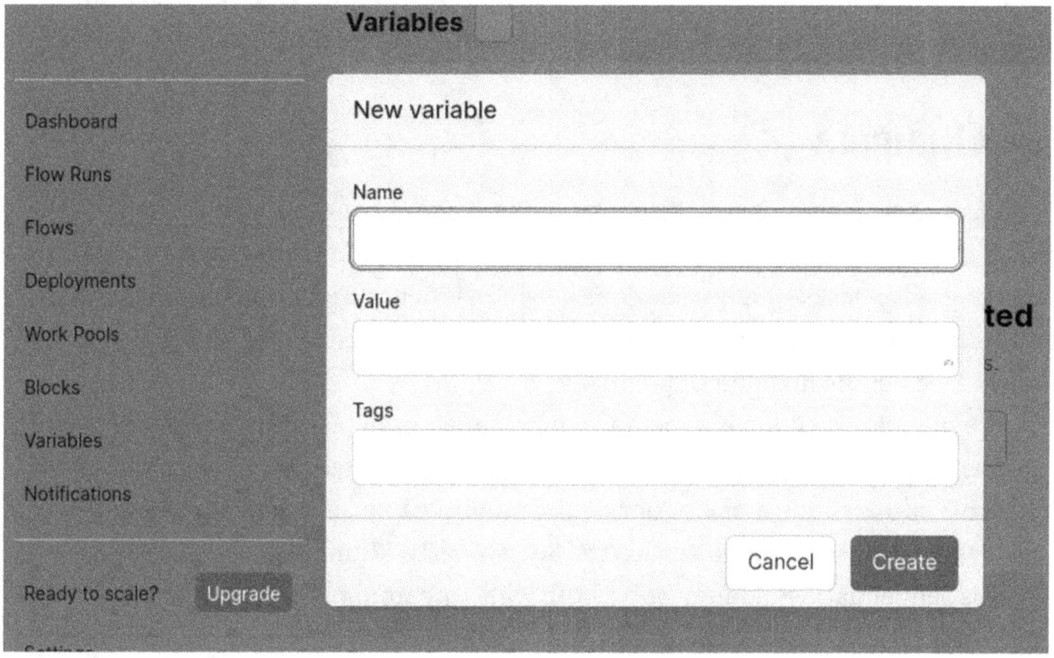

Figure 13-20. Creating a new variable in the Prefect server

Here's how to access Prefect variables using the Prefect API:

```
from prefect import variables

VariableFromPrefect = variables.get('prefect_var_1')
print(VariableFromPrefect)

# value contained in prefect_var_1
```

Variables in .yaml Files

Variables can be referenced in project YAML files. The syntax for referencing a YAML file in Prefect is "{{ variable name }}". Here is an example that Prefect documentation provides, where a variable is passed to specify a branch for a git repo in a deployment pull step:

```
pull:
- prefect.deployments.steps.git_clone:
    repository: https://github.com/PrefectHQ/hello-projects.git
    branch: "{{ prefect.variables.deployment_branch }}"
```

The abovementioned variable will be evaluated at runtime. One can also change or modify the variables without having to update a deployment directly.

Task Runners

In Prefect, tasks can be run in different ways. They are concurrent, parallel, and distributed ways of executing a task. Task runners are not required to be specified for executing tasks. Tasks simply execute as a Python function and produce appropriate returns for that function. By using default settings, when a task is called within a flow, Prefect executes the function sequentially.

A sequential execution is a method where functions or tasks are executed one after the other, only after the previous task or function is completed. This is the simplest model of task execution and it performs tasks one at a time, in a specific order. Prefect provides a built-in task runner, called the SequentialTaskRunner.

This SequentialTaskRunner works both with synchronous and asynchronous functions in Python. Here is an example:

```
from prefect import flow, task
from prefect.task_runners import SequentialTaskRunner

@task
def task1():
        print("I am task1")

@task
def task2():
        task1.fn()
        print("I am task2")

# using submit enables to create Prefect Future that helps return both
state and result of the task
@flow (task_runner=SequentialTaskRunner())
```

```
def flow1():
    task1.submit()
    task2.submit()

if __name__ == "__main__":
    flow1()
```

Prefect also provides ConcurrentTaskRunner, another built-in task runner.

ConcurrentTaskRunner executes tasks in a concurrent way, where multiple functions or tasks may start, run, and complete without having to wait for another process to complete. The execution of one task does not block the execution of another task in the flow. Here is an example:

```
from prefect import flow, task
from prefect.task_runners import ConcurrentTaskRunner
import time

@task
def stop_at_floor(floor):
    print(f"elevator moving to floor {floor}")
    time.sleep(floor)
    print(f"elevator stops on floor {floor}")

@flow(task_runner=ConcurrentTaskRunner())
def elevator():
    for floor in range(10, 0, -1):
        stop_at_floor.submit(floor)

if __name__ == "__main__":
    elevator()
```

In addition, Prefect also comes with DaskTaskRunner, a parallel task runner. It requires the Python package "prefect-dask" to be installed. The DaskTaskRunner submits the tasks to a distributed scheduler. For any given flow run, a temporary Dask cluster is created. Here is an example:

```
import dask
from prefect import flow, task
from prefect_dask.task_runners import DaskTaskRunner
```

```
@task
def show(x):
    print(x)

# Create a `LocalCluster` with some resource annotations
# Annotations are abstract in dask and not inferred from your system.
# Here, we claim that our system has 1 GPU and 1 process available per worker
@flow(
    task_runner=DaskTaskRunner(
        cluster_kwargs={"n_workers": 1, "resources": {"GPU": 1, "process": 1}}
    )
)
def my_flow():
    with dask.annotate(resources={'GPU': 1}):
        future = show(0)  # this task requires 1 GPU resource on a worker

    with dask.annotate(resources={'process': 1}):
        # These tasks each require 1 process on a worker; because we've
        # specified that our cluster has 1 process per worker and 1 worker,
        # these tasks will run sequentially
        future = show(1)
        future = show(2)
        future = show(3)
if __name__ == "__main__":
    my_flow()
```

Conclusion

In this chapter, we looked at Prefect, another open source workflow management and orchestration tool. Prefect offers a flexible and convenient approach to workflow orchestration that addresses many of the challenges of an engineering team. Prefect is Python native, and the concepts of flows and tasks and the idea of simply using a decorator to create a workflow are convenient. So far, we looked at the components of the workflow orchestrator, followed by setup and installation. We gained an appreciation for flows and flow instances, tasks, and task instances. We further looked into results, persisting results, and various artifacts. We also discussed states, state change hooks, and variables. We looked at blocks for secrets management; there is also integration available with HashiCorp Vault, for secrets management. It may be worth noting that the choice of a workflow orchestration solution may depend on several factors (current infrastructure, team's expertise, requirements of the project, etc.). Overall, you have gained higher-level skills by learning Airflow and Prefect, enabling you to build scalable and more efficient data pipelines.

CHAPTER 14

Getting Started with Big Data and Cloud Computing

Introduction

The landscape of technology in general and data processing have transformed significantly in the past decade. In the past decade alone, the evolution of cloud computing revolutionized the concept of building data pipelines in the cloud. This chapter will serve as an introduction or preamble to engineering data pipelines using major cloud technologies. We will focus on the early adopters of cloud computing, namely, Amazon, Google, and Microsoft, and their cloud computing stack. In this chapter, we will discuss how cloud computing is packaged and delivered, along with some technologies and their underlying principles. Although many of these are now automated at present, it is essential to have an understanding of these concepts.

By the end of the chapter, you will learn

- Essential networking concepts for cloud engineering
- Foundations of big data processing and Hadoop ecosystem
- Adoption of Hadoop and Spark and growth in cloud computing adoption
- Cloud computing deployment models
- Architecture and principles of cloud computing

- Cloud computing service models
- Some of the services cloud computing offers

Background of Cloud Computing

The earliest version of what we refer to as remote computing started as client–server computing. Client–server computing involved two personal computers where one computer requests for service and another computer responds with the requested information. Traditionally, accessing the Internet or email followed the model of client–server interaction. There has also been peer-to-peer computing, where all the computers in a connected network are identical and interact with each other. There have also been various protocols that supported accessing one computer from the other. As the operating system and graphics evolved, the ability to access the entire GUI of the remote computing and perform various operations became possible. As the Internet and World Wide Web took off and grew faster, companies started serving larger groups of audiences. These companies needed better infrastructure that could scale up in order to serve their customers. It may be safe to begin a cloud computing journey from here on. Let us look at a couple of terms:

> Capital expenditure: Capital expenditure refers to expenses to acquire physical equipment so it provides benefit to the company. This is often referred to as CAPEX. A good example of incurring capital expenditure is to buy blade servers for the company's IT infrastructure.
>
> Operational expenditure: Operational expenditure refers to the expenses incurred by a company in day-to-day operations to generate revenue. These are often referred to as OPEX. Examples include renting office space, paying electricity bills, maintenance and repair costs for various equipment, etc.

Networking Concepts for Cloud Computing

It is important to understand networking plays an important role in IT infrastructure. Let us review some networking concepts and terminologies that may benefit your learning journey.

IP address

IP address stands for Internet protocol address. An IP address is a way servers are identified on the Internet. IP version 4 defined the IP address as a 32-bit number. However, there is increasing adoption of IP version 6, which defines the address as a 128-bit number.

DNS

DNS stands for Domain Name System. It can be seen as an address book but only for the Internet. Instead of remembering the complex IP address digits and characters, you can see this as a string of words, an easily remembered domain name that can be used to access the website. DNS is a database where the IP addresses are mapped with the domain names.

Ports

Ports are logical entities that serve as gates to the servers. In order to enter a server, not only you need to have the appropriate IP address but also the correct port. Port numbers range from 0 to 65535. Imagine this as an airport with 65,535 gates. In order to board your flight, you have to reach the correct airport and access the correct gate to get to your flight. For instance, port 1521 is used by Oracle database, port 22 is used by SSH, etc.

Firewalls

A firewall is like a gated system where only certain IP addresses are given access. Think of it more like an invite-only launch event that allows members who are in the white list.

Virtual Private Cloud

A Virtual Private Cloud is a virtualized network environment provided by the cloud computing vendor. It can be seen as a small gated space within your own public cloud environment, where you get isolation and control over your cloud resources.

This is different from a Virtual Private Network, which provides secure and encrypted connection over private networks so you can access private services from remote locations.

Virtualization

Virtualization is the process of running a virtual instance of an operating system whose hardware is isolated from the operating system. So the computer hardware may have certain specifications. By adopting virtualization, we are able to provision only a subset of the entire capabilities that bare metal hardware possesses, and so we are able to provision multiple operating systems within a same computer. A hypervisor software that sits between computer hardware and the operating system manages how much hardware resources to allocate to each system. The operating system that a hypervisor spins off is referred to as a virtual machine.

Introduction to Big Data

Organizations saw value in obtaining insights from data for better decision making. Data started growing on a massive scale in the last decade. There was a challenge and need for technical systems to accommodate this growth and still be able to gain insights from it. This is where the concept of big data was conceived. Big data technologies are basically created to meet the needs of massively scalable systems capable of parallel processing and delivering insights.

Hadoop

The most widely accepted system in the big data space is Hadoop. Hadoop is an ecosystem of tools and technologies that supports processing and delivering insights using big data. After its first release in the early 2000s, Hadoop was open-sourced and contributed by people all over the world. Hadoop has its roots in distributed computing. Hadoop enables the creation of a distributed computing platform that runs on a large number of consumer-grade machines. The Hadoop ecosystem consists of a cluster that has a name node and one or more worker nodes, wherein the name node would be responsible for assigning and coordinating tasks and the worker nodes would carry out the tasks that are assigned to them.

The Hadoop ecosystem has its own file system called Hadoop Distributed File System (HDFS). HDFS supports the MapReduce paradigm, which enables parallel computation on big data. Hadoop enables writing quick and easy MapReduce jobs using a scripting language called Pig Latin. Pig Latin is helpful in writing complex data transformations and useful in exploratory data analysis with big datasets. Furthermore, Hadoop also supports Hive, a data warehousing solution on top of Hadoop's file system. The Hive tool provides a SQL-like interface for querying data stored in Hadoop's file system. Hive translates this SQL-like syntax into MapReduce jobs and enables distributed computation on the data.

Spark

The biggest leap in the evolution of big data systems happened with Apache Spark. Apache Spark is an analytics engine that supports big data processing and possesses the ability to run in-memory computations. The idea of performing in-memory computations enabled Apache Spark to be 100 times faster than MapReduce processing on disk. Furthermore, Spark supports multiple programming languages like Java, Scala, Python, and R offering exhaustive analytics capabilities that include SQL processing, real-time stream processing, machine learning pipeline development, and graph processing as well.

Spark played a significant role in the shaping and evolution of modern data platforms like Delta Lake architecture, data lake house, etc. Delta Lake is an open source storage layer that runs on top of data lakes and is basically built on Spark; Delta Lake provides various capabilities like ACID transactions on data lakes and integrating streaming and batch data processing.

The growth of big data, leading to greater need for big data systems and their closely associated tools and technologies, certainly contributed to increased adoption of cloud computing technologies. As organizations realized gathering analytics from vast amounts of data to gain competitive intelligence, among other insights, it was a relatively easier choice to migrate to the cloud. Cloud computing systems, regardless of vendors, offer compute, memory, and storage at scale. It is much quicker to provision and set up a data processing cluster; even better, cloud computing vendors offer to charge only for the time you consume their resources. This is called serverless computing and you will see more of this in later sections and chapters.

Introduction to Cloud Computing

Cloud computing is the concept of renting and consuming computer resources that are located remotely. Cloud computing is a business term and can be seen as a buzzword. Cloud computing enables an organization to move away from incurring capital expenditure and focus on the operational expenditure. An organization, instead of buying several blade servers and rack servers to stand up their own IT infrastructure, can provision the same configuration within a cloud and pay operational expenses based on their level of consumption.

One of the biggest and clear advantages is the pace or speed it takes to set up a basic infrastructure. In the traditional model, computers need to be procured and set up along with storage, memory and networking that need to be configured, and to house that physical infrastructure in a suitable office space is required. All of this may take longer than anticipated. With cloud computing, it would take not more than an hour to provision various computes, storage, memory, and other components related to the IT infrastructure. This provides a significant advantage of time to market the product for an organization.

Second, you can go global with cloud computing. The fact that you can easily scale your app and sometimes the cloud computing vendor dynamically adds computing resources to meet the demand enables you to access new markets. As mentioned earlier, you no longer have to invest in dedicated physical servers and take several weeks to set up and get up and running.

Security is another major advantage in cloud computing. Major cloud vendors, to the best of my knowledge, do seem to offer various privacy and security tools and methods like access control, encryption at rest and in motion, bringing your own encryption key so you are driving the security, and multifactor authorization, to name a few. You still need a security expert though. However, the fact that you can avail all these services well integrated with the cloud vendor offerings and not having to procure each one of these separately is an advantage of adopting cloud computing.

Cloud Computing Deployment Models

Cloud computing can be deployed in various methods depending upon the use case and the nature of the industry that the customer is in. Let us look at them in detail.

Public Cloud

Public cloud is the most common method of deploying cloud computing for a customer. It is also considered as the most economical option. In the public cloud, the services are delivered over the Internet connection. And so, the customer just needs to have a secured workstation and a strong enough Internet bandwidth to be able to access services rendered by cloud computing vendors, seamlessly. It is also the easiest to set up and may take not more than a few minutes to create an account and begin to utilize the cloud services.

Private Cloud

Private cloud is where an organizational customer has the exclusive rights and control of the IT infrastructure. This is similar to running a traditional IT infrastructure. You can have a third-party cloud vendor provide and manage the exclusive IT infrastructure, or it can be the case that the organization has deployed their own exclusive cloud that is managed by them. You can still avail of the more recent technology, automations, virtualization, etc. and reap the benefits of cloud computing. However, the cost may be higher when compared with public cloud offerings.

Hybrid Cloud

As the name suggests, it is a mix of both private and public cloud concepts. In hybrid cloud, some resources are hosted on-premises, while some are hosted in the public cloud. This can be seen more often where organizations stay on-premise with certain applications and move to the cloud for hosting certain other applications. And so, organizations can keep their data within their own premise while procuring and utilizing certain services in the cloud.

Community Cloud

A community cloud is a variant of the public cloud, except that the IT infrastructure is shared between several organizations from a specific community or industry. They can either be managed internally or externally by a third-party vendor. Community cloud is a collaborative effort. It provides better security when compared with public cloud. Examples include banking clients and healthcare clients, where specific regulations and compliances are to be met, which are set by their respective governing bodies.

Government Cloud

Government cloud is made for government bodies and entities; they are a sub-variant of community cloud except that they are made to cater specific compliance and regulatory requirements. This would involve cloud computing vendors providing extra layers of security and policies that would comply towards specific requirements that government entities may have.

Multi-cloud

Multi-cloud is where an organization is leveraging more than one cloud vendor to host their applications and infrastructure. The advantage can be that you are not tied to a single cloud vendor and so you can easily move between vendors. It can also be advantageous from the cost point of view. An offering at one cloud vendor may be less costly compared with an offering at another cloud vendor. The downside is that it has a high operational complexity.

Cloud Architecture Concepts

Let us review some of the concepts that may directly or indirectly touch on designing data pipelines of engineering and machine learning projects. You may have seen or read the seven essential "-ilities" or the ten "-ilities" of software architecture. In this section, we will look at such similar concepts to keep in mind, when designing the data architecture for a data engineering or machine learning project. The volume, veracity, and variety of data have gradually grown over the past several years, and so that is the motivation behind discussing the following concepts.

Scalability

Every data pipeline is meant to process and/or transport data. The amount of computational resources consumed is directly proportional to the volume of data that is processed. As the data pipeline processes more data, the entire pipeline gets overworked. Scalability is the process of handling the stress caused by increasing in usage while ensuring your data pipeline effectively performs its functions smoothly and consistently. You have to have a rough idea about the maximum workload your pipeline can process and whether you may encounter any memory leaks or stack overflow. Scalability can be achieved by two approaches, namely, horizontal scaling and vertical scaling. Horizontal scaling is adding more compute nodes to a cluster that can handle an increase in the processing traffic. Vertical scaling is increasing the capacity of a given compute node to meet the traffic.

Elasticity

Elasticity refers to the ability to adapt to the change in workloads of a given data pipeline. Elasticity is the idea of dynamically resizing the underlying computational resources to meet the data processing needs and also to reduce the utilization of computational resources. This is different from scalability as scalability focuses on scaling up during peak consumption period, whereas elasticity refers to the elastic nature of the data processing system to dynamically scale up or down given the need to minimize the resource utilization. A good example is the idea of serverless computing offered by the cloud vendors. Serverless is where you do not pay for the underlying computer but rather you pay only for the units of work completed within the execution time. In addition, elasticity is also that feature where you can add or dynamically provision compute and storage during runtime.

High Availability

High availability is defined as the ability of the data processing system to be available continuously without intervention for a specified amount of time. The idea is to eliminate the single point of failure and be made available anytime when the need for data processing arises. High availability can be measured by uptime percentage during a given point in time. Even in case of system maintenance or failure of the underlying computing environment, the ability to make the data processing system to be available is the idea of high availability.

Fault Tolerance

Fault tolerance is the idea of the data processing system to be available in the event of actual failure of the underlying computing environment. It is slightly different from the idea of high availability although they are kind of two sides of the same coin. Both are aiming to make the system available without any interruptions. The way you can achieve fault tolerance is by adding compute nodes with the same data processing system that is present in a different region. For instance, if your data processing pipeline is served from one geographical region, you can have one more instance of a system that performs the same task in a different geographical region. This way if one region fails, the other region can continue to make itself available till the primary region resumes serving its customers. Many distributed computing clusters have the concept of a main node managing communications and worker nodes performing tasks. Fault tolerance is the idea of having backup main nodes readily available.

Disaster Recovery

Disasters mean macro-level disasters, such as earthquakes and floods that cause disruption in everyday services and other human-related events like diseases, pandemic or epidemic, or some other event that causes prolonged interruption to services. They can be intentional or unintentional. Disaster recovery is the idea of recovering the data management systems, data processing, and machine learning systems, restoring the functionality after a disaster event happens. The idea here is to have a backup site for the data management, processing, and machine learning systems. In this context, disaster recovery is the ability of cloud computing vendors to offer this as a service so the businesses continue to function as is.

Caching

Caching is not exactly an architectural principle; however, it is an important aspect of architectural design. Caching is a data storage concept, where commonly used data during computations are stored in a place. The idea is to reuse previously computed or processed data so we do not have to run the same computation again and again. The cached data is usually stored in RAM, or random access memory. Caching improves speed as it fetches the result from RAM compared with performing the same operation again using compute and storage. From the CPU architecture point of view, there may be several other caching layers, which are beyond the scope of cloud data architecture. There are several types of caching data in memory; some of them are API response caching, Content Delivery Network caching, etc. Our focus is on database caching. When you have a database, there may be few queries that are expensive; even after optimizations, it requires heavy resource utilization. By obtaining the results of the queries from cache instead of the source, we minimize the latency and increase the overall performance of the data processing system.

CHAPTER 14 GETTING STARTED WITH BIG DATA AND CLOUD COMPUTING

Cloud Computing Vendors

There are three major cloud computing vendors currently in the industry, Amazon, Google, and Microsoft. Amazon put cloud computing in light, back in the year 2006, when it started offering Amazon Web Services. Google launched their cloud computing platform called Google Cloud Platform. In 2010, Microsoft launched their cloud computing platform called Azure. In addition to these three major cloud vendors, there are also several other companies offering cloud computing services. Some of them are Oracle, DigitalOcean, IBM, and a few regional providers, catering to specific nations.

Cloud Service Models

Once we have deployed the type of cloud computing model, we can proceed to choose the appropriate service model. The way cloud computing vendors offer their services is broadly classified into three major categories, and we will discuss them in greater detail.

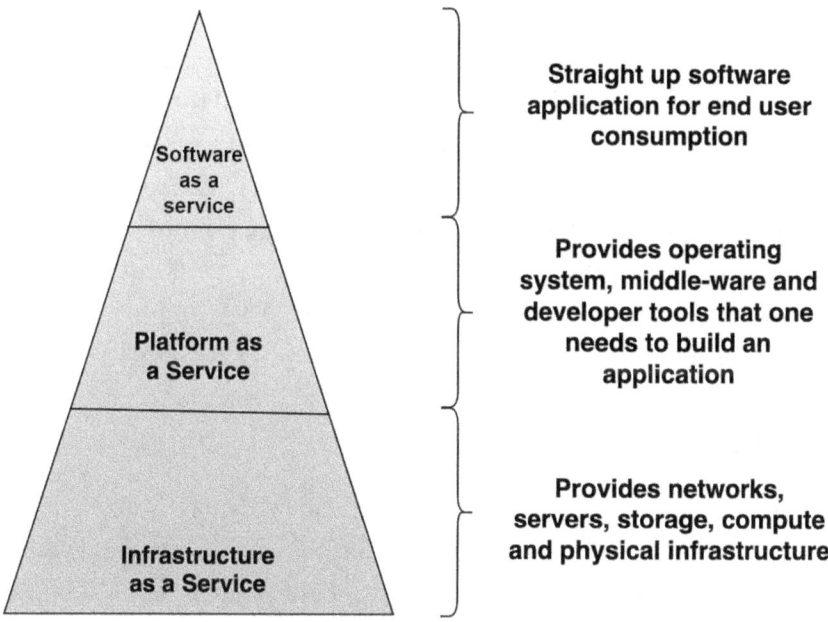

Figure 14-1. *A pyramid describing cloud service models*

461

Infrastructure as a Service

This is where the cloud computing vendor provides the computing infrastructure as a service. The idea is the cloud vendor provides you the service and you can use it to run applications, perform computations, etc. To run applications, you need infrastructure. The infrastructure consists of computer hardware like processors, memory, operating systems, networking, storage, etc. Typically the cloud vendor will provide you a server, storage, networking, and a virtual machine with an operating system. It is up to you to decide which software to install to perform the tasks you require. This model is for administrators who are looking to set up and manage the infrastructure.

Examples of Infrastructure as a Service include Microsoft Azure, Amazon Web Services, and Google Cloud Platform, to name a few.

Platform as a Service

Platform as a Service is where the cloud computing vendor provides you all the infrastructure and the environment to develop your application or work with the application. Imagine you were able to rent out a multi-core processor with memory and storage configured, an operating system installed with all the updates and patches done, and the software development installed and managed so you can jump right into developing your code or deploying your code. Platform as a Service is very helpful to developers who are looking to develop something from the ground up without having to worry about administrative efforts.

Examples of Platform as a Service include Salesforce CRM, Tableau dashboards, or a subscription-based database as a service (Postgres instance on AWS RDS).

Software as a Service

Software as a Service is something you see all over you every day. You check your free email account provided by Google by entering their domain address and your user id and password or your social media provider like Instagram or Facebook; you are using it for free of charge. Basically, these companies provide you with everything from hardware, operating system, memory, storage, and software application (Gmail, Dropbox, etc.). As an end user you would access their service, consume it to your heart's content, and, perhaps, return to it another time.

Examples of Software as a Service may include Zoom video conferencing, Microsoft Office 365 (Word, Excel, PowerPoint), Dropbox online storage, etc.

Cloud Computing Services

Identity and Access Management

Identity and access management is a business process of enabling the right users to access the right datasets at the right time. Access management enables security for data assets and controlling access to various data assets to various constituents. It is a continuous process that ensures only authorized users and services can access data assets. Amazon offers AWS Identity and Access Management and AWS Lake Formation to manage access (including fine-grained access control policies) to various resources including to its data lake environment. Microsoft offers Azure Active Directory and role-based access control services that enable access control to various services. Google provides Cloud IAM service to provide fine-grained access control over its resources and data assets.

Compute

Compute means computing processing power (like a motherboard), memory, storage, and network to connect to it. This is the most basic cloud service that one can offer. In 2006, Amazon started their Amazon Web Services and provided the EC2 as one of their earliest offerings. You have the option to pick and choose the kind of microprocessor that may offer various numbers of cores, the amount of physical memory you need, and the kind of networking connection you wish.

Storage

Cloud storage is basically storing your data (files, backups, pictures, videos, audios, etc.) in a third-party storage provider that can be accessed via the network. Cloud storage offers security and cost-effective storage management. Cloud storage has multiple types of storage offerings deepening upon the need and use case of the customer. These storage systems can autoscale wherever they need to be and can be highly available to be accessed over a network.

Object Storage

Object storage is a specialized type of file system storage that can be used to store any kind of files. The object storage manages data as objects, as opposed to file systems where data is managed as files. Each data is an object and comes with metadata and a unique identifier that is associated with the single data. Object storage is used in a variety of applications including cold storage of data (where data is infrequently accessed) and archives of data.

Examples of object storage are Red Hat Ceph, MinIO, Cloudflare R2, Azure Blob Storage (if you are in the Microsoft ecosystem), etc.

Databases

One of the most common offerings is the relational database management systems. Cloud vendors offer databases as a service in various options. You have the option of renting out a computing space from a cloud vendor with a high-speed network and installing a database by yourself; or you can rent out a preinstalled database of the latest version in a matter of a few clicks. The vendor may provide support with some of the database administration and maintenance tasks if you choose the fully configured route. With databases, you can autoscale, programmatically interact, and build data pipelines, in addition to loading, querying, and generating SQL reports.

NoSQL

In addition to relational databases, cloud computing vendors also provide NoSQL database solutions. NoSQL stands for Not Only SQL, meaning that these are database management systems that are designed to handle a variety of data models and data structures. The idea here is NoSQL database systems complement and enable efficient storage of data that may not fit into the cookie-cutter relational data models. However, NoSQL can never replace the traditional relational database systems, the data integrity they impose using various normalization forms, and the effectiveness relational databases deliver in transactional systems. NoSQL systems allow data to be inserted without having to predefine a schema.

Schema on Write vs. Schema on Read

The difference between SQL and NoSQL database systems is that relational database management systems impose a schema on the incoming data. The SQL database systems do not accept incoming data if the incoming data does not adhere to the schema standards. The NoSQL database systems, on the other hand, are schemaless; the structure of the data is interpreted when it is being read from the NoSQL database. This is why, traditionally, RDBMSs use "Schema on Write," whereas NoSQL database systems use "Schema on Read."

> **Note** NoSQL database systems are generally schemaless; however, some NoSQL database systems offer a schema enforcement option, which helps with data consistency. Various types of NoSQL data systems (document, key–value, graph, columnar) tend to handle schemas differently. MongoDB does allow for schema validation (which is like entering into the SQL arena); few RDBMSs have begun to offer support for JSON data, and PostgreSQL even lets you write a complex query on unstructured data. The line between SQL and NoSQL is getting blurry to say the least; however, it is good to understand that NoSQL has better support and a faster query engine over JSON data and unstructured data, and RDBMSs still champion transactional and structured relational data..

NoSQL has been in existence since the early 2010s. Today, there are various NoSQL offerings that major cloud vendors do seem to offer. Let us look at each one briefly.

Document Databases

Document databases are data systems that are designed for managing semistructured data. A good example is JSON data, where data is captured in JSON documents. These document databases have flexible data models and they are not defined prior to loading. Even though the JSON documents may have complex structure and nested structures within each document, they will always be consistent—as long as they come from a single source. And so, the document databases will accommodate the incoming structure of these documents. A famous example of a document database is MongoDB. It is a fast, lightweight, simple-to-use database that comes with a querying language that can be used to treat the JSON documents as relational data and write queries on top of them.

Column-Oriented Databases

The other oldest form of NoSQL database system is the column-oriented databases. Please note that column-oriented databases are different from column-store databases. Column stores are still relational database systems that are used for OLAP processing and data warehousing purposes. They are extremely fast and efficient for generating complex analytics reports. Column-oriented databases are NoSQL systems where the schema evolves with the incoming data. Column-store databases mostly have dimensional data models that are defined prior to loading data. For instance, Postgres is a column-store database; so is Google BigQuery.

Column-oriented databases are also referred to as wide-column databases. It is where the columns for a given set of data can be added at any time. Google Bigtable and Apache Cassandra are two great service offerings of wide-column databases. Microsoft's Cosmos DB and Amazon's DynamoDB are equally competitive service offerings for wide-column database systems.

Key–Value Stores

Key-value stores are types of NoSQL database systems where the data is stored as a collection of key-value pairs (very similar to dictionary data types in Python). The data model is very simple and straightforward, where the keys are identifiers and can be used to retrieve the corresponding value. Hence, these systems are incredibly fast for retrieval, as it does not involve scanning a few tables and matching certain filters and conditions. Good examples are Amazon DynamoDB, Redis, Azure Cosmos DB, etc.

Graph Databases

Graph databases are another type of NoSQL databases, where the underlying data model is composed of nodes and edges. Nodes are entities that contain various information like persons, names, or other discrete objects. Edges are connections between nodes that describe the relationship between two connecting nodes.

Let us look at a graphical illustration of how two entities may share a relationship and how there can be various types of relationships between entities.

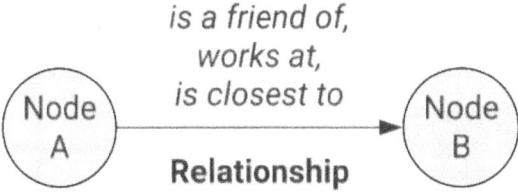

Figure 14-2. Example of a relationship between two nodes

Some good examples of graph databases are Neo4j, Amazon Neptune, Azure Cosmos DB, etc.

Time Series Databases

Time series databases are another variant of NoSQL database systems, where the data model is basically the data value indexed over a timestamp. This type of data exists in various industries—for instance, price of a stock during trading time or monitoring the temperature using an IoT sensor. These databases are designed to be write-heavy while being efficient in retrieval of aggregated metrics over time intervals, time range–based reports, etc. Good examples of such databases are Google Bigtable, Azure Time Series Insights, and Amazon Timestream, to name a few.

Vector Databases

Vector databases are newer in the NoSQL database systems, where they are designed to store, index, and fast search vector data. Given a query vector, the vector database would perform a fast search to retrieve the vector that has the most similarity with the given vector. This is also known as similarity search, which is very effective in image recognition and other artificial intelligence environments. Here, vectors are simply an array of numbers; depending upon the context, these arrays may carry certain meaning. Once a vector is given, one can specify the type of measure that is relied on to obtain the nearest neighbor vector that resembles the same. The similarity measure can be Euclidean, Tanimoto, or Manhattan distance. All these vectors are indexed for faster similarity search and retrieval.

Data Warehouses

Data warehouses are columnar-style relational database management systems that are used for analytical processing. These data stores leverage dimensional data models and column stores in order to retrieve data faster and be able to generate analytic insights. Major cloud vendors provide data warehouse software, access control, and some key functionalities out of the box, so you can easily navigate to standing up your data warehouse instance. The time taken to provision a data warehouse and go live is "relatively" faster but not exponentially faster. Amazon's AWS Redshift, Google's BigQuery, and Azure's Synapse Analytics are good examples of relational data warehouse environments.

Data Lakes

Data lakes are a centralized repository for storing structured, semistructured, and unstructured data. Data lake environments enable data storage similar to data warehouses but without the need for defining predefined schemas. You can ingest data from various sources, of multiple types, and can interact with and query data using various analytical tools like Spark or distributed computing environments like Hadoop. A data lake environment is ideal if you have big data or are attempting to integrate relational and unstructured data to create analytical datasets. Data lakes are a good option to look at for developing machine learning pipelines. Good examples of data lake environments are Amazon's Lake Formation, Azure Data Lake Storage Gen2, and Google's Bigtable.

Data Warehouses vs. Data Lakes

Data warehouses house structured schema and load processed data in a schema that is already defined. Data lakes, on the other hand, store unstructured or semistructured data without having the need to enforce schema. Data lakes are used by data scientists and machine learning engineers for running various types of advanced analytics, whereas data warehouses are targeted at data analysts and developers. Data warehouses continue to provide the value they always have generated. It may be slightly expensive to stand up a warehouse environment and you may see the value only over time through effective practices. Data lakes are relatively less expensive, as the modern cloud offerings provide managed object storage solutions that can be used as data lakes; however, data lakes require more processing power based on the growth rate of data.

For instance, a healthcare company data warehouse may be designed for retrieval of patient information in a structured format, like a dimensional schema. A healthcare company data lake may consist of structured patient data, medical images, reports of various tests, videos of patients' health tests, etc.

Real-Time/Streaming Processing Service

As we see, most businesses and applications generate lots of real-time data as users work on these applications every day. There is opportunity if one can quickly analyze the data to identify trends or happening events over a given time period on the given day. This is where streaming ETL pipelines come in. One can leverage Apache Kafka to design producers and consumers and use Apache Flink to create streaming data analytics pipelines. Amazon offers AWS MSK, which is managed streaming for Apache Kafka, and AWS Kinesis, which offers Apache Flink. Google also offers Kafka and Flink through Google Cloud, along with Google Pub/Sub service. Microsoft also offers these services in the name Azure Stream Analytics.

Serverless Functions

Serverless means that you have the option of consuming service and pay only for the resources that have been utilized. Serverless in a major cloud means that one can run functions that interact with one or more components and set up tasks to run without having to provision the underlying computer instance. Amazon offers a serverless compute service called AWS Lambda, Google Cloud Platform has Cloud Functions, and Microsoft has Azure Functions—all three offer serverless compute services.

Data Integration Services

Data integration is the process of bringing data from various sources to provide business the value and insights so they can make faster and better decisions. Data integration can integrate various types of data and can enable generating simple reports and running complex analytics. The most common methodology of data integration is the idea of ETL pipelines, which represent extract, transform, and load tasks; there also is the concept of ELT, which would load the data into a modern data platform and proceed to perform transformation on it. Google Cloud provides Cloud Data Fusion, a data integration service, whereas Amazon provides AWS Glue and Microsoft provides Azure Data Factory. All of these services are fully managed by their respective cloud vendors.

Continuous Integration Services

Continuous integration (CI) is the software development strategy that automates integration of code development and changes from various members of the team into the main build. When developers program new changes and developments, the CI process would run the code into automated test cases, perform code review where required, and integrate them into the main build. This is also referred to as continuous development (CD). The term "CI/CD" is synonymous with this process. Amazon offers AWS CodePipeline, Microsoft offers Azure Pipelines, and Google has GCP Cloud Build. In addition to continuous integration, these tools also support continuous deployment. Continuous deployment is where the software is automatically deployed into a production environment.

Containerization

Containers mimic virtual machines; they host an application or service. Unlike virtual machines they consume much lower resources. When compared to the idea of serverless, containerization enables an app or a service to be always available. Docker is the well-known container software that helps containerize the application in the form of container images. A container image is an executable software that contains everything from code, dependencies, system tools and libraries, etc. needed to run the application independently. A container registry is where one or more container images are stored and distributed. Amazon provides AWS Elastic Container Registry, Microsoft provides Azure Container Registry, and Google offers Container Registry as container registry services.

Data Governance

Data governance is the combination of tools, processes, and practices where data, throughout the lifecycle of it, is managed safely and in a compliant manner. It is important to have data governance practice in every organization as data has become the most valuable asset an organization possesses. Major cloud vendors like Amazon, Google, and Microsoft all provide various services that can be used to establish and run data governance practice. Let us look at some of them in greater detail here.

Data Catalog

A data catalog is a centralized repository for storing and managing metadata about data stored in the cloud. It also serves as a system where users can discover data about various data sources and understand how these sources can be used. Amazon offers AWS Glue Data Catalog that is a centralized repository for storing and managing data. Microsoft has Azure Data Catalog, a fully managed cloud data cataloging service that serves as a knowledge hub about various data assets in Microsoft Azure. Google Cloud Platform provides Google Cloud Data Catalog, a scalable data catalog and metadata management service that is used to discover all the data sources in the cloud.

Compliance and Data Protection

Compliance is the process of adhering to the rules and regulations that have been set by an appropriate body. Data protection is the process of identifying and protecting personally identifiable information that an organization may currently have. For instance, in the healthcare industry, there is HIPAA compliance that ensures that the personally identifiable information is protected from fraud and theft. Amazon offers AWS Artifact that generates audit and compliance reports that are helpful with various regulatory agencies and policies. Microsoft offers Azure Policy, which offers a drill-down report to assess compliance and enforce standards. Google offers Security Command Center that helps monitor compliance and other security standards.

Data Lifecycle Management

Data lifecycle management is a process of establishing policies and processes to manage the data throughout its lifecycle. From the time new data is generated till the data moves to long-term archival, data is separated into phases based on different criteria. Data moves through various phases that are defined as per the criteria. Microsoft, Google, and Amazon offer lifecycle management storage policies in their object storage, namely, Azure Blob Storage, Google Cloud Storage, and AWS S3, respectively. These policies enable automation of moving the data objects in the object storage to various storage classes depending upon the conditions that have been defined. You can define conditions based on which these data objects are automatically deleted.

Machine Learning

All the major cloud vendors offer a machine learning platform as a service where you can build machine learning models, train these models, and be able to connect with other cloud services for deployment or reporting the results. Amazon offers AWS SageMaker, Microsoft offers Azure ML, and Google offers Vertex AI. These services offer a complete lifecycle of machine learning models and have the option to either program your own custom machine learning model or use low code to build a machine learning model.

Conclusion

Cloud computing has revolutionized the technology landscape. The amount of flexibility and cost-effective procurement of services has enabled engineering teams to deliver more value. We have seen how the rise in big data paved the way for various evolutions of tools and technologies. You can now provision a Hadoop cluster in a relatively shorter amount of time, compared with a decade ago. However, it is important to know how we evolved so we appreciate the magnitude of sophistication in the technology and data systems we have. Though there may be reasons an organization may or may not choose to go cloud, it remains evident that cloud computing vendors offer ease of use and latest technologies, tools, and processes to gain insights from data.

CHAPTER 15

Engineering Data Pipelines Using Amazon Web Services

Introduction

Cloud computing is a concept of obtaining the necessary IT infrastructure for development and deployment of data and software services. By providing compute power, memory, storage, and various software installed on top of physical machines, an organization can get up to speed faster than before. This way an organization eliminates the capital expenses for buying physical equipment as cloud computing providers offer pay-as-you-go pricing models for their services.

In this chapter we will look at Amazon Web Services in detail. Amazon is one of the early adopters of cloud computing and one of the first few companies to provide IT infrastructure services to businesses, as early as 2006. Cloud computing was referred to as web services. Today, Amazon offers more than 200 products in cloud computing including but not limited to compute, storage, analytics, developer tools, etc.

Let us look at the AWS service offerings in the context of developing data and machine learning pipelines in detail. AWS classifies the countries where it is hosting its data centers into three regions, also known as partitions. They are standard regions, China regions, and US GovCloud regions. AWS suggests choosing a region that is geographically close to the physical presence to minimize latency.

CHAPTER 15 ENGINEERING DATA PIPELINES USING AMAZON WEB SERVICES

As you embark on your journey exploring Amazon Web Services, please may I request you to be mindful of managing the cloud resources, especially when you are done using them. In the case of traditional infrastructure, you may switch off a computer when you are done. However, in the case of cloud infrastructure, where you often procure a plan that is based on a pay-as-you-go model, you will be billed for your resources even if they are not being used. Signing off AWS does not mean powering off the services you have procured. This is particularly important for virtual machines, databases, and any other compute-intensive resources. Please, always, make it a habit to review your active resources and shut down these specific resources. If you are no longer planning on using a service, you may wish to delete the same. And so, you are only paying for the services you actively use and keep your costs under control. I also would like to thank AWS for providing free credits for new users and having actively supported new learners in bringing their cost down.

By the end of the chapter, you will learn

- Fundamentals of Amazon Web Services and the AWS management console
- How to set up an AWS account and configure various parameters
- AWS's object storage, database, and warehouse solutions
- Data processing and analytics tools
- How to build, train, and deploy machine learning models using AWS SageMaker

AWS Console Overview

One can access various cloud services by utilizing the AWS management console. It is a simple graphical user interface that enables access to more than 200 AWS services that can be hosted in more than 190 countries with simpler pricing.

Setting Up an AWS Account

To sign up for a new account, visit `https://aws.amazon.com/` and click "Create an AWS Account." Once you get the following screen, follow the prompt and enter the necessary information to create an account:

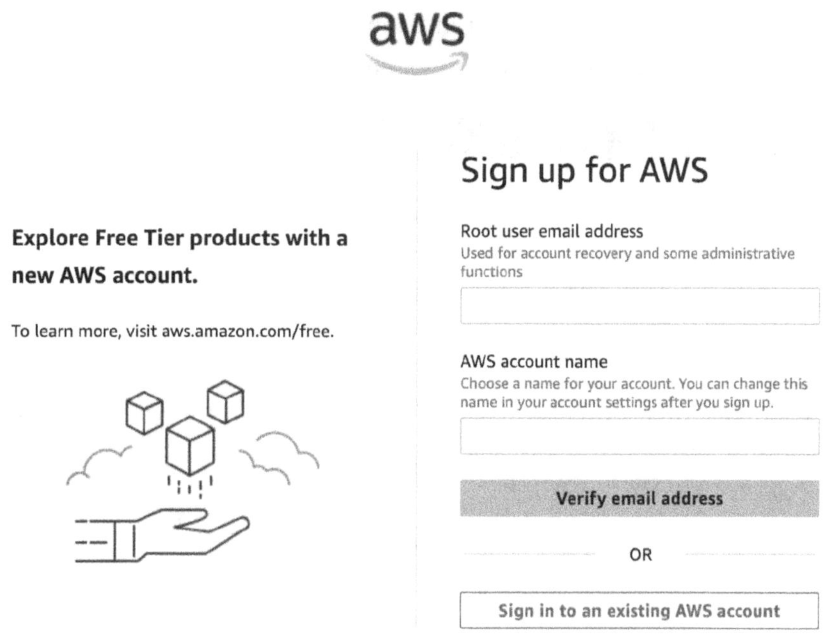

CHAPTER 15   ENGINEERING DATA PIPELINES USING AMAZON WEB SERVICES

Figure 15-1. AWS console

It may be possible to provision and access services directly from your root account, but I recommend you create an admin account, to provision and access AWS services. To accomplish that, let us utilize AWS Organizations service. AWS Organizations is an account management service that enables one to manage multiple accounts. The process is that, within your organization (the root account), you would set up organizational units. Organizational units would basically be various departments within an organization. This is followed by creating and assigning service control policies to various accounts in various organizational units. For our illustration purposes, we will create two organizational units, namely, "dev" and "prod." We will invite the team members for our organizational units.

Note   To simulate team members, creating temporary email addresses and adding them to your organizational units is an option.

CHAPTER 15   ENGINEERING DATA PIPELINES USING AMAZON WEB SERVICES

Once you have completed the registration and have signed in, you would get the AWS management console page. Here is how that looks:

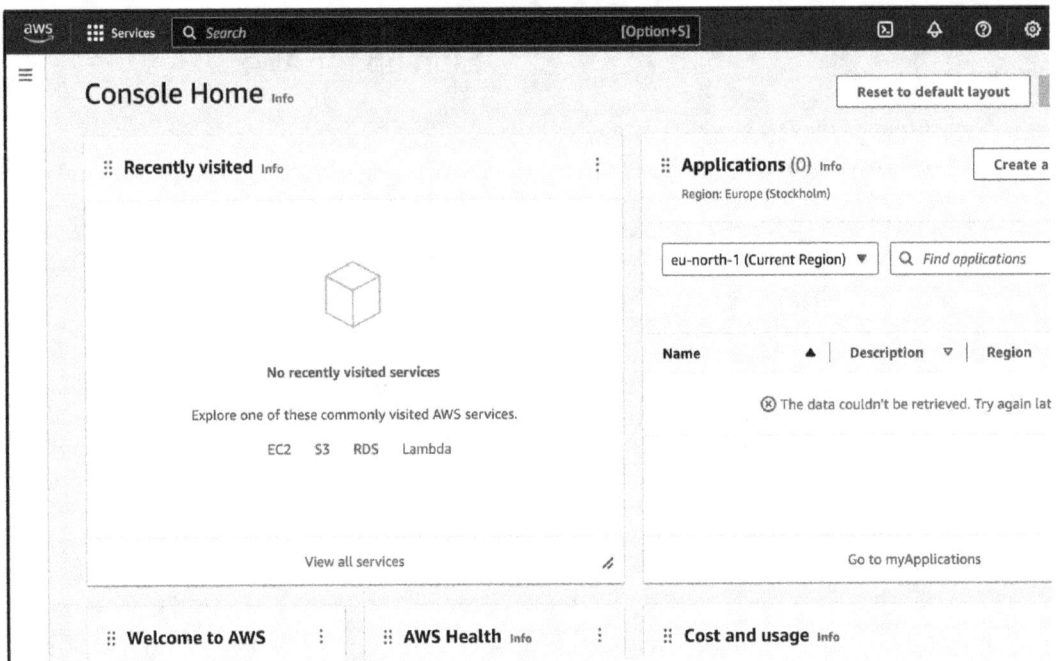

Figure 15-2.   Home page of the AWS console

CHAPTER 15 ENGINEERING DATA PIPELINES USING AMAZON WEB SERVICES

Navigate your cursor to the search section, and enter "AWS Organizations" and click "Create a new organization." You will get the following screen:

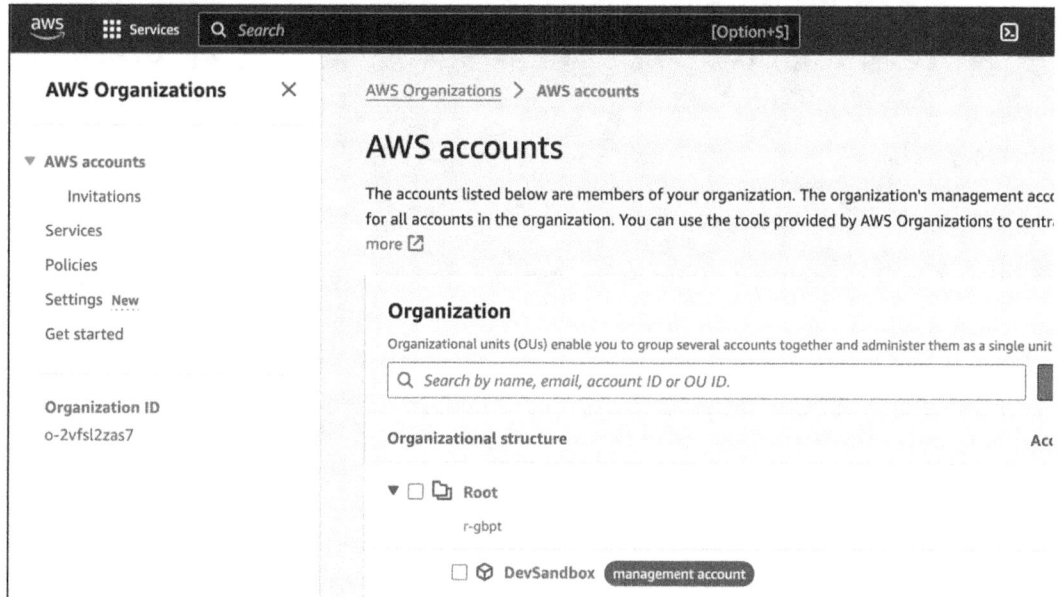

Figure 15-3. *Creating a new organization under AWS Organizations service*

The next step is to create an organizational unit. You will notice the organizational structure mentioned in the center of the screen. Make sure to check the "Root" and click Actions, followed by creating an organizational unit.

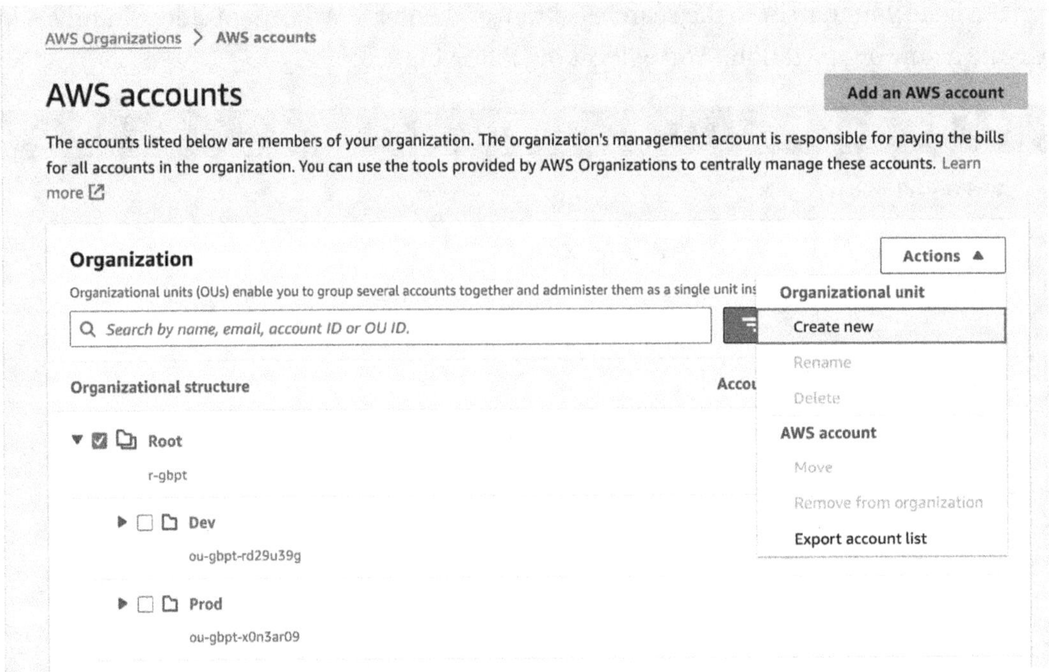

CHAPTER 15   ENGINEERING DATA PIPELINES USING AMAZON WEB SERVICES

Figure 15-4. Creating a new organizational unit

This is followed by creation of service control policies. A service control policy is a type of an organizational policy that manages permissioning within the organization. These policies enable controlled access to members within the organization. We will now enable the service control policy by clicking the appropriate button, as shown here:

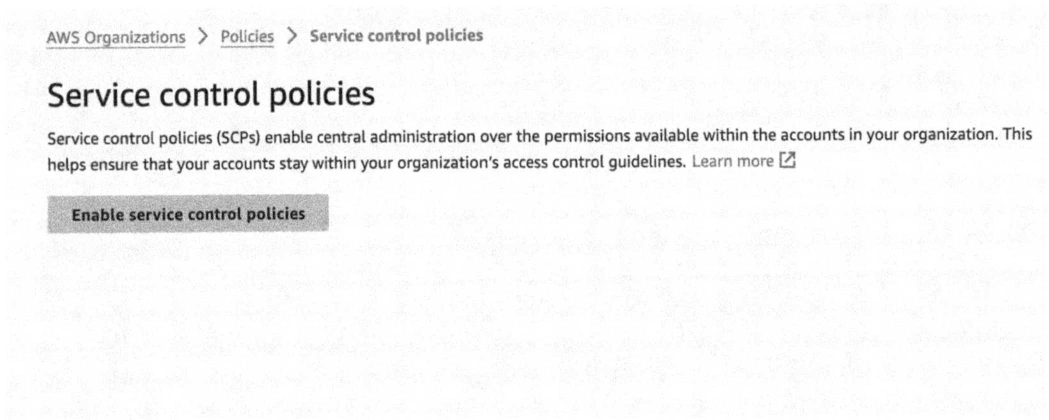

Figure 15-5. Enabling service control policies

CHAPTER 15   ENGINEERING DATA PIPELINES USING AMAZON WEB SERVICES

Now that we have service control policies enabled, let us create a new policy for our organization by clicking "Create policy":

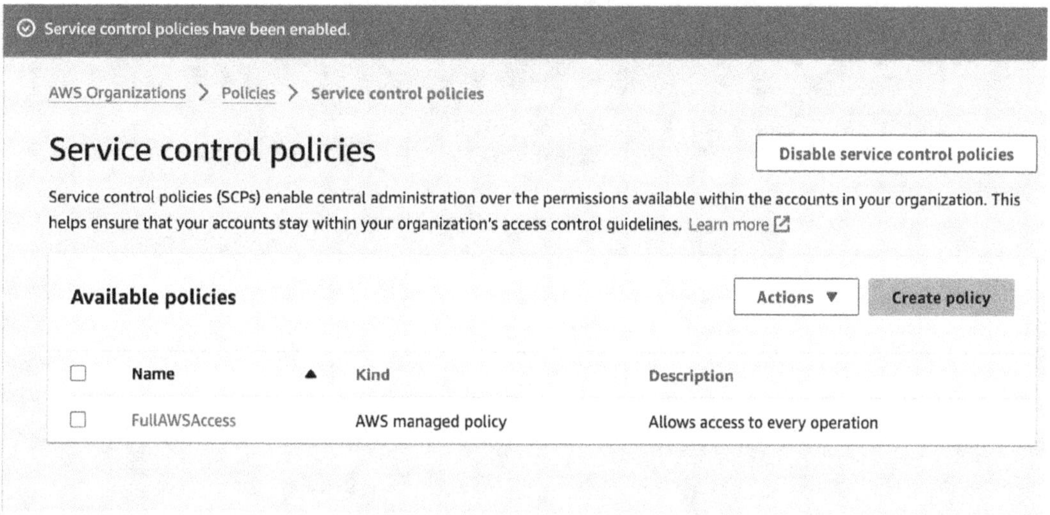

Figure 15-6. *Creating a new service control policy*

You will have the option of selecting one of many AWS services and choosing the appropriate privileges to the policy, along with resources as well. Once you have chosen the appropriate policy, choose "Save changes." AWS will automatically create a JSON file based on what we have chosen here. Here is how policy creation would look:

CHAPTER 15   ENGINEERING DATA PIPELINES USING AMAZON WEB SERVICES

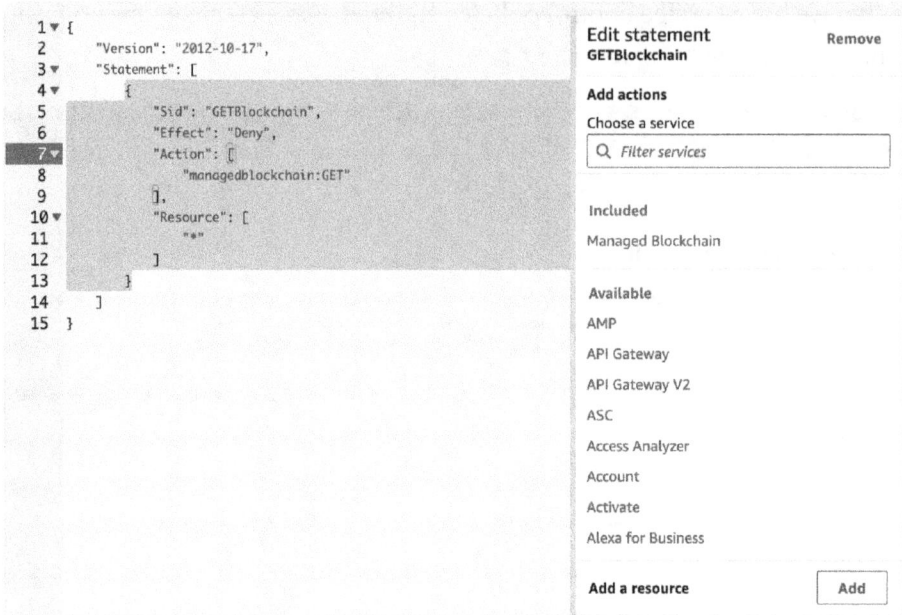

Figure 15-7. *Configuration of a new service control policy*

Once these policies are created, it is time to attach a policy with an account. To accomplish that, go to the main policies page, choose the appropriate policy, and click "Actions" followed by "Attach Policy," attaching the policy to the appropriate organizational unit.

CHAPTER 15   ENGINEERING DATA PIPELINES USING AMAZON WEB SERVICES

Here is how that looks:

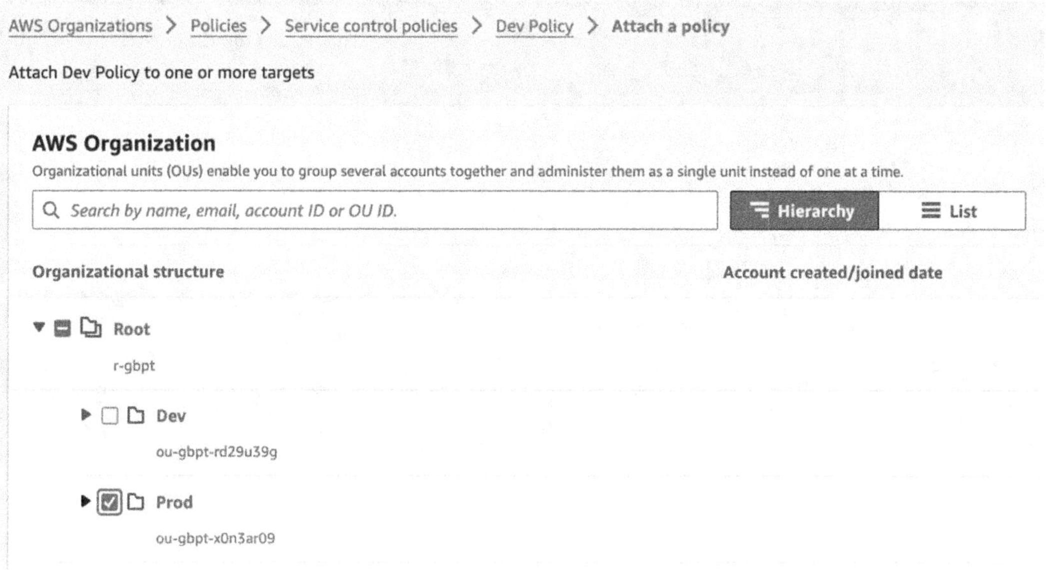

Figure 15-8. Attaching the new service control policy to an AWS organizational unit

Once the respective policies have been attached, you can sign in to the user's account to see if the policies have been enforced. Now visit the management console and look for "IAM Identity Center." Please ensure that you have logged in with the account used to create the AWS organization. AWS IAM Identity Center provides a single sign-on (SSO) service within AWS. You need to enable AWS IAM Identity Center. Here is how the page looks:

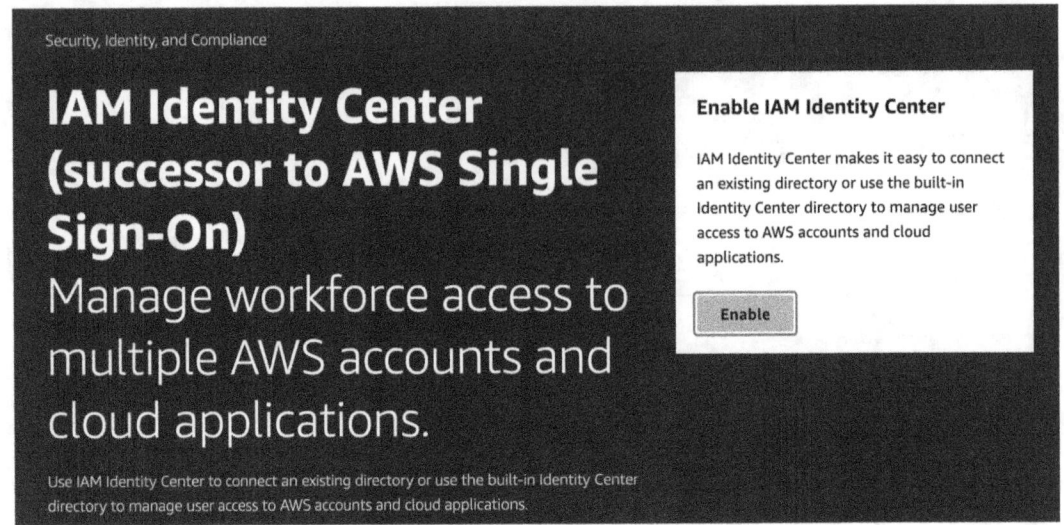

Figure 15-9. Enabling IAM Identity Center

> **Note** A single sign-on service is such that it enables centralized authentication and authorization to access and utilize a number of services, programs, and software using only one login credential. Commonly referred to as SSO, you log in once, and you would be granted access to various services.

SSOs are considered secure, can be auditable as they provide log trails, and do help organizations meet regulatory compliance requirements. Famous names include SAML, OAuth, Okta, etc.

CHAPTER 15 ENGINEERING DATA PIPELINES USING AMAZON WEB SERVICES

Once you have enabled the identity center, you need to create permission sets. Permission sets are templates for creating policies of identity and access management. Permission sets enable granting access to various services for groups and users within the organization. Here is how that looks:

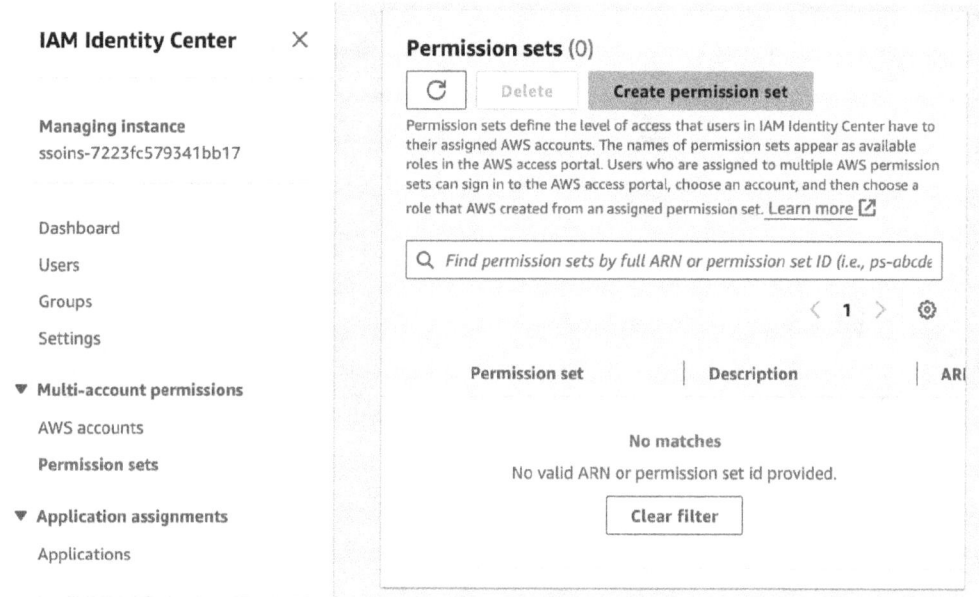

Figure 15-10. Home page of IAM Identity Center

Now we will create a new permission set. Let us try to work with a permission set that is predefined by Amazon. For our illustration, we have chosen the "DatabaseAdministrator" permission set.

CHAPTER 15 ENGINEERING DATA PIPELINES USING AMAZON WEB SERVICES

Here is how that looks:

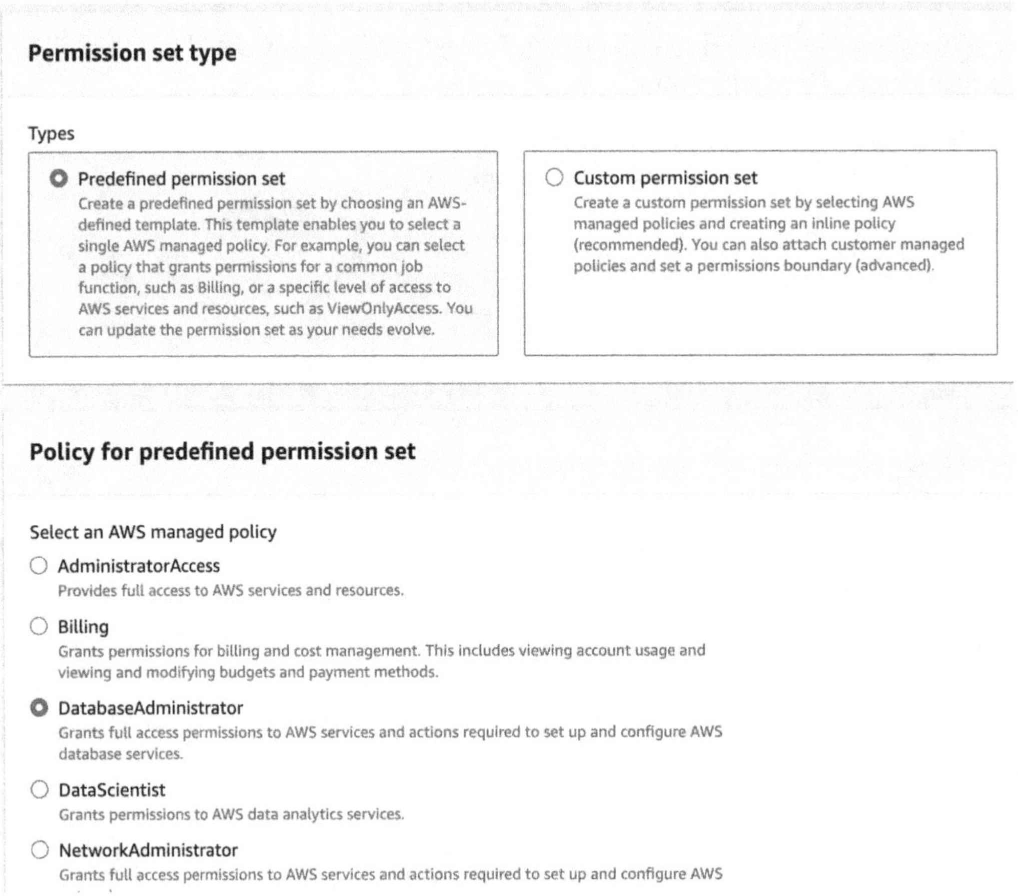

Permission set type

Types

- ◉ **Predefined permission set**
  Create a predefined permission set by choosing an AWS-defined template. This template enables you to select a single AWS managed policy. For example, you can select a policy that grants permissions for a common job function, such as Billing, or a specific level of access to AWS services and resources, such as ViewOnlyAccess. You can update the permission set as your needs evolve.

- ○ **Custom permission set**
  Create a custom permission set by selecting AWS managed policies and creating an inline policy (recommended). You can also attach customer managed policies and set a permissions boundary (advanced).

Policy for predefined permission set

Select an AWS managed policy

○ AdministratorAccess
Provides full access to AWS services and resources.

○ Billing
Grants permissions for billing and cost management. This includes viewing account usage and viewing and modifying budgets and payment methods.

◉ DatabaseAdministrator
Grants full access permissions to AWS services and actions required to set up and configure AWS database services.

○ DataScientist
Grants permissions to AWS data analytics services.

○ NetworkAdministrator
Grants full access permissions to AWS services and actions required to set up and configure AWS

Figure 15-11. *Choosing the predefined permission set*

CHAPTER 15 ENGINEERING DATA PIPELINES USING AMAZON WEB SERVICES

This will enable you to enter further details about the permission set. Once this step is successful, you will obtain the following screen:

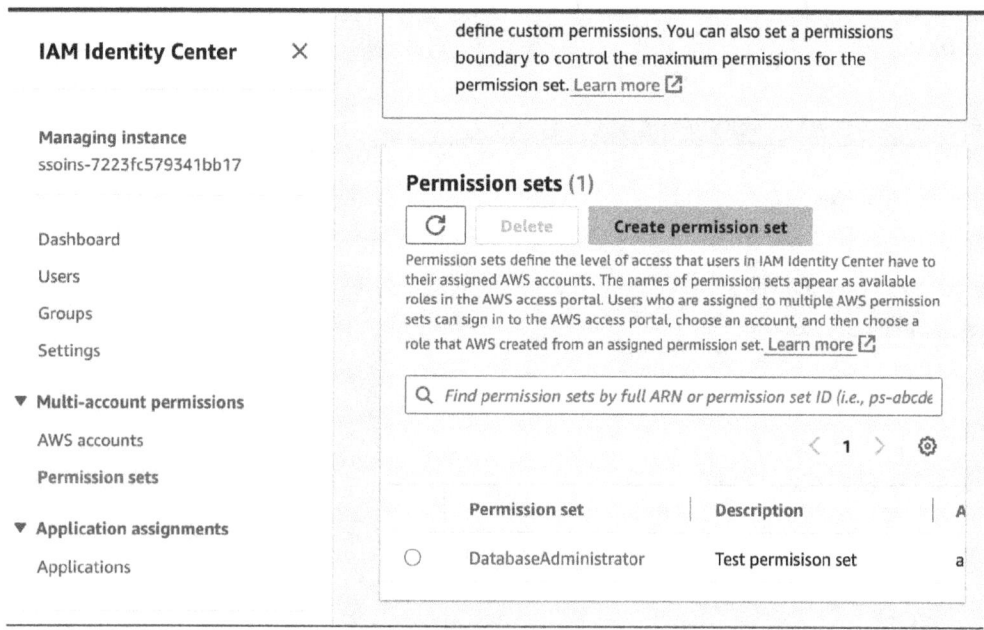

Figure 15-12. Creation of a new permission set

CHAPTER 15   ENGINEERING DATA PIPELINES USING AMAZON WEB SERVICES

Now we will create a user. The idea is that we will create users and assign various users with various permission sets. You can also assign users to groups and assign permission sets to the groups as well. It is an optional step though. Here is how that looks:

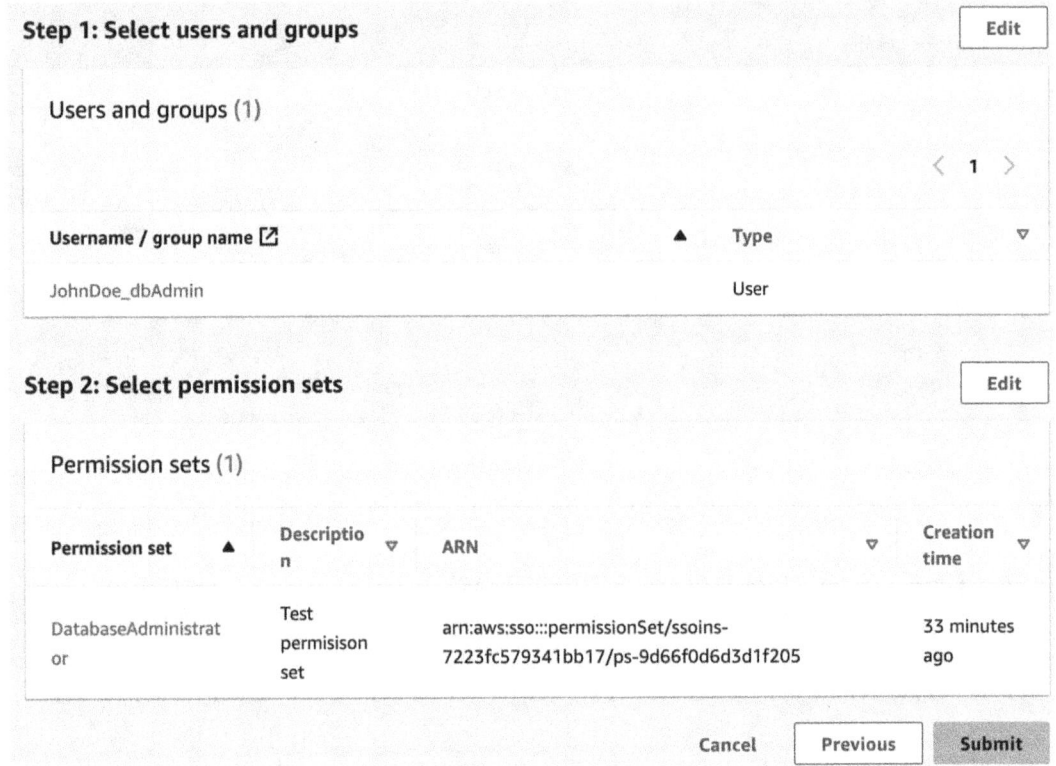

Figure 15-13. Creation of a new user within IAM Identity Center

Now, we need to connect the users, permission sets, and organizations; revisit AWS Organizations, select the appropriate organization, and select the user and the permission sets to the respective user. Here is how AWS Organizations looks like:

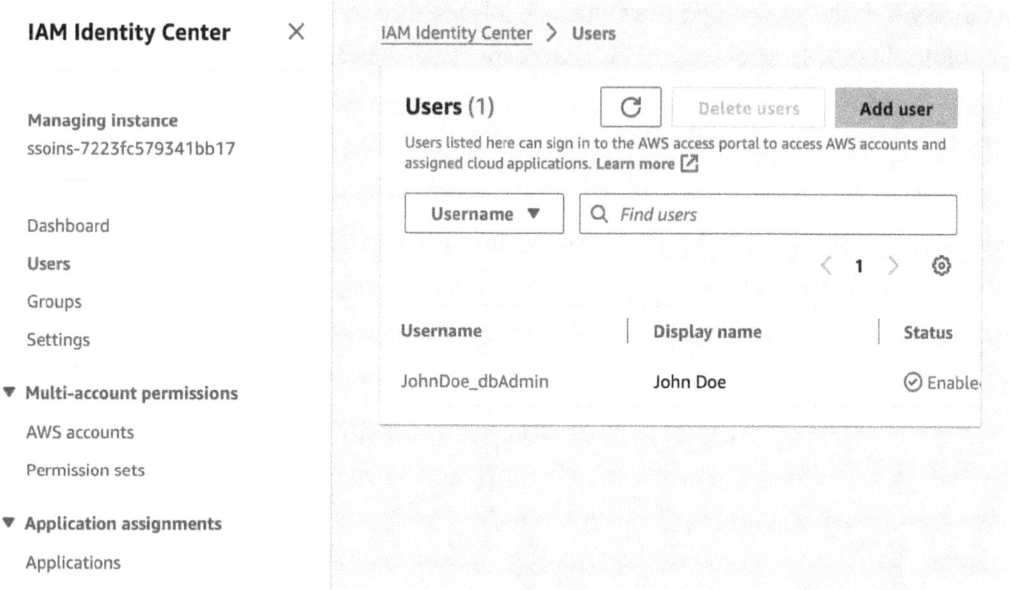

CHAPTER 15   ENGINEERING DATA PIPELINES USING AMAZON WEB SERVICES

Figure 15-14. Connecting new users, permission sets, and organizations

Click Assign users or groups, and choose the user we created. Upon clicking Next, choose the permission sets for the respective user for the respective organization. Click Submit and wait a few seconds for AWS to create the permission sets for the user for the organization. Here is how that final screen looks:

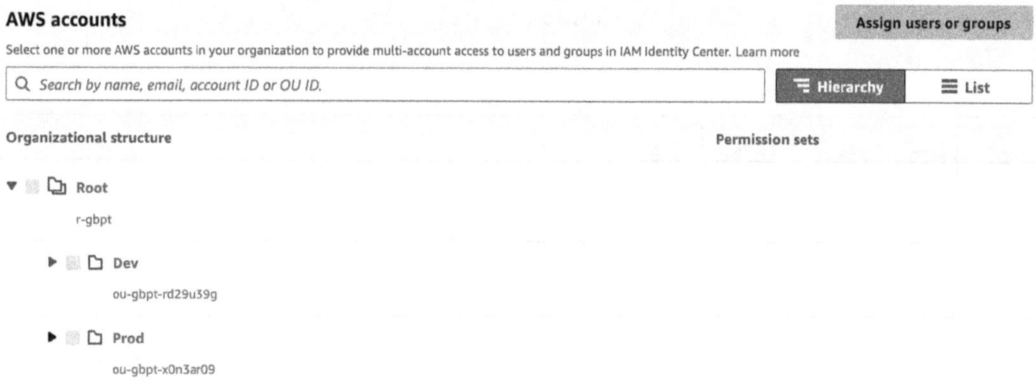

Figure 15-15. Assigning users to organizations

487

Once you have completed the steps, click "Dashboard" in AWS IAM Identity Center and obtain the AWS access portal for that respective user. Here is how that looks:

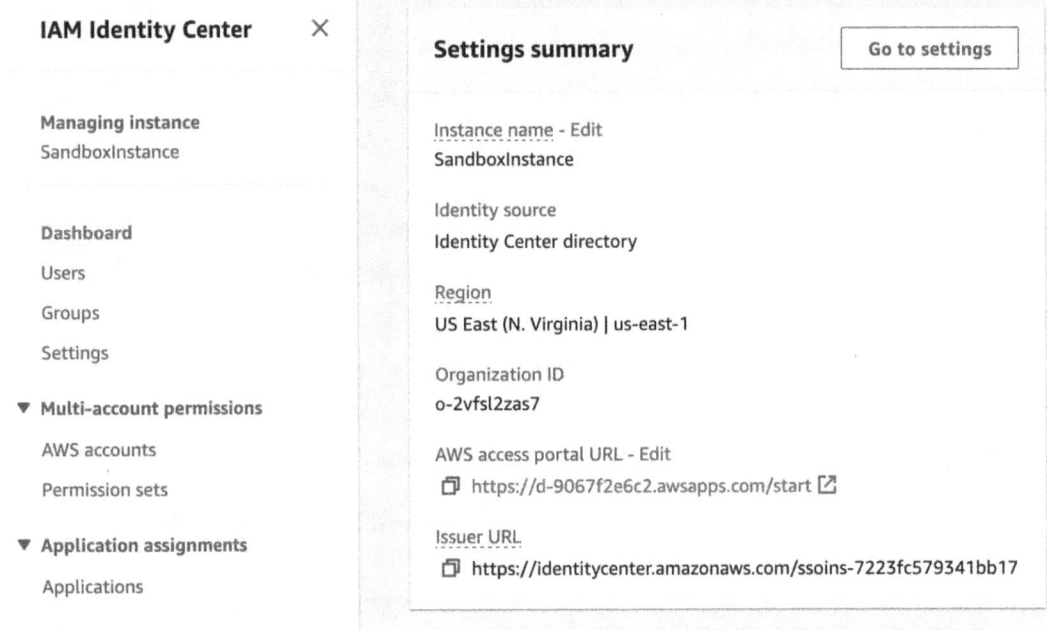

Figure 15-16. *Obtaining the access portal for the respective user*

You can use the AWS access portal URL to log in. Please use the user credentials we have set up using AWS IAM Identity Center and not the root credentials. It may also be beneficial to bookmark the link.

Installing the AWS CLI

The AWS CLI is the command line interface, available for major operating systems like Windows, Linux, and MacOS. The instructions are detailed in the AWS documentation website.

Upon setting up the AWS organization, it is time to set up what is called a "single sign-on" solution.

To configure the command line interface, using AWS SDK or REST API to connect to a S3 bucket, we need to set up security credentials to obtain access keys and secrets. To create an access key and passcode, simply visit the Identity and Access Management portal, locate the security credentials section, and follow instructions to create access keys and secrets. These would appear to be a long string of integers and characters.

CHAPTER 15   ENGINEERING DATA PIPELINES USING AMAZON WEB SERVICES

To configure the command line interface, enter

```
aws configure
```

You would be required to enter the following:

```
AWS Access Key ID [None]:
AWS Secret Access Key [None]:
Default region name [None]:
Default output format [None]:
```

And supply the required credentials to proceed.

Once you have configured, you can issue commands like listing buckets, moving objects between buckets, and so on. Here is a command that lists all buckets within your S3 service:

```
aws s3 ls
```

Here is a command to list all files within a given bucket:

```
aws s3 ls s3://<name-of-your-bucket>
```

AWS S3

AWS S3 stands for simple secure storage, which is an object storage service. S3 allows for retention of unstructured data. S3 is considered to be optimized for use cases where you read many times and write once. S3 stores data in the form of objects, as compared with file systems where data is stored as files. It is a great option to store photos, videos, and similar content. AWS provides a programmatic interface to allow apps to perform create, update, read, and delete operations on the data stored on the S3 object storage.

Amazon offers a variety of storage types within their S3 service. Amazon S3 Standard is for data that has a high volume of reads and is an ideal option for applications that seek high throughput and demand low latency. Amazon S3 IA (Infrequent Access) is for data that is not accessed and not read as much; ideal applications would include business recovery and disaster planning, long-term backups, etc. Amazon S3 Glacier is for data that may never be accessed but requires to be held; applications are end-of-lifecycle data, backup that is kept for regulatory and compliance purposes, etc.

> **Note** Let us learn a couple of terms. In the context of cloud object storage, we have two terms that we use in context to access patterns and storage requirements. These terms are "hot data" and "cold data," respectively.
>
> **Hot data** is frequently accessed or used by one or more users at the same time. The data needs to be readily available and is in constant demand.
>
> **Cold data** is accessed rarely; data includes periodic backups and historical data. It may be wise to provide a cost-effective storage rather than performance optimized (as compared with hot data).

AWS supports a global bucket system. To create a global bucket, it must be named uniquely within a given partition. There is also an option to create a public bucket that is accessible to everyone. Here is how to create an AWS S3 bucket:

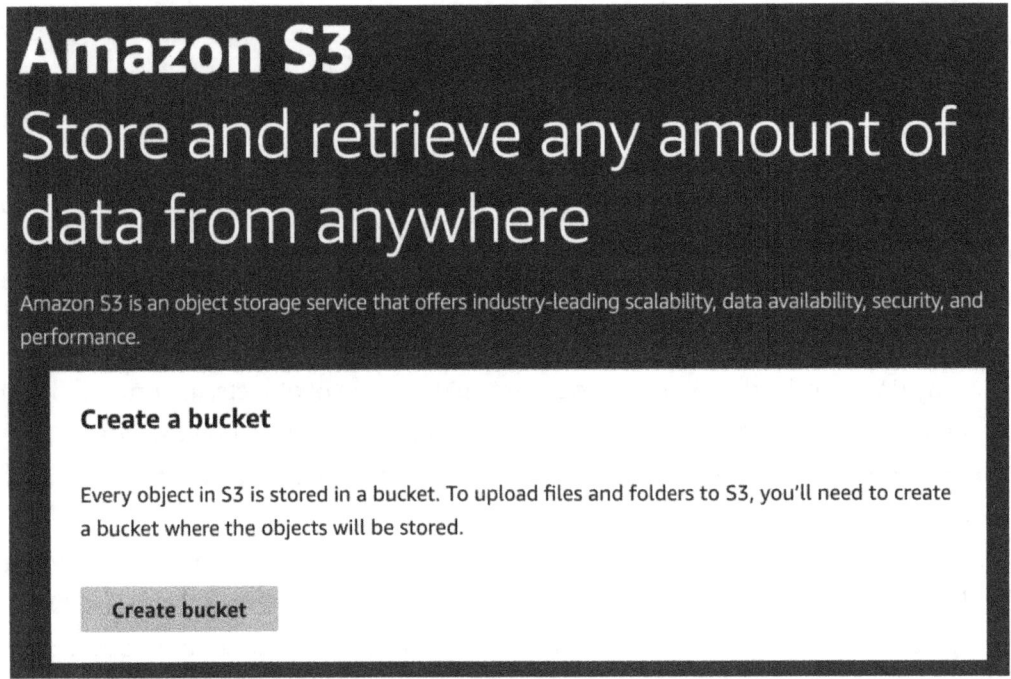

Figure 15-17. *Home page of AWS S3 service*

CHAPTER 15 ENGINEERING DATA PIPELINES USING AMAZON WEB SERVICES

Click Create bucket and you will be presented with a bucket creation page. While it is best to leave the options preselected as is, here are some options you wish to consider choosing that work best for you:

- AWS region: Choose a region that is geographically close to your location.

- Bucket type: If you are new to AWS, it may be better to choose general-purpose buckets.

- Bucket name: In addition to following AWS guidelines for naming buckets, follow a naming convention that includes your organization and department codes, project type, and whether is it a development sandbox or a production bucket; it would help you greatly when you have multiple buckets to sift through.

Here is how that looks:

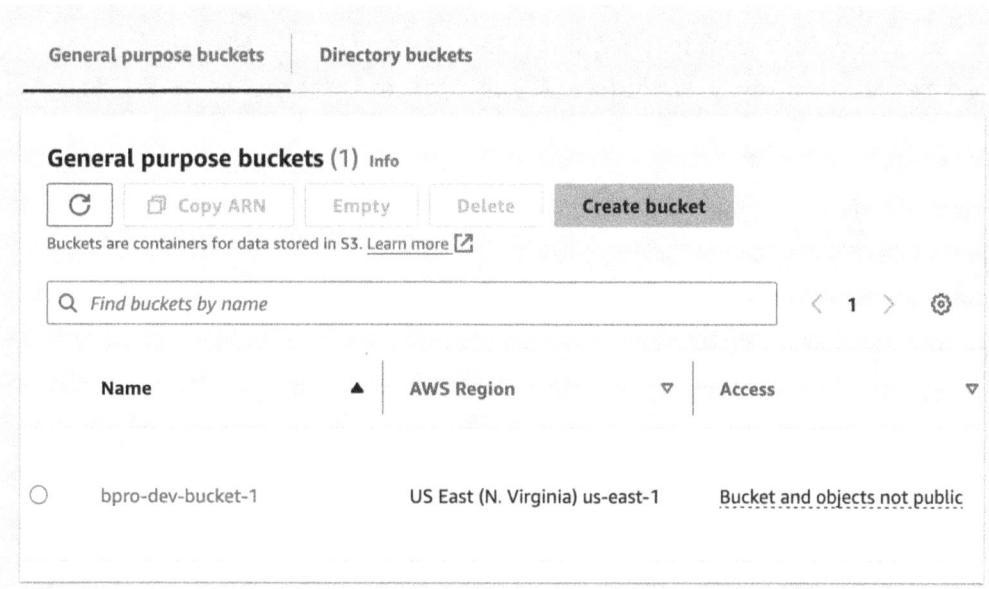

Figure 15-18. Creation of a new bucket

Uploading Files

Here is an example of uploading objects (files) in a given bucket using the AWS command line interface:

```
aws s3api put-object \
    --bucket <your-bucket-name> \
    --body /Users/me/<path-to-your-object-here> \
    --key destination-file-name
```

Here is how to upload a file using AWS SDK:

```python
import boto3
import configparser

config = configparser.ConfigParser()
path = '/Users/pk/Documents/de/aws/s3/.aws/credentials.cfg'
config.read(path)
aws_access_key_id = config.get('aws','access_key')
aws_secret_access_key = config.get('aws','access_secret')

aws_region = 'us-east-1'

session = boto3.Session(
   aws_access_key_id=aws_access_key_id,
   aws_secret_access_key=aws_secret_access_key,
   region_name=aws_region
)

# create a S3 client
s3_client = session.client('s3')

file_to_upload = '/Users/pk/Documents/abcd.csv'
bucket_name = 'bpro-dev-bucket-1'

try:
   s3_client.upload_file(
        file_to_upload,
        bucket_name,
        'uploaded_abcd.csv'
     )
```

```
    print(f"The file {file_to_upload} has been uploaded successfully \
          in {bucket_name} S3 bucket")
except Exception as e:
    print(f"Exception {e} has occurred while uploading \
          {file_to_upload}")
```

In the preceding example we are storing the AWS credentials in a configuration file and calling the required access key and access secret from the said file. A good security practice is to create a hidden folder and store the credentials. Hidden folders will provide some level of security, but keep in mind anyone with root access to your development environment can see the credentials.

When you work with version control systems like git, hidden folders by default will not be included when you add files to the repository. It is also essential to have a .gitignore file within your git repository that dictates which files to ignore.

Here is a very basic .gitignore file:

```
# Ignore sensitive or confidential information (e.g., credentials)
secretsconfig.json
topsecretpasswords.txt
credentials.cfg
```

Don't forget to place the .gitignore file in the root of the directory and commit the file so others who are forking it may also benefit. Please ensure that ignoring certain files according to .gitignore won't affect the code in any way.

AWS Data Systems

AWS provides a number of data system options for users. In this section, you will learn the most common ones or most widely adopted services, if you will.

Amazon RDS

RDS stands for relational database service. As its name goes, Amazon RDS provides relational databases with several options for database engines. Some of them include but are not limited to MySQL, PostgreSQL, Oracle, MS SQL Server, etc. AWS RDS is easier to set up within a few minutes and easier to operate and can scale up when necessary. AWS will manage the common database administration tasks so you may not have to worry about maintenance.

AWS RDS is a primary option for those who are looking to migrate their databases to the cloud from the on-premise environment. By choosing the appropriate database engine for their migration, one may be able to achieve faster speeds and reliability during transfer of relational database systems.

Let us look at creating a new database instance. You may initiate the process by searching for "RDS" in the AWS console, opening the appropriate page, and selecting "Create a database." You have the option of creating a new database either by customizing individual options for availability, security, maintenance, etc. or choosing the AWS-recommended configurations for your use case.

You can also specify the nature of this instance, whether you are in need of a highly available production database instance, or a development sandbox with limited performance might do it for you. Here is how this may look for you:

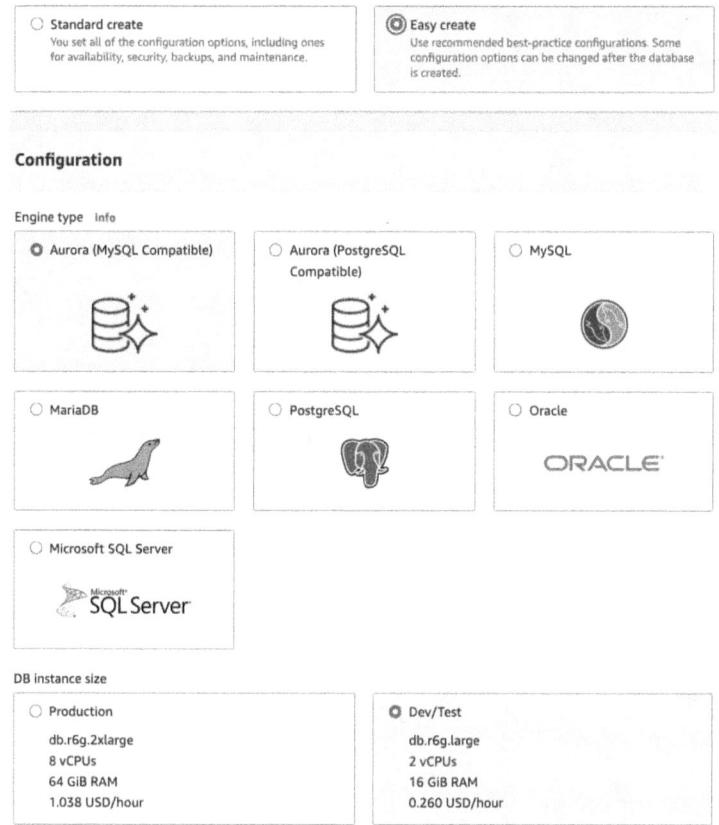

Figure 15-19. *Creating a new database instance in AWS RDS*

In addition, make sure to add an inbound network rule in the security configuration by enabling all incoming IPs to talk through TCP/IP. You will find this option under the VPC tab. Here is how that may look for you:

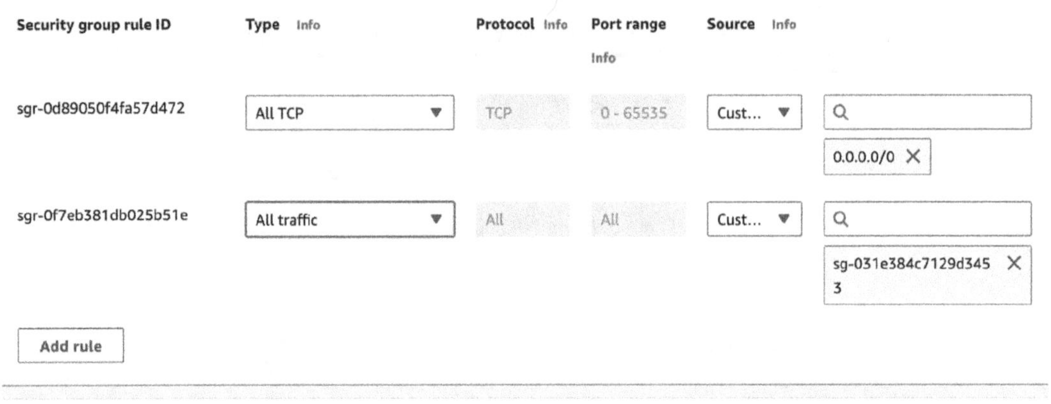

CHAPTER 15   ENGINEERING DATA PIPELINES USING AMAZON WEB SERVICES

Security group rule ID	Type	Protocol	Port range	Source	
sgr-0d89050f4fa57d472	All TCP	TCP	0 - 65535	Cust...	0.0.0.0/0 ✕
sgr-0f7eb381db025b51e	All traffic	All	All	Cust...	sg-031e384c7129d3453 ✕

Add rule

Figure 15-20. *Enabling all incoming IPs to talk through TCP/IP (for learning and development purposes only)*

Here is a Python application that connects to AWS RDS and uploads data to a table:

```
import csv
import psycopg2
import configparser

# CONFIG
config = configparser.ConfigParser()

# reading the configuration file
config.read('aws_rds_credentials.cfg')

dbname = config.get("rds", "database_name")
user = config.get("rds", "username")
password = config.get("rds", "password" )
host = config.get("rds", "endpoint" )
port = config.get("rds", "port" )

# Connect to your RDS instance
conn = psycopg2.connect(
    dbname=dbname,
    user=user,
    password=password,
    host=host,
    port=port
)
```

```
cur = conn.cursor()

# Path to your CSV file
csv_file_path = "/Users/<your-path-here>/iris.csv"

try:
    with open(csv_file_path, 'r') as f:
        reader = csv.reader(f)
        next(reader)  # Skip the header row
        for row in reader:
            cur.execute(
                "INSERT INTO users (id, name, age)
                 VALUES (%s, %s, %s)",
                row
            )
    conn.commit()
    print("Data uploaded successfully")
except:
    print("error loading data into AWS RDS")
finally:
    cur.close()
    conn.close()
```

Amazon Redshift

Redshift is a massively scalable cloud data warehouse. It allows you to load petabytes of data, execute complex analytical queries to retrieve data using SQL, and connect business intelligence applications to render reports and dashboards. You do not have to pay when the data warehouse environment is in idle state as charges occur only when you are actively interacting with Redshift by passing a query or loading data. Amazon Redshift is based on open standard PostgreSQL; however, it has been customized to support online analytical processing applications and other read-heavy applications.

Amazon Redshift is a column-store-like data warehouse that is scalable and fully managed by AWS. Storing data in a column store reduces the number of input–output operations on Redshift and optimizes the performance of analytical queries. Furthermore, you can also specify the data in Redshift to be compressed so that it further reduces storage requirements and even optimizes the query. When you execute a query against compressed data, the data is uncompressed and gets run.

Amazon Redshift instances are spun off in clusters. A cluster has a leader node and compute nodes. The leader node manages communications, both with client applications and the compute nodes. For an analytical query supplied by the user, the leader node creates an execution plan and lists the execution steps necessary to return the result. The leader node compiles the query and assigns a portion of the execution to the compute node. The leader node would assign a query to a compute node that has the underlying table mentioned in the query.

The compute nodes are themselves partitioned into slices. A portion of physical memory and storage is allocated to each slice within a compute node. So, when the leader node distributes the work to compute nodes, the work really goes to a given slice within a given compute node. The compute node would then process the task in parallel within the cluster. A cluster can consist of one or more databases where the data is stored in compute nodes.

Let us try to set up a new Amazon Redshift cluster. To begin, we have to create a workgroup. A workgroup in Redshift is a collection of compute resources that also include VPC configuration, subnet groups, security groups, and an endpoint. Let us visit the AWS management console and search for Redshift. Once we are on the main page, let us choose to create a workgroup. You will be directed to a page where you can specify the RPUs for the workgroup. RPUs are Redshift Processing Units, a measure of available computing resources. You may also choose the VPC configuration and subnet groups. Here is how that may look for you:

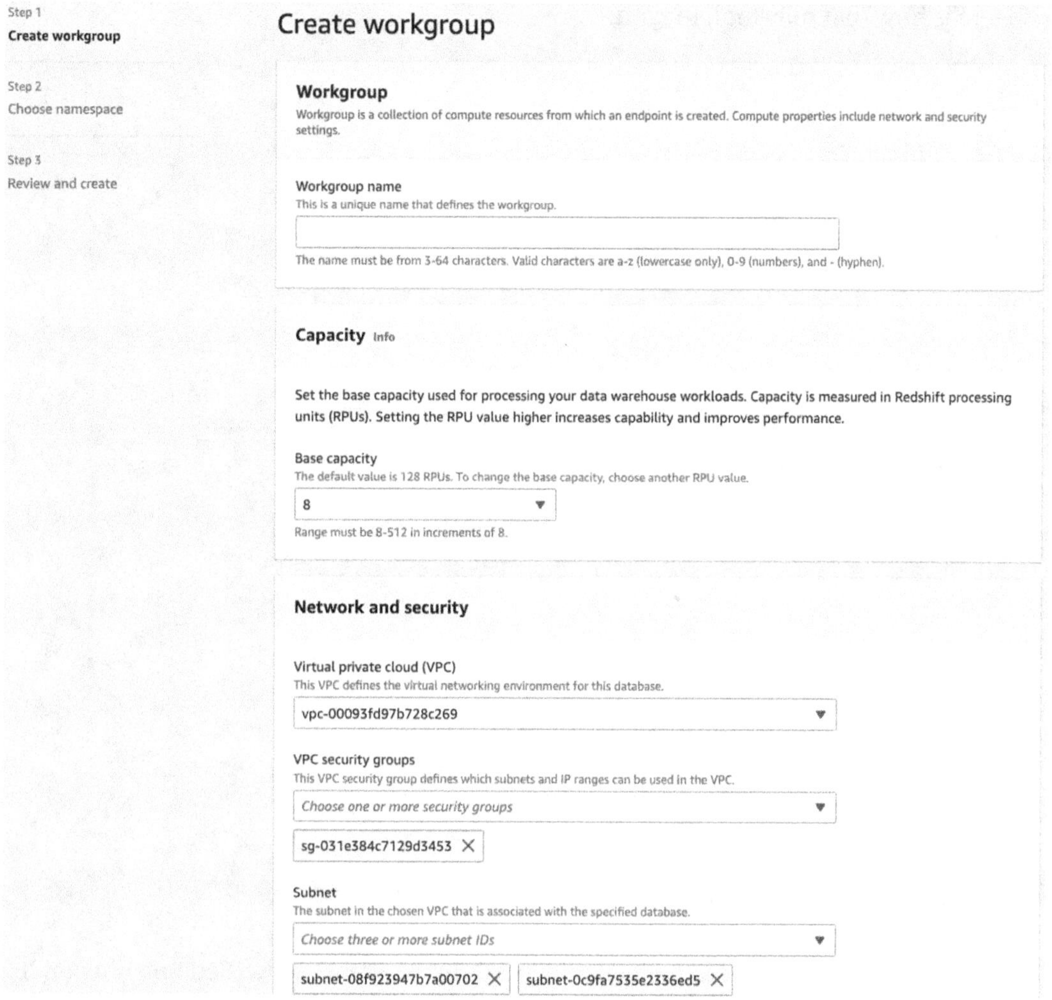

CHAPTER 15   ENGINEERING DATA PIPELINES USING AMAZON WEB SERVICES

Figure 15-21. Creating workgroups in Amazon Redshift

Once you have specified the various parameters, you may wish to navigate to the provisioned cluster dashboard to create a new cluster. When creating a new cluster, you may specify the name of the cluster, the number of nodes you wish to have on the cluster, and the type of compute, RAM, and storage you wish to obtain for your data warehouse cluster.

Here is how that may look for you:

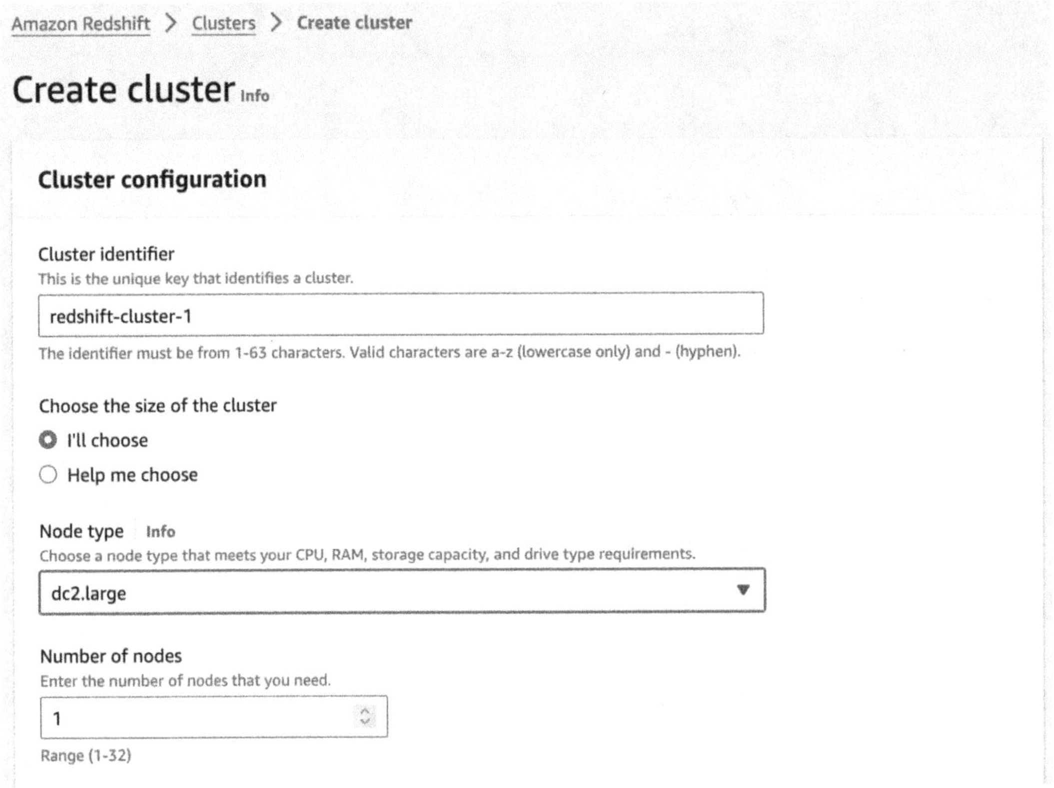

Figure 15-22. Creating a new cluster within Amazon Redshift

You may also specify the database credentials when creating the Redshift cluster. If you wish to get started on querying the Redshift cluster using sample data, you may click Load sample data. In addition, you can also choose to create a new role or associate an existing IAM role with their cluster.

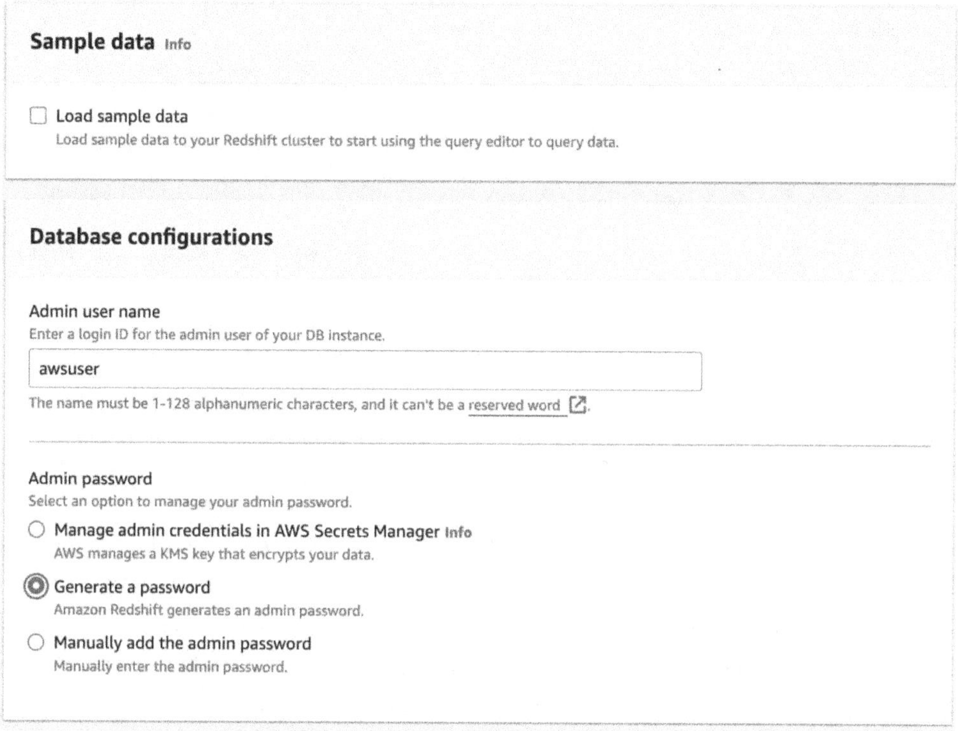

CHAPTER 15 ENGINEERING DATA PIPELINES USING AMAZON WEB SERVICES

Figure 15-23. Configuring the database credentials for the Redshift cluster

Once you have chosen other options, then you are ready to create a new Redshift cluster. After you click Create the new cluster, you may have to wait a few minutes for the cluster to be created. Once you have the cluster ready and loaded with sample data, you may wish to click the query editor on the cluster dashboard page. We have the option of choosing query editor version 2, which is a much cleaner interface. Here is how the query editor may look:

CHAPTER 15   ENGINEERING DATA PIPELINES USING AMAZON WEB SERVICES

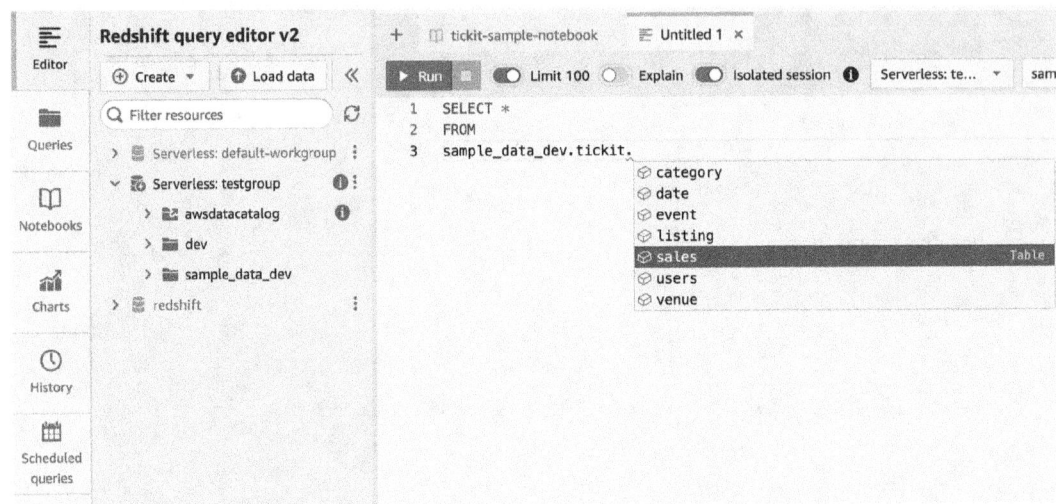

Figure 15-24. Redshift query editor 2 with a new interface

You can write queries, access various databases, and also explore AWS Glue Data Catalog from your query editor. Furthermore, you can create views, stored procedures, and functions within the query editor against the data in the Redshift cluster. In the top-left corner, you have options to create a new database or table using the graphical user interface or load data directly from one of your S3 buckets. Let us look at how we may load the data from a S3 bucket to Redshift using Python. Here is a sample code that picks up a flat file from a S3 bucket and loads it into a Redshift database:

```
import boto3
import configparser

config = configparser.ConfigParser()

path = '/Users/<your-path-here>/credentials.cfg'

config.read(path)

aws_access_key_id = config.get('aws','access_key')
aws_secret_access_key = config.get('aws','access_secret')
aws_region = 'us-east-1'

s3_bucket_name = 'your-bucket'
s3_key = 'key'
```

```python
table_to_be_copied = 'nba_shots.csv'

redshift_cluster_identifier = config.get('redshift','cluster_id')
redshift_database = config.get('redshift','r_dbname')
redshift_user = config.get('redshift','r_username')
redshift_password = config.get('redshift','r_password')
redshift_copy_role_arn = config.get('redshift','redshift_arn')

s3 = boto3.client(
    's3',
    aws_access_key_id=aws_access_key_id,
    aws_secret_access_key=aws_secret_access_key,
    region_name=aws_region
)

redshift = boto3.client(
    'redshift',
    aws_access_key_id=aws_access_key_id,
    aws_secret_access_key=aws_secret_access_key,
    region_name=aws_region
)

copy_query = f"""
COPY {table_to_be_copied}
FROM 's3://{s3_bucket_name}/{s3_key}'
IAM_ROLE '{redshift_copy_role_arn}'
FORMAT AS CSV;
"""

def execute_redshift_query(query):
    try:
        redshift_data = boto3.client(
            'redshift-data',
            aws_access_key_id=aws_access_key_id,
            aws_secret_access_key=aws_secret_access_key,
            region_name=aws_region
        )
```

```
        response = redshift_data.execute_statement(
            ClusterIdentifier=redshift_cluster_identifier,
            Database=redshift_database,
            DbUser=redshift_user,
            Sql=query,
            WithEvent=True
        )
        return response
    except Exception as e:
        print(f"Error executing Redshift query: {e}")
        return None
response = execute_redshift_query(copy_query)
print(response)
```

Amazon Athena

Amazon Athena is an interactive querying service to analyze the data in the object storage AWS S3 directly using standard SQL syntax. With Amazon Athena, you also have the option to run data analytics using Apache Spark, directly on the S3 buckets. There is no need to set up the underlying Spark infrastructure or set up a database connection driver. Amazon will take care of all these for you. Amazon Athena is a serverless service, and so you will be charged only for what you use, by the number of requests.

As you have the ability to query the buckets in your object storage directly, you can now analyze the unstructured, semistructured, and structured data within the buckets. The data can be of any formats, including but not limited to CSV, JSON, Parquet, and Apache ORC. All of these can be analyzed using Athena. By using Amazon Athena, you can query various kinds of data within S3; therefore, you can use S3 as your data lake, with Athena as a query layer for the data lake. The queries will automatically be parallelized and executed, offering natively faster performance.

To get started with Amazon Athena, search for Athena in the AWS management console and choose an option to query your S3 bucket data (either SQL or Spark). Here is how that may look for you:

CHAPTER 15 ENGINEERING DATA PIPELINES USING AMAZON WEB SERVICES

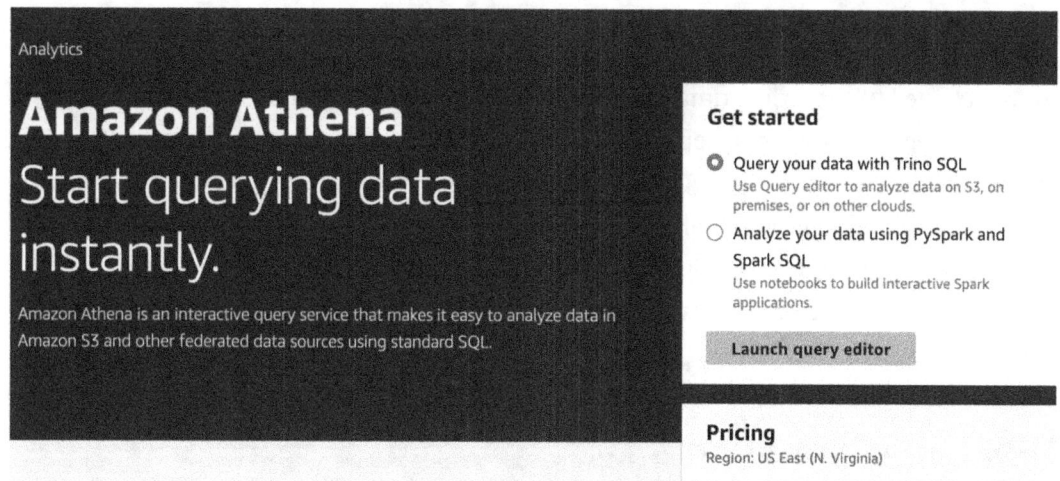

Figure 15-25. *Home page of AWS Athena*

You can also use Spark SQL on top of the objects in your S3 bucket, by interacting with Jupyter Notebook. Here is how that may look for you:

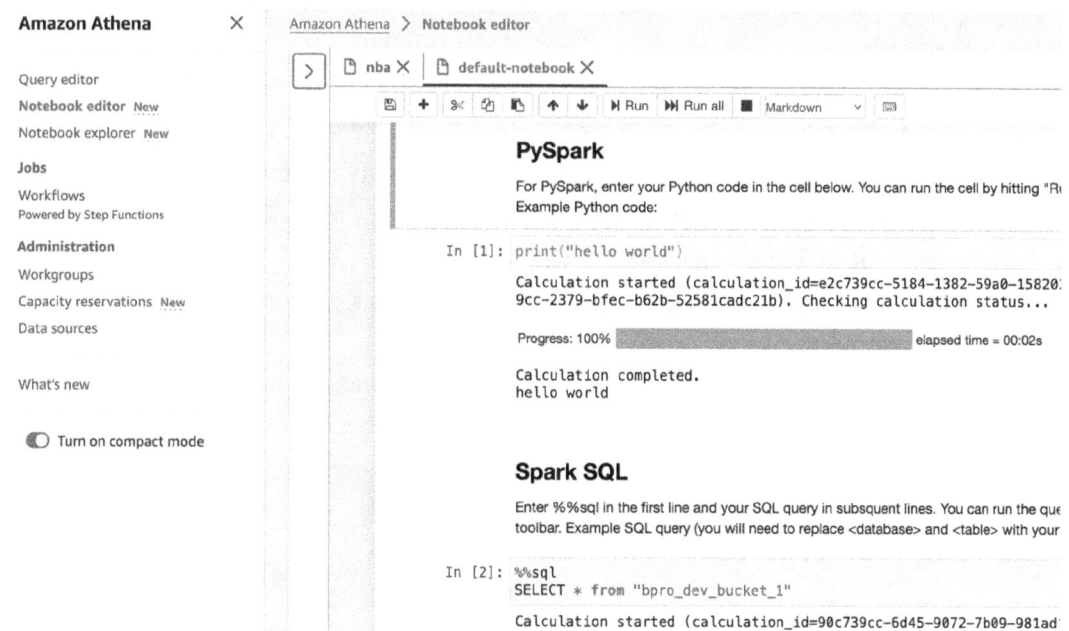

Figure 15-26. *AWS Athena—Jupyter Notebook workspace*

Amazon Glue

Amazon Glue is a serverless data integration tool. You can use Glue to discover data sources, integrate data, and prepare datasets for data analytics, machine learning, and application development. Amazon Glue has Glue Studio, which lets you prepare your data integration jobs visually, where you can drag and drop components into a workspace. Once you have designed your jobs, you can run and monitor these jobs directly from the Glue environment. You can trigger a job based on a scheduled time or an event.

AWS Lake Formation

AWS Lake Formation is a data lake service. A data lake is where both structured and unstructured data can coexist and be queried and analyzed. It simplifies building and managing the data lakes. Instead of multiyear projects, AWS Lake Formation can get you on production much faster. AWS Lake Formation provides a data cataloging tool that captures metadata about various datasets so that you can search and retrieve a given dataset effectively.

AWS Lake Formation stores data in S3 buckets, regardless of whether it is structured or unstructured data. One of the key aspects of creating data lakes with Lake Formation is that you list various data sources regardless of their data types and also their security and access policy that will be applied to the data lake. Let us look at a simple example of using Lake Formation with Athena and Glue.

To get started, let us create a new bucket in the object storage:

CHAPTER 15   ENGINEERING DATA PIPELINES USING AMAZON WEB SERVICES

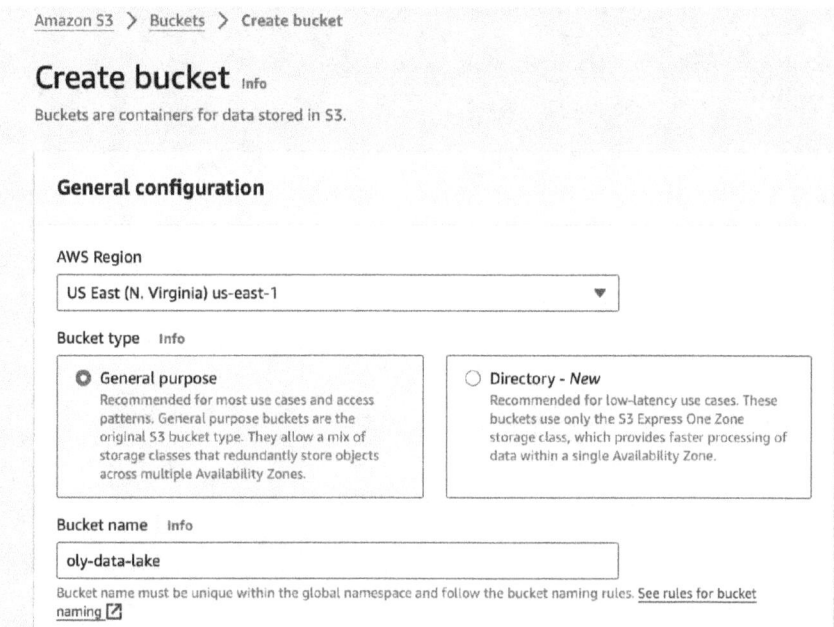

Figure 15-27. *Creating a new S3 bucket for a Lake Formation project*

Let us create two folders, one for the source data and the other for running Glue-based reports within the data lake bucket. Here is how that may look for you:

CHAPTER 15   ENGINEERING DATA PIPELINES USING AMAZON WEB SERVICES

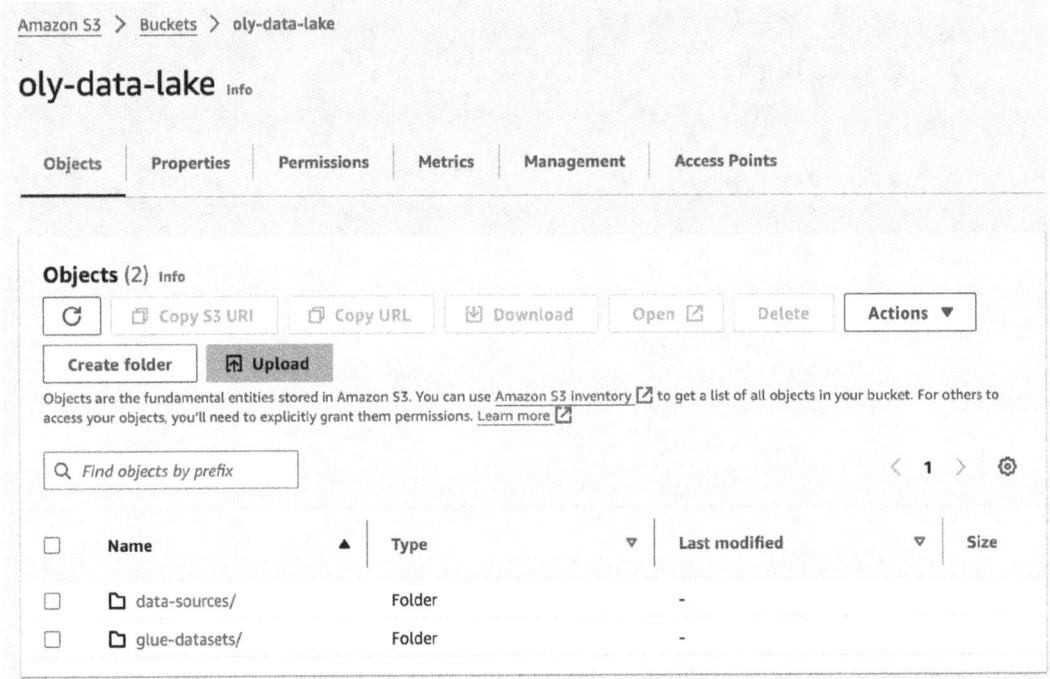

Figure 15-28. *Creating folders within the object storage*

Let us now configure the data lake. We will navigate to the AWS management console and search for Lake Formation. We will create a database within AWS Lake Formation and configure our data sources folder from the S3 bucket. Create a database, assign a name, and choose the data source location for the database. Here is how that may look for you:

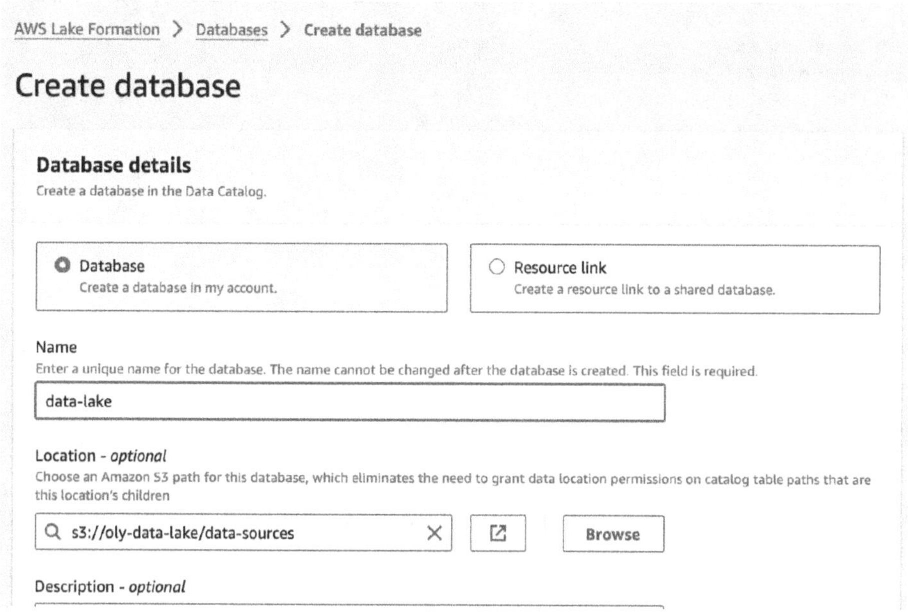

CHAPTER 15   ENGINEERING DATA PIPELINES USING AMAZON WEB SERVICES

Figure 15-29. Creating a new database in Lake Formation

Now let us configure AWS Glue, so we can automatically catalog the tables in the data sources bucket. Before that, we need to create an IAM role within AWS that AWS Glue will use to create a crawler. A crawler in AWS Glue is a program that can scrape or obtain data from multiple sources. First, let us look for IAM within the AWS management console, navigate to the service, and click Roles, to create a role. Here is how that may look:

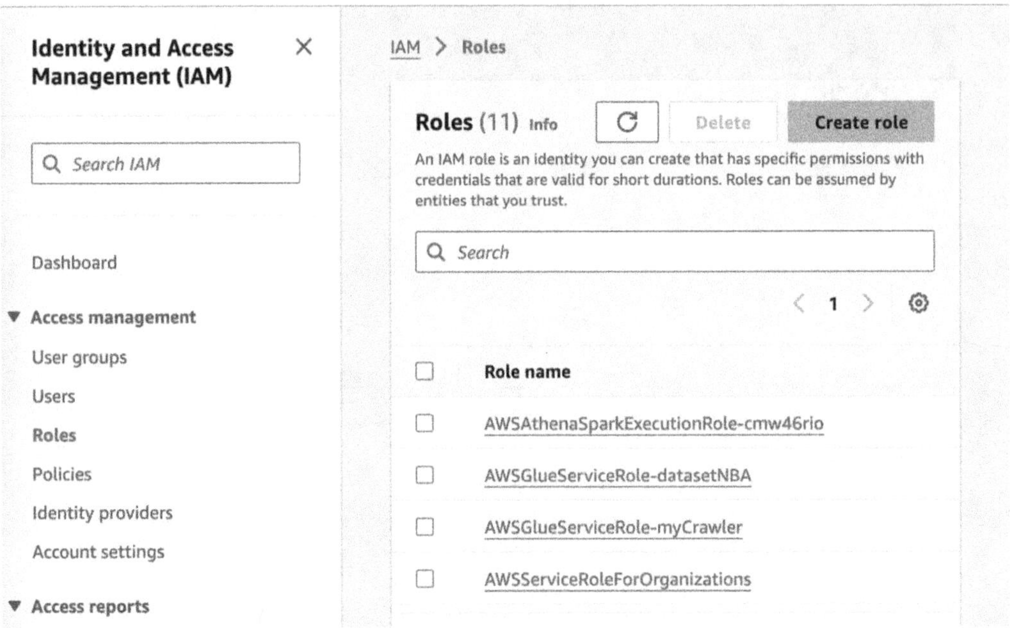

Figure 15-30. Creating a new role in IAM

For the trusted entity type, choose "AWS service," and for the use case, click the drop-down and select "Glue." Here is how that may look:

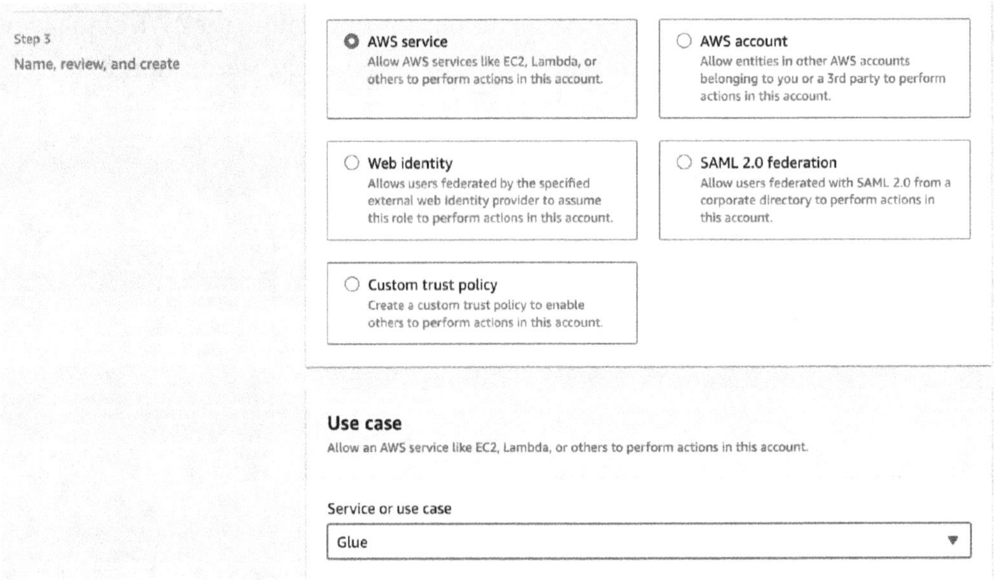

Figure 15-31. Choosing the trusted entity type within IAM

CHAPTER 15 ENGINEERING DATA PIPELINES USING AMAZON WEB SERVICES

Here are the following permissions you would need for this IAM role; choose an appropriate name and create the IAM role:

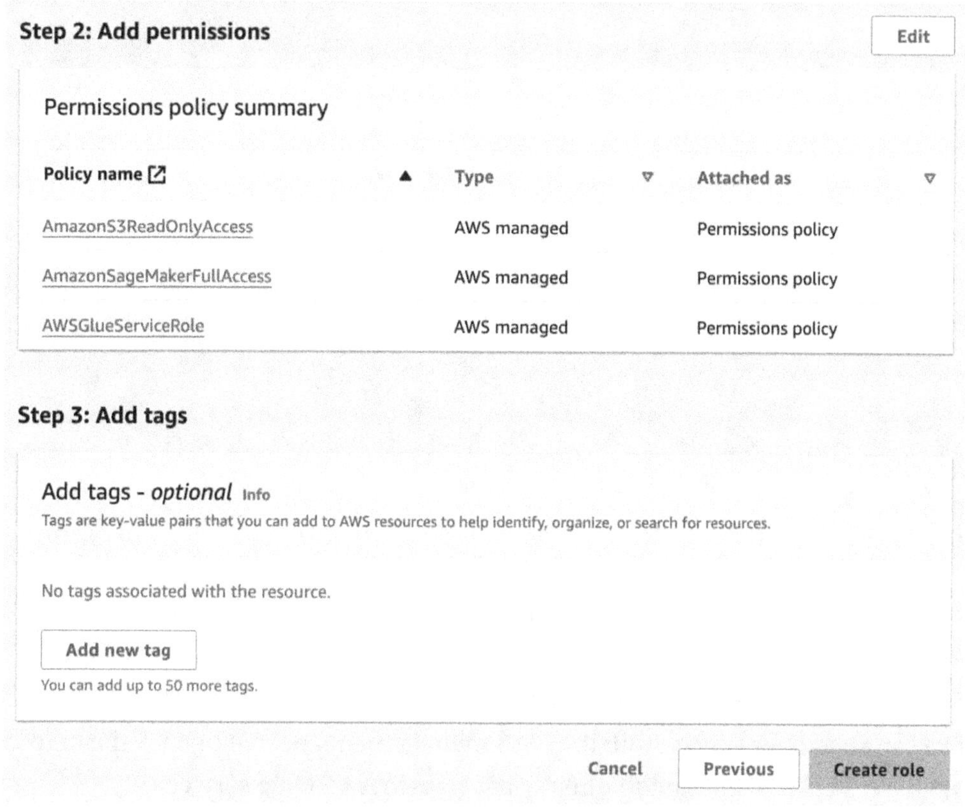

Figure 15-32. Adding permissions for the new IAM role

Now, we will grant permission for this newly created role to create a catalog in our Lake Formation. Navigate to AWS Lake Formation, choose our data lake, and grant permission from Actions:

Here is how this may look for you:

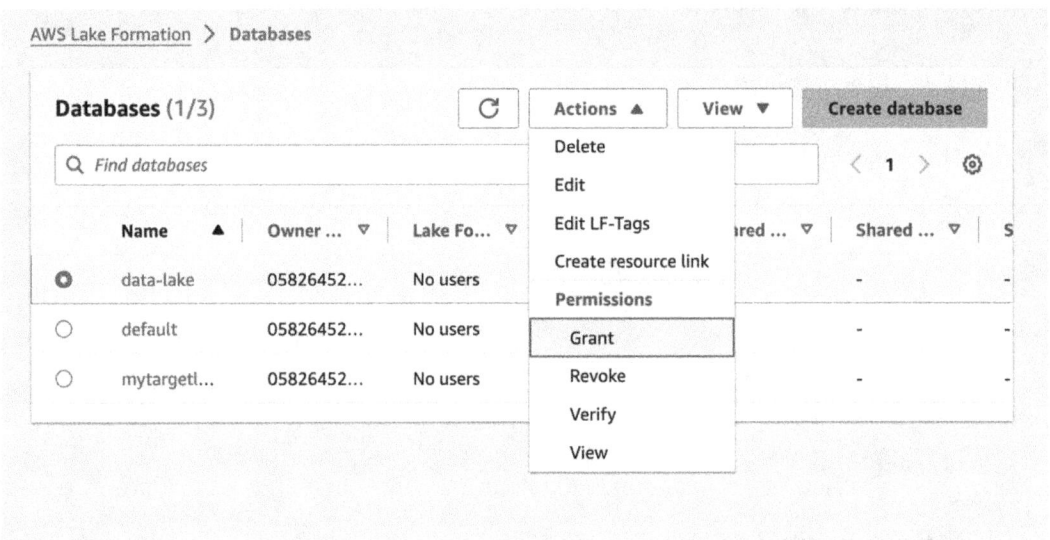

Figure 15-33. *Grant permission for the newly created role in Lake Formation*

Select the newly created IAM role and select the data lake under the named data catalog resources. Within the database permissions, provide the create table access option as well. Let us now locate the AWS Glue service.

You can search for Glue within the AWS management console or click the crawler option in AWS Lake Formation to directly navigate to AWS Glue service.

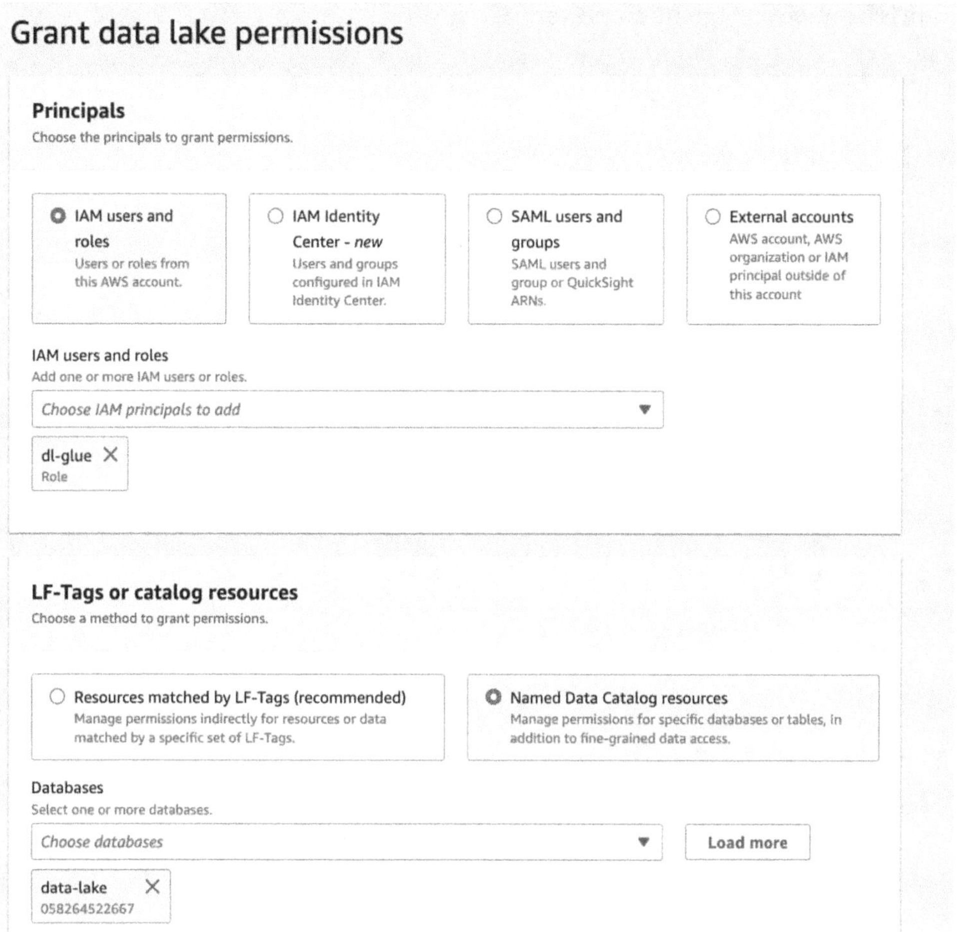

CHAPTER 15   ENGINEERING DATA PIPELINES USING AMAZON WEB SERVICES

Figure 15-34. *Choosing the IAM user within the permissions page*

We are going to create a new crawler, where we will specify the "data sources" folder from the data lake bucket as the input, assign the newly created "dl-glue" IAM role, and choose the "data lake" bucket for the target database.

CHAPTER 15 ENGINEERING DATA PIPELINES USING AMAZON WEB SERVICES

Here is how we can create a crawler:

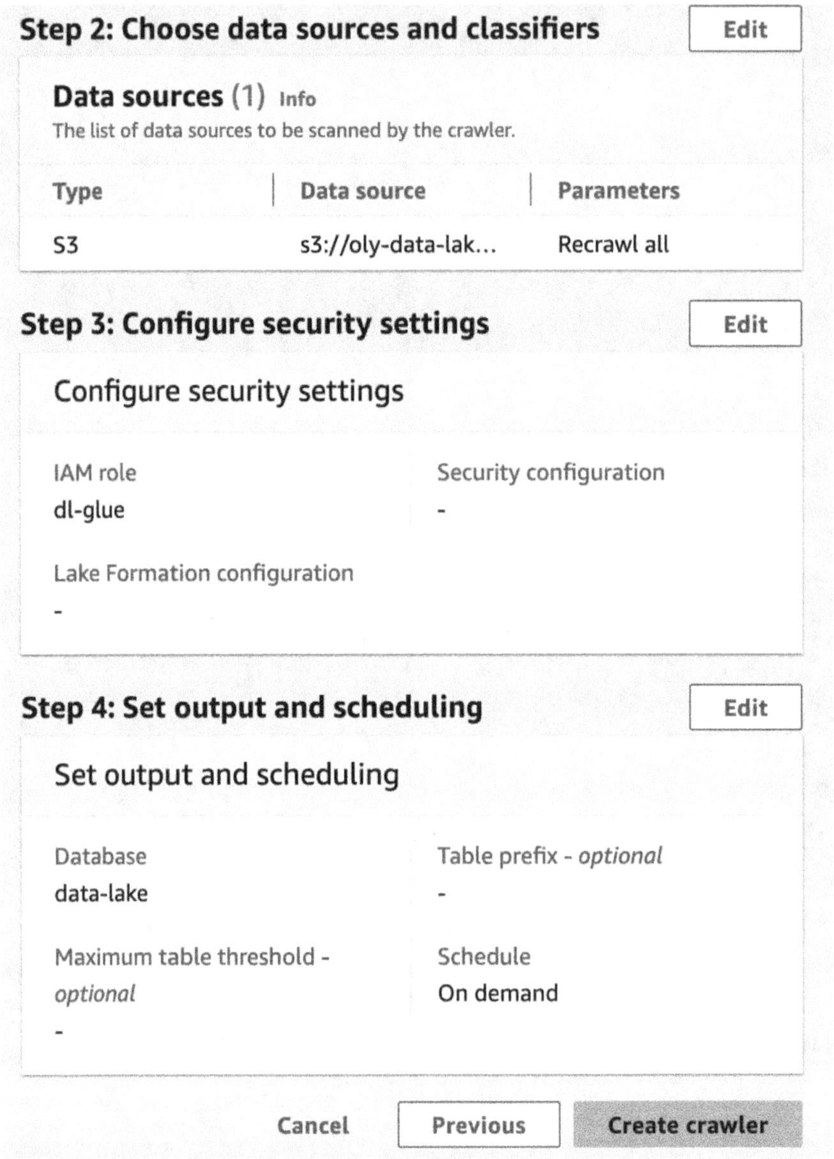

Step 2: Choose data sources and classifiers — Edit

Data sources (1) Info
The list of data sources to be scanned by the crawler.

Type	Data source	Parameters
S3	s3://oly-data-lak…	Recrawl all

Step 3: Configure security settings — Edit

Configure security settings

IAM role
dl-glue

Security configuration
-

Lake Formation configuration
-

Step 4: Set output and scheduling — Edit

Set output and scheduling

Database
data-lake

Table prefix - *optional*
-

Maximum table threshold - *optional*
-

Schedule
On demand

Cancel | Previous | **Create crawler**

Figure 15-35. Creating a new crawler with specified sources and IAM roles

Once you have created the crawler (may take a couple of minutes), you can see your newly created crawler in AWS Glue.

Choose the crawler and click Run. You may have to wait a few more minutes for the crawler to run.

CHAPTER 15   ENGINEERING DATA PIPELINES USING AMAZON WEB SERVICES

During this process, the AWS Glue crawler would infer the structure of the input dataset from the data sources folder and store the metadata in the data catalog.

Here is how that may look:

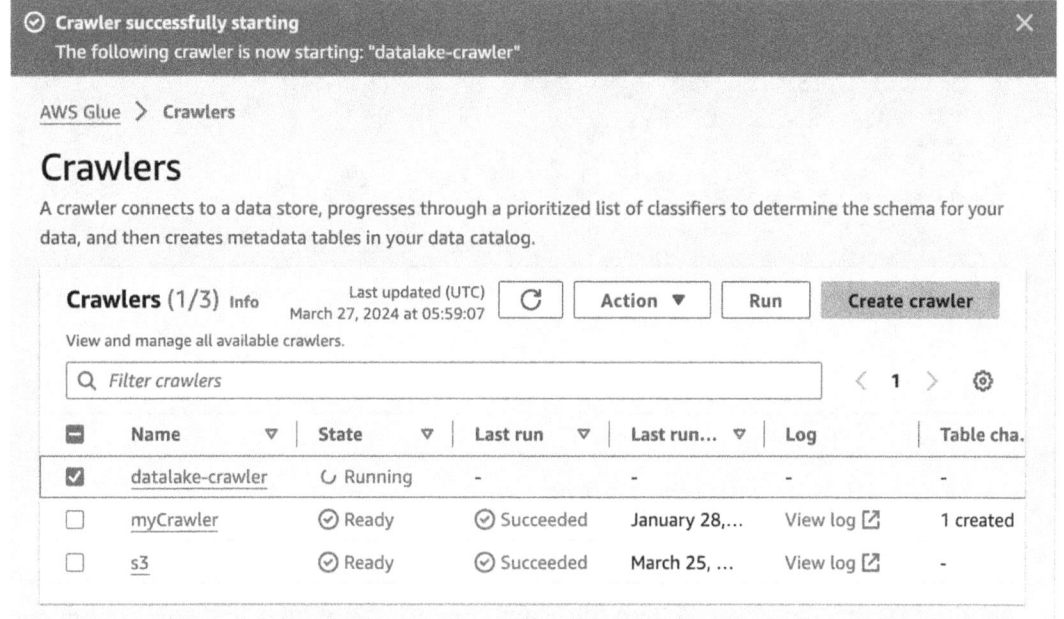

Figure 15-36. Initialization of a new crawler in AWS Glue

So far, we have successfully set up AWS Lake Formation, pointed the data sources at the AWS S3 bucket, created the necessary service account using AWS IAM, configured and ran the AWS Glue crawler, and populated the data catalog. We will now use AWS Athena to query the data to extract insights. Visit the AWS management console to navigate to AWS Athena service. Select the option of querying the data with Trino SQL and click Launch the query editor. Here is how that may look:

515

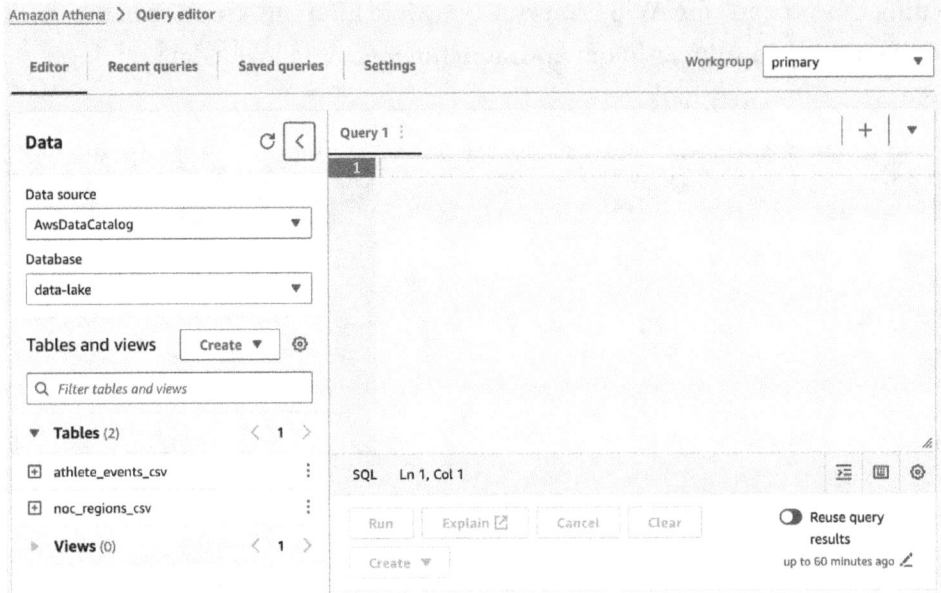

Figure 15-37. *AWS Athena query editor*

Let us now set up an output folder where we can store results of the query. At the top, click the Settings option and then the Manage settings option. Click the Browse S3 option, where you navigate to the data lake S3 bucket and select the Output datasets option. Save the settings.

Now, we are ready to write queries into AWS Athena and extract the insights from the dataset. Let us write a simple query:

```
select
    distinct year, season, city
from
    events_csv
where
    year < 2020
order by year desc
```

Here is how that looks in the editor:

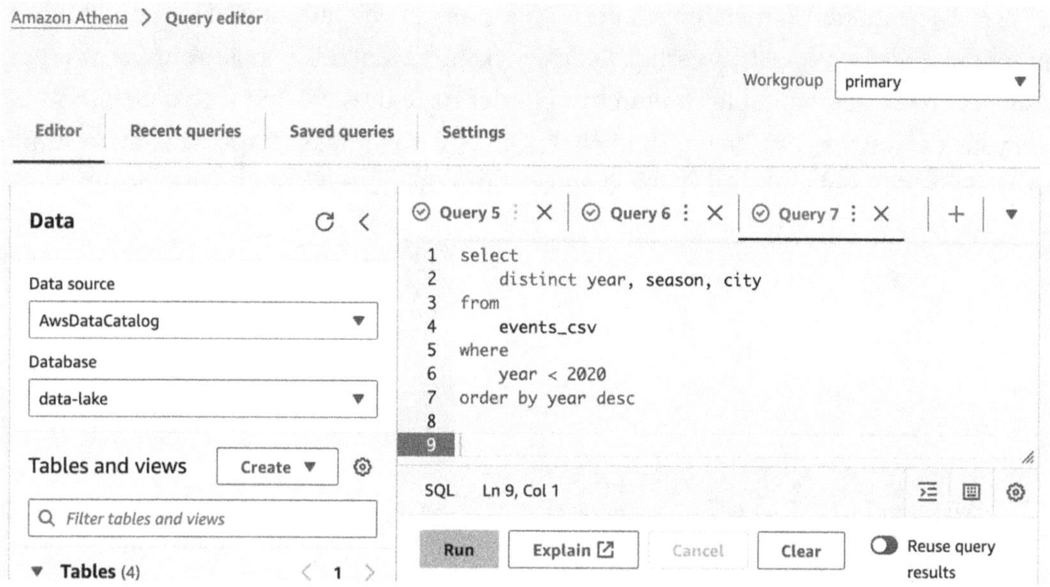

CHAPTER 15   ENGINEERING DATA PIPELINES USING AMAZON WEB SERVICES

Figure 15-38. Querying the data lake using AWS Athena

The query would yield the following results:

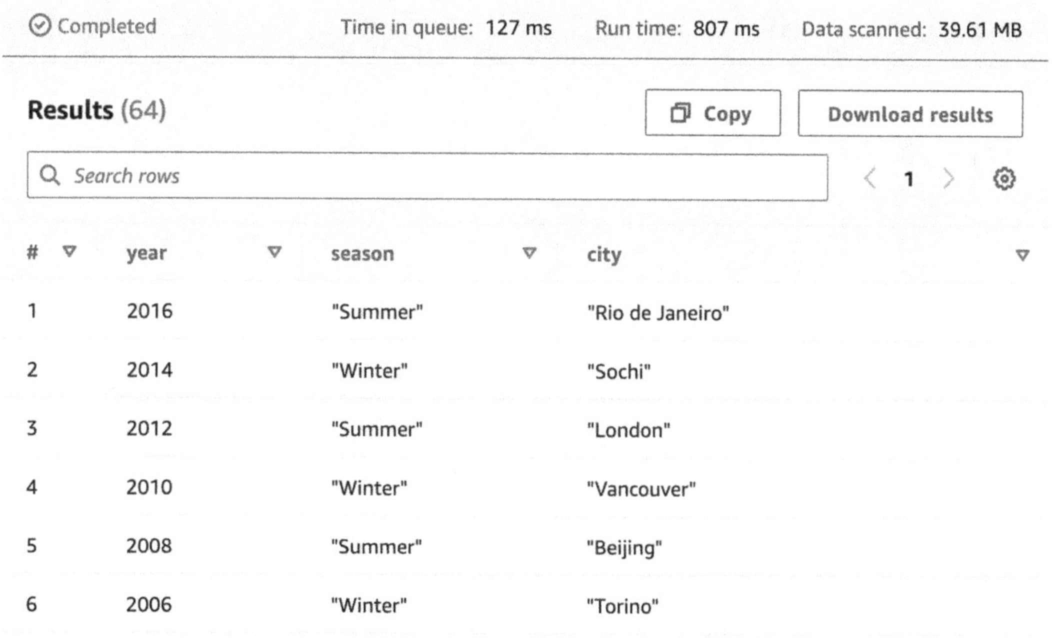

#	year	season	city
1	2016	"Summer"	"Rio de Janeiro"
2	2014	"Winter"	"Sochi"
3	2012	"Summer"	"London"
4	2010	"Winter"	"Vancouver"
5	2008	"Summer"	"Beijing"
6	2006	"Winter"	"Torino"

Figure 15-39. Retrieval of results in AWS Athena

CHAPTER 15   ENGINEERING DATA PIPELINES USING AMAZON WEB SERVICES

As we mentioned earlier, the results of this query are already stored in the S3 bucket under the folder we specified earlier. Let us navigate to that folder to look at the query results. You will see a complex hierarchy of folder structures created with date, month, and year. Locate the CSV file type and click "Query with S3 Select." You will have a simple query already populated, and so just choose to run the query to obtain the results.

Here is how that may look:

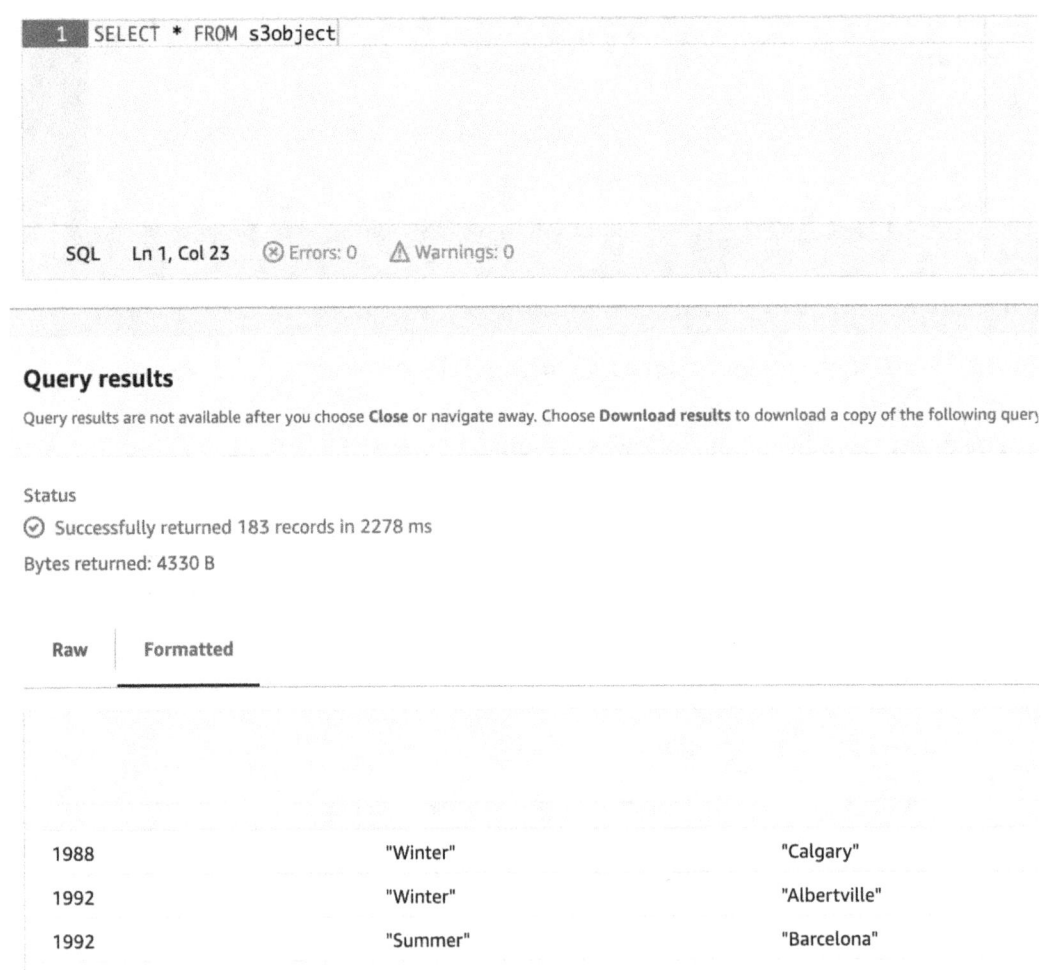

Figure 15-40.   Query results in a folder within the S3 bucket

CHAPTER 15   ENGINEERING DATA PIPELINES USING AMAZON WEB SERVICES

AWS SageMaker

Amazon SageMaker is a fully managed machine learning service by AWS. It enables you to build, train, test, and deploy machine learning models at scale. You have the option of developing a no-code or low-code machine learning model as well. AWS SageMaker supports industry-leading machine learning toolkits like sk-learn, TensorFlow, PyTorch, Hugging Face, etc.

AWS SageMaker comes in a few variants. AWS SageMaker Studio comes with a fully functioning ML integrated development environment. It comes with a Jupyter notebook, which gives you control over every step of the model being built. AWS SageMaker Studio Lab is a user-friendly, simplified version of AWS SageMaker Studio, but it comes with limited hardware support. You would get 16 gigabytes of physical memory and 15 gigabytes of storage, and it cannot be extended, compared with SageMaker Studio where you can specify larger compute and greater memory. There is also SageMaker Canvas, which is a low-code environment for developing predictive models.

To setup AWS SageMaker Studio, get to the service page, and choose the option "Set up SageMaker Domain." Here is how that may look:

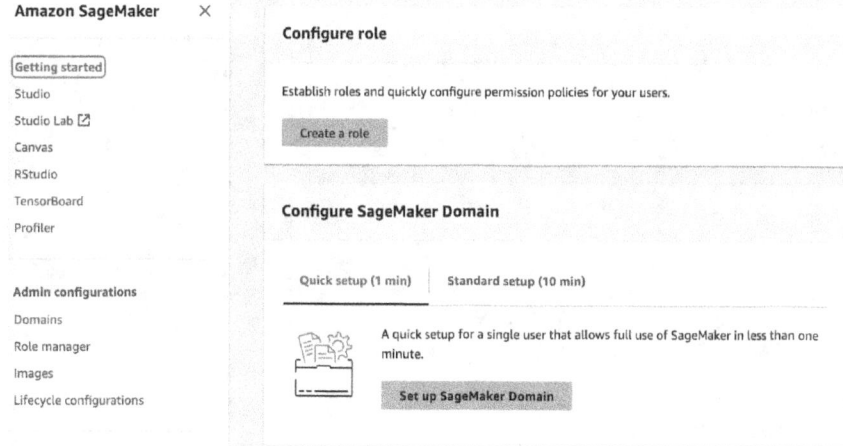

Figure 15-41. Setting up a SageMaker Domain

Once the SageMaker Domain is created for you, you can proceed to launch SageMaker Studio from the Domain. Here is how that may look:

CHAPTER 15   ENGINEERING DATA PIPELINES USING AMAZON WEB SERVICES

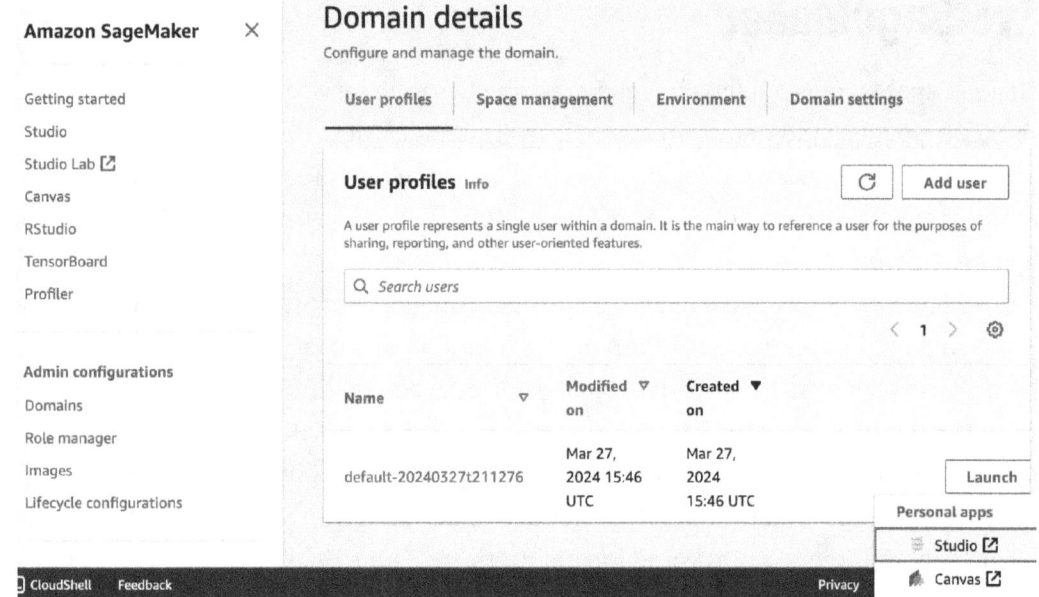

Figure 15-42. Launching AWS SageMaker Studio

Here is how SageMaker may look for you:

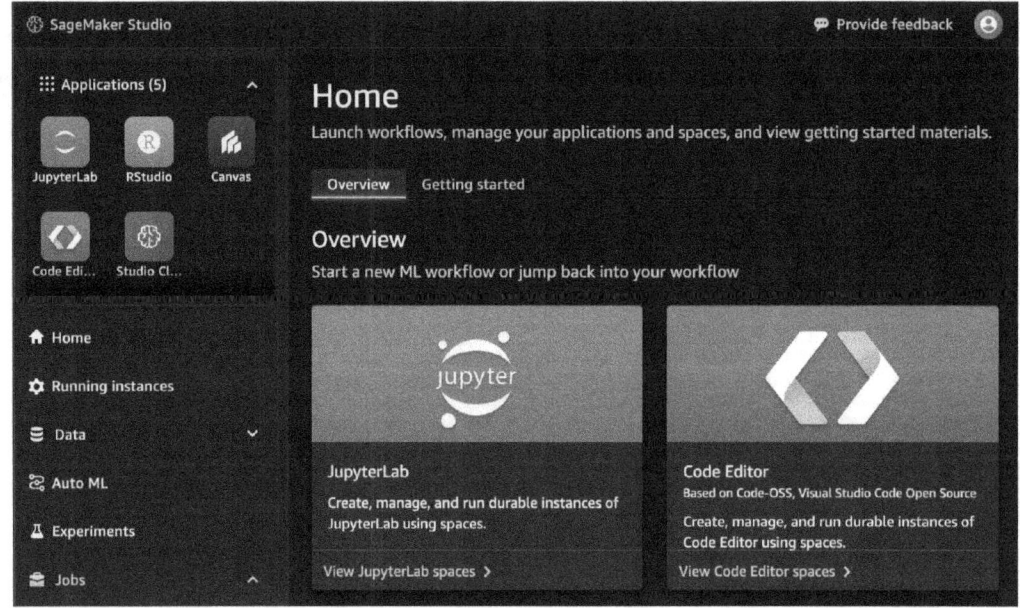

Figure 15-43. Home page of AWS SageMaker

CHAPTER 15   ENGINEERING DATA PIPELINES USING AMAZON WEB SERVICES

Let us try to build and deploy a machine learning model using Amazon SageMaker. We will build a simple machine learning model, host it on an instance, create an API that can be used to interact with the model, and create a lambda function that would access the endpoint.

To get started, let us open JupyterLab and choose the new JupyterLab space. You will provide a name for the JupyterLab Instance and proceed to choose the underlying hardware. My suggestion is to use the instance "m1.t3.medium" as you get 250 free hours of compute when you start using SageMaker initially.

The Image option is basically a Docker image appropriately configured for machine learning tasks.

Here is how the settings may look, for instance:

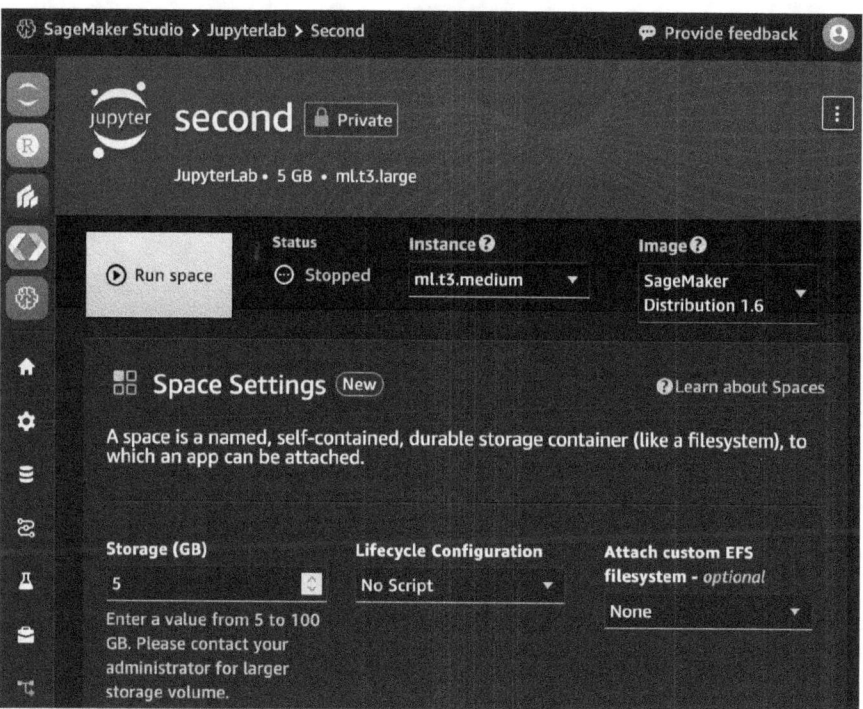

Figure 15-44. *Creating and configuring a new Jupyter space on AWS SageMaker*

For our simple illustration, we will choose the titanic dataset, where the outcome is the likelihood of a passenger to survive, in case of a ship colliding with the iceberg. For our illustration purposes, we are choosing a simpler dataset and have removed a few columns as well. We are importing necessary libraries, preprocessing the dataset,

CHAPTER 15   ENGINEERING DATA PIPELINES USING AMAZON WEB SERVICES

running the SageMaker's linear learner model, and creating an endpoint that can be hosted on another instance. We are using Amazon SageMaker's linear methods to obtain the model.

Here is the code:

```
import sagemaker
from sagemaker.sklearn.model import SKLearnModel
from sagemaker import get_execution_role
import numpy as np
import pandas as pd

df = pd.read_csv("titanic.csv")

df = df.dropna(
    subset=["Age"]
)

df = df.drop(
    ['Name','Sex','Ticket','Cabin','Embarked'],
    axis=1
)

df.to_csv("titanic_subset.csv")

rawdata = np.genfromtxt(
        "titanic_subset.csv",
        delimiter=',',
        skip_header=1
    )

train = rawdata[:int(len(rawdata) * 0.8)]
test = rawdata[int(len(rawdata) * 0.8):]

Xtr = train[:, :-1]
Ytr = train[:, -1]
Xts = test[:, :-1]
Yts = test[:, -1]
```

```
linear = sagemaker.LinearLearner(
        role = get_execution_role(),
        instance_count = 1,
        instance_type = 'ml.m5.large',
        predictor_type='regressor',
        sagemaker_session=sagemaker.Session()
    )

train_data_records = linear.record_set(
   Xtr.astype(np.float32),
   labels=Ytr.astype(np.float32),
   channel='train'
)

linear.fit(train_data_records)

predictor = linear.deploy(
      initial_instance_count=1,
      instance_type='ml.t3.medium'
   )
```

Let us run this code on a JupyterLab notebook and make a note of the endpoint that SageMaker generates. We will host this endpoint on another SageMaker instance. We will then have a lambda function that would call the hosted API. Lambda functions usually have a certain IAM role associated to work with other services.

Let us create a new IAM role that can be used to interact with machine learning models. Let us find IAM in the AWS management console, locate the Identity and Access Management option on the left, and click Roles; create a new role and assign full access to Lambda, SageMaker, and CloudWatch services. Here is how it may look for you:

CHAPTER 15 ENGINEERING DATA PIPELINES USING AMAZON WEB SERVICES

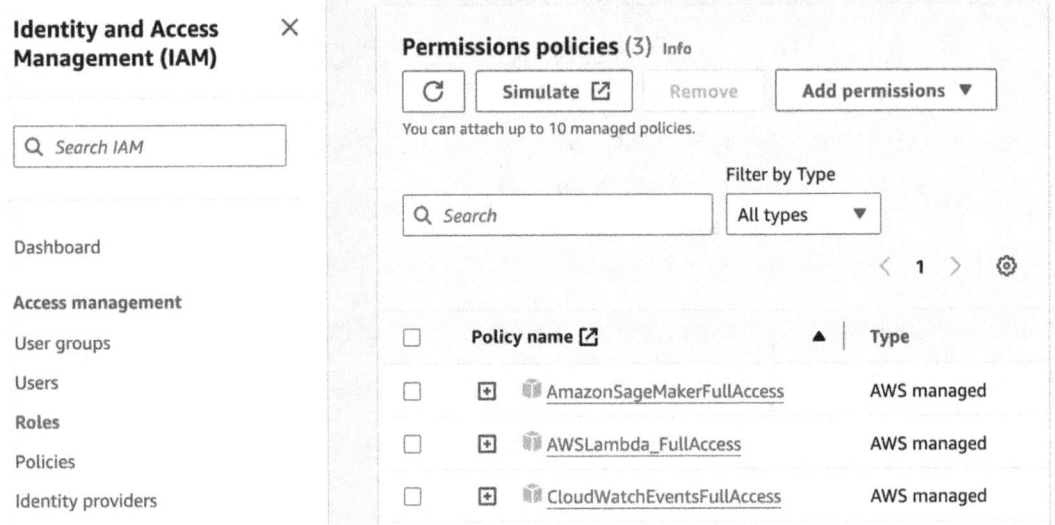

Figure 15-45. *Creating a new IAM role for ML*

Let us now create a lambda function. Navigate to the AWS management console and search for "Lambda" service. Choose to create a function, select the Author from scratch option, choose Python as the runtime configuration, and leave the architecture selection as is.

Here is how the Create function page in AWS Lambda looks like:

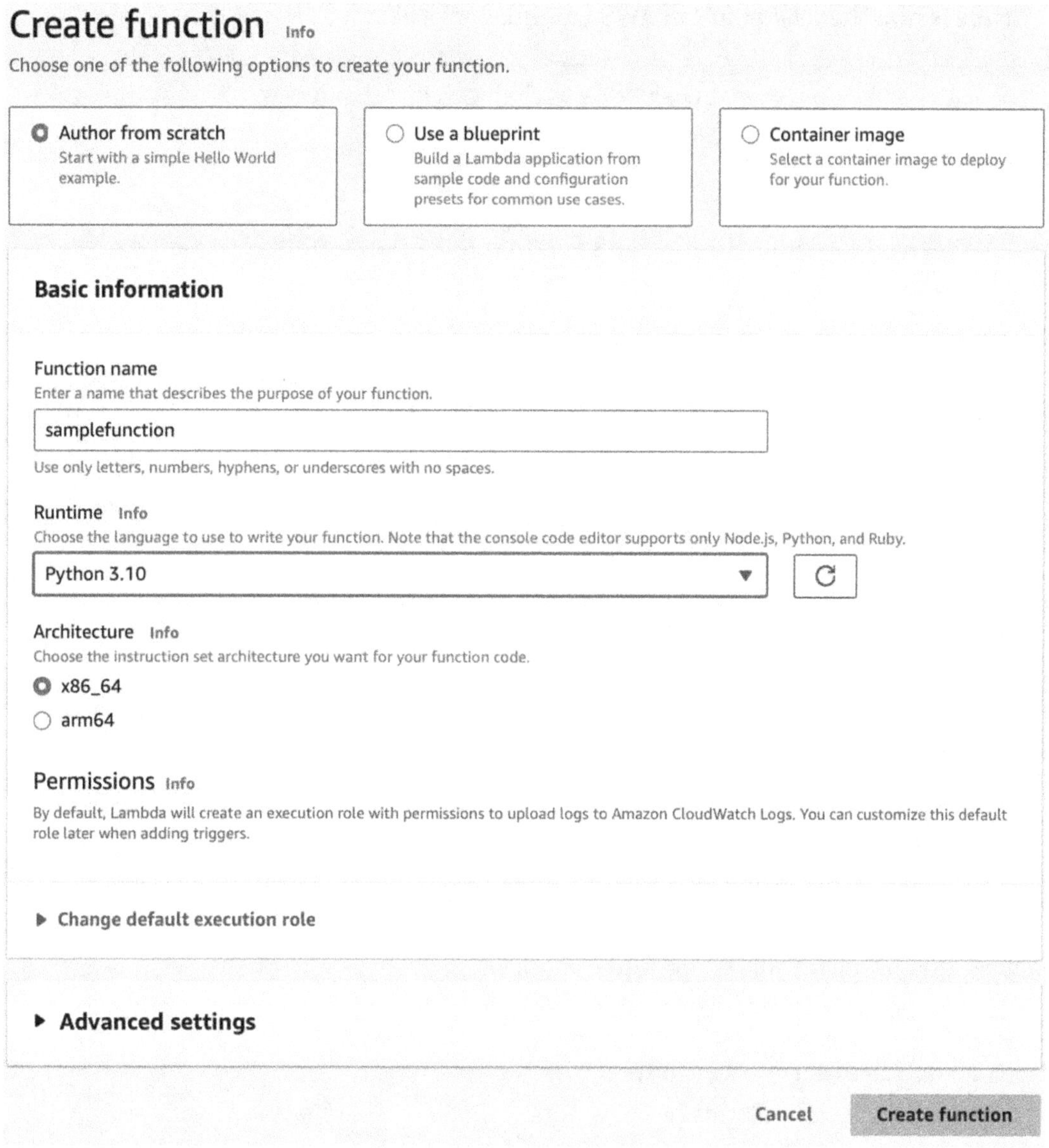

Create function

Choose one of the following options to create your function.

- ● **Author from scratch** — Start with a simple Hello World example.
- ○ **Use a blueprint** — Build a Lambda application from sample code and configuration presets for common use cases.
- ○ **Container image** — Select a container image to deploy for your function.

Basic information

Function name
Enter a name that describes the purpose of your function.

```
samplefunction
```
Use only letters, numbers, hyphens, or underscores with no spaces.

Runtime Info
Choose the language to use to write your function. Note that the console code editor supports only Node.js, Python, and Ruby.

```
Python 3.10
```

Architecture Info
Choose the instruction set architecture you want for your function code.
- ● x86_64
- ○ arm64

Permissions Info
By default, Lambda will create an execution role with permissions to upload logs to Amazon CloudWatch Logs. You can customize this default role later when adding triggers.

▶ Change default execution role

▶ Advanced settings

Cancel **Create function**

Figure 15-46. Creating a new lambda function in AWS Lambda

Ensure that you change the default execution role by using an existing role, and select the IAM role you created from the drop-down option.

The AWS lambda function comes with starter code, where the lambda function has two parameters, one that takes an event that triggers the function and another containing runtime parameters.

Here is how the dashboard in AWS Lambda looks like:

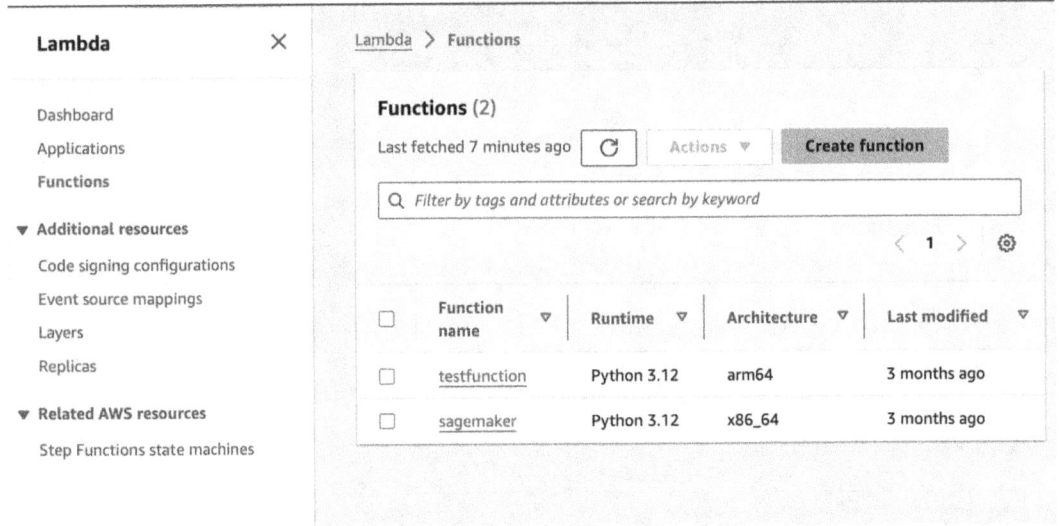

Figure 15-47. Dashboard page of AWS Lambda

We parse the JSON string to obtain the actual data that will be sent to SageMaker for prediction. We then request the SageMaker endpoint and return the response. This is again followed by JSON parsing of the response and returning the prediction score.

Here is the lambda function:

```
import json, os, boto3

endpoint = os.environ['endpoint']
runtime = boto3.client('runtime.sagemaker')

def lambda_handler(event, context):
    event_data = json.loads(
        json.dumps(event)
    )
    testdata = event_data['data']
    response = runtime.invoke_endpoint(
        EndpointName=endpoint,
        ContentType='text/csv',
        Body=testdata
    )
```

```
result = json.loads(
            response['Body'] \
            .read() \
            .decode()
        )
return int(result['predictions'][0]['score'])
```

In addition, you may wish to set the environment variable for this lambda function by navigating to the Configuration option and choosing Environment variables. Here is how that may look:

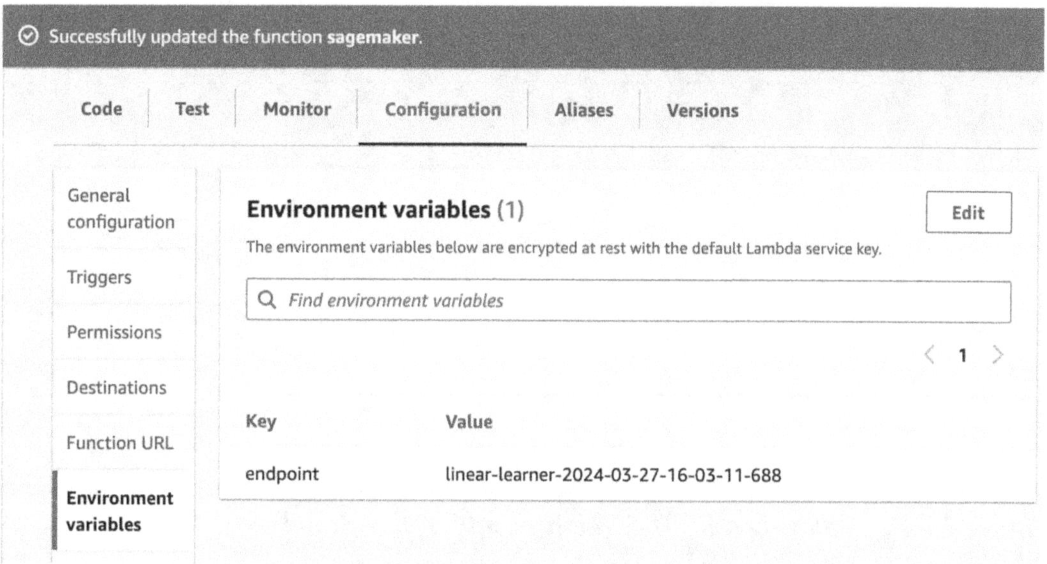

Figure 15-48. Setting the environment variables for the lambda function

Let us now deploy this function. From the AWS management console, navigate to the API Gateway service and choose to create an API. Choose to build a REST API that will provide you more control over the request and response. Enter a name for the new API, leave the endpoint type as is, and choose to create an API.

Here is how this may look:

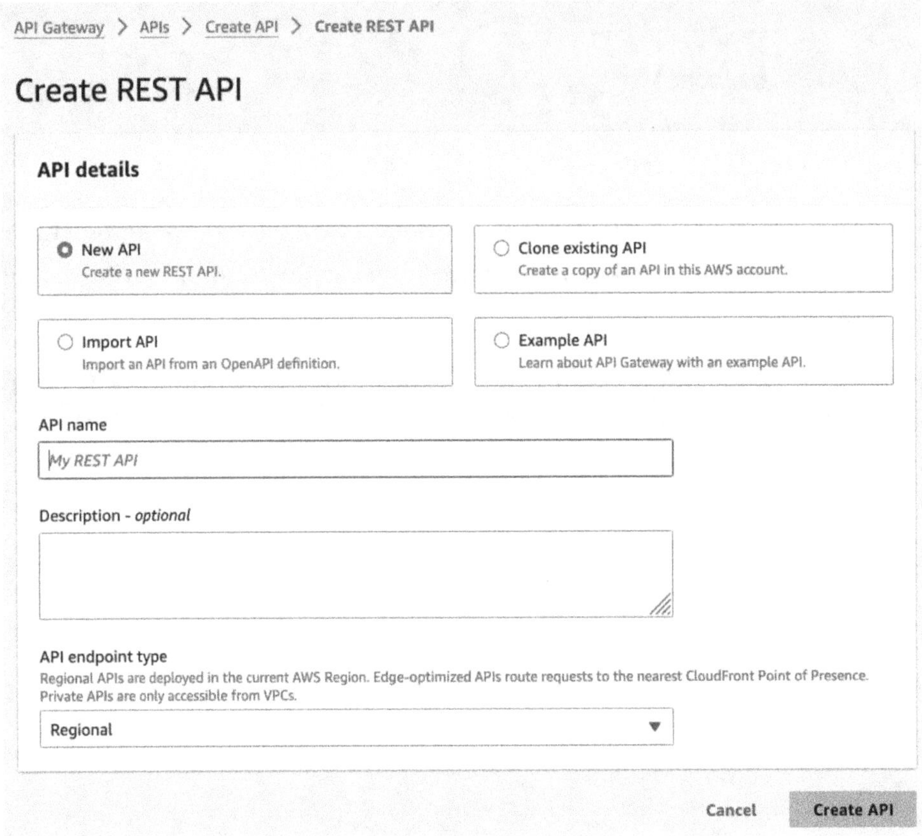

Figure 15-49. Creating a REST API endpoint for our model

Since we will be sending data to the API to obtain a prediction score, we will create a POST method that would post the data points to the machine learning API and return a prediction score. And so, choose to create a method, select POST as method type, and select the lambda function as the integration method.

From the Method type drop-down, please choose the POST method and not the PUT method.

CHAPTER 15   ENGINEERING DATA PIPELINES USING AMAZON WEB SERVICES

Here is how that may look:

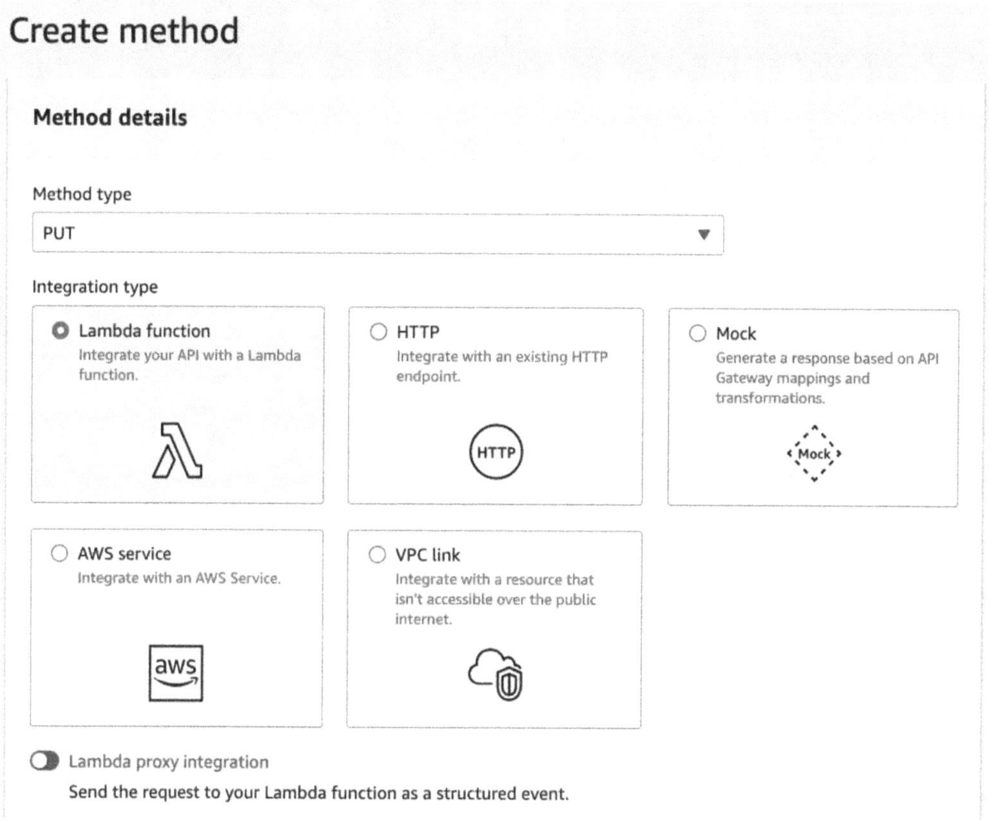

Figure 15-50. Creating a POST method

Once you have created this POST method, you can test the API functionality to see if it is working as expected. On API Gateway, navigate to the Test option and enter the data to be tested within the Request Body section and click Test. You will get the response code and the response that the API returns.

Once everything looks good, we can proceed to deploy the API that is publicly accessible. In the top-right corner, choose the "Deploy API" option, choose "New stage" for the Stage drop-down, and enter a name for the stage. You could either mention Development, Testing, Production, or Alpha, Beta, etc.

Here is how that may look for you:

Figure 15-51. Deploying the API to be publicly accessible

Conclusion

So far, I hope you gained a comprehensive view of Amazon Web Services, focusing on its data engineering and machine learning capabilities. We have covered a wide range of services starting from simple object storage to hosting your own machine learning models on the cloud. We created IAM roles for secured access to services; stood up databases with RDS, data warehouses with Amazon Redshift, data lakes with Lake Formation, and machine learning models with AWS SageMaker; and also gained experience in working with AWS Lambda, AWS Athena, and AWS Glue.

The examples provided here may seem to use comparatively generous IAM role privileges. This is only for illustration purposes. In production, you should practice the principle of least privilege, granting only the specific permissions required for a given task. Overall the topics we covered may serve you well in your projects.

CHAPTER 16

Engineering Data Pipelines Using Google Cloud Platform

Introduction

In this chapter, we will look at Google Cloud Platform, one of the early adopters of cloud computing. We will look at exploring various services of GCP, focusing on data engineering and machine learning capabilities. We will start by understanding core concepts of Google Cloud Platform and looking at key services. This is followed by a detailed look at the various data system services offered by Google Cloud Platform. We will finally look at Google Vertex AI, a fully managed machine learning platform for building, training, and deploying machine learning models within one service.

By the end of the chapter, you will learn

- Fundamentals of Google Cloud Platform
- How to set up Google CLI for managing resources
- Google's object storage and compute engine for virtual machines and data systems
- Google's data processing tool Dataproc for running Spark and Hadoop workloads
- Developing, training, and deploying machine learning models using Google Vertex AI

CHAPTER 16 ENGINEERING DATA PIPELINES USING GOOGLE CLOUD PLATFORM

As you embark on your journey exploring Google Cloud Platform, please may I request you to be mindful of managing the cloud resources, especially when you are done using them. In the case of traditional infrastructure, you may switch off a computer when you are done. However, in the case of cloud infrastructure, where you often procure a plan that is based on a pay-as-you-go model, you will be billed for your resources even if they are not being used. Signing off GCP does not mean powering off the services you have procured. This is particularly important for virtual machines, databases, and any other compute-intensive resources. Please, always, make it a habit to review your active resources and shut down these specific resources. If you are no longer planning on using a service, you may delete the same. And so, you are only paying for the services you actively use and keep your costs under control. I also would like to thank GCP for providing free credits for someone who is a new user and having actively supported new learners in bringing their cost down.

Google Cloud Platform

Google Cloud Platform is a set of cloud computing services offered by Google. Google started providing cloud computing as early as 2008, when they provided App Engine, a platform for hosting and managing web applications where the data centers were completely managed by Google. Google has data centers located all over the world. They are designated as regions and zones. Each region has a collection of zones. These zones are isolated from each other. Google Cloud provides various services that could either be hardware or software products or resources.

Google Cloud has the concept of projects. If you are attempting to use a Google Cloud service, then it must be part of a project. Google defines a project as an entity made up of settings, permissions, and other metadata that describe your application. Every time you create a new project, Google Cloud would provide you a project name, project id, and project number. You can create multiple projects and be tied to the same account for billing purposes.

In AWS, we witnessed the concept of resource hierarchy where we have a root account that enables us to create organizational units through AWS Organizations and create AWS accounts from there on and provision various services within such accounts. In GCP, the hierarchy starts off with an organization resource that is comparable to an AWS root account.

The organization resource exercises control over billing, access control lists, and managing various resources. An organizational node can have one or more folders, each representing some kind of a sub-organization or a department within the organization.

CHAPTER 16 ENGINEERING DATA PIPELINES USING GOOGLE CLOUD PLATFORM

Each folder can have one or more projects. The project represents the lowest level of hierarchy in an organization. Any given project is where a virtual machine or a storage instance can be provisioned.

For illustration purposes, let us set up folders as part of GCP setup. You are going to need a Google account to begin with. I would suggest you start off by creating a sandbox folder and create your projects there. You may set up a development folder upon gaining momentum and test your use cases there. Google Cloud can be accessed and interacted through a graphical user interface, command line interface, and client libraries that offer API to manage resources and services.

Set Up a GCP Account

Let us get started by visiting https://cloud.google.com. Click Sign in and follow the on-screen instructions for creating a new Google account or log in to an existing Google account. Enter the details as required and proceed to get started. We want to get started with the Google Cloud console page (similar to the AWS management console).

Here is how the Google console page may look like:

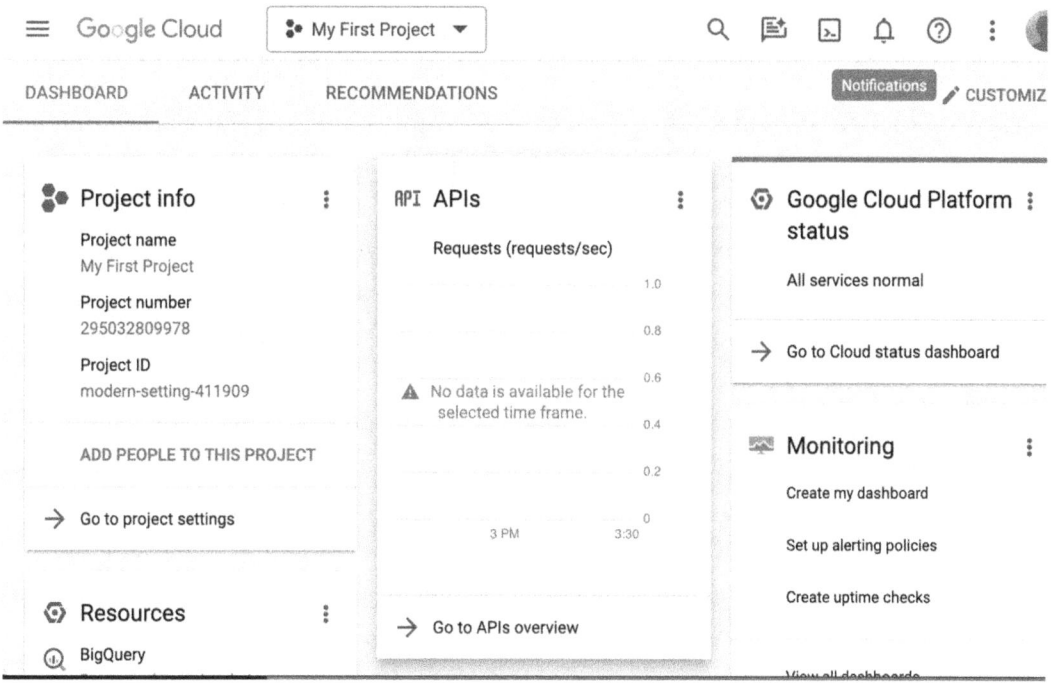

Figure 16-1. Google Cloud console page

CHAPTER 16　ENGINEERING DATA PIPELINES USING GOOGLE CLOUD PLATFORM

Google Cloud Storage

Google's Cloud Storage is an object storage service. Cloud Storage stores data as a unique object or a blob. Unlike relational database management systems, you do not need to define a schema to use Cloud Storage. You can store data files, images, videos, audios, and so on. These files are referred to as objects. These objects are placed in a container called buckets. These objects can be referenced by a unique key in the form of a URL. There are many types and applications of data storage with Cloud Storage; you can start off with storing the contents of your site, the data file that collects data in the form of relational tables, periodic archives, videos and interviews you host on your website, documents, etc. These files are considered as atomic, meaning if you attempt to access an object, you are accessing an entire object.

As far as data security is concerned, you can utilize Google IAM and Cloud Storage's access control list mechanism for authentication and authorization. In Cloud Storage, individual objects have their own access control list as well. The object storage service naming must be unique as Cloud Storage has a global namespace. Once you create an object service with a certain name, you cannot modify the name at a later point. You can, however, rename an object within a bucket.

Google's Cloud Storage comes with four different classes of storage, namely, standard storage, near line storage, cold line storage, and archival storage. The standard storage option is for data that is frequently accessed and stored only for a brief amount of time. The near line storage is for storing data that is less frequently accessed with a slightly lower availability, where data is stored for at least 30 days. The cold line storage comes with lower availability and is for storing infrequently accessed data where data is stored for at least 90 days. The archival storage is for disaster recovery and long-term backups where the data is not accessed as much and the service may not be as available as the other three storage classes.

Cloud Storage provides object lifecycle management configurations where you can define and set criteria for a given bucket. For instance, you can change the storage class of an object depending upon a certain time threshold. You can specify the number of days a certain object or a bucket should be retained. You can enable object versioning, where there exist multiple variants of the same object. However, this would significantly increase the storage size and costs incurred, and so it is advised to exercise caution.

CHAPTER 16   ENGINEERING DATA PIPELINES USING GOOGLE CLOUD PLATFORM

Google Cloud CLI

Google Cloud has a command line interface that can be used to create and manage Google Cloud resources. Depending upon the operating system, the installation may vary. Here are some notes on setting up on an Ubuntu/Debian machine:

1. First off, run the update command and install any dependencies your system may need:

    ```
    sudo apt update
    sudo apt-get install apt-transport-https ca-certificates gnupg curl
    ```

2. Import the Google Cloud CLI public key and add the GCloud CLI distribution as a package source:

    ```
    curl https://packages.cloud.google.com/apt/doc/apt-key.gpg | sudo gpg --dearmor -o /usr/share/keyrings/cloud.google.gpg
    ```

    ```
    echo "deb [signed-by=/usr/share/keyrings/cloud.google.gpg] https://packages.cloud.google.com/apt cloud-sdk main" | sudo tee -a /etc/apt/sources.list.d/google-cloud-sdk.list
    ```

3. Run the update command again and install GCloud CLI:

    ```
    sudo apt-get update && sudo apt-get install google-cloud-cli
    ```

4. Run Google Cloud CLI to get started:

    ```
    gcloud init
    ```

You will have to authenticate in a browser by entering your credentials and allowing the Google Cloud SDK to access your account.

Here is how that may look:

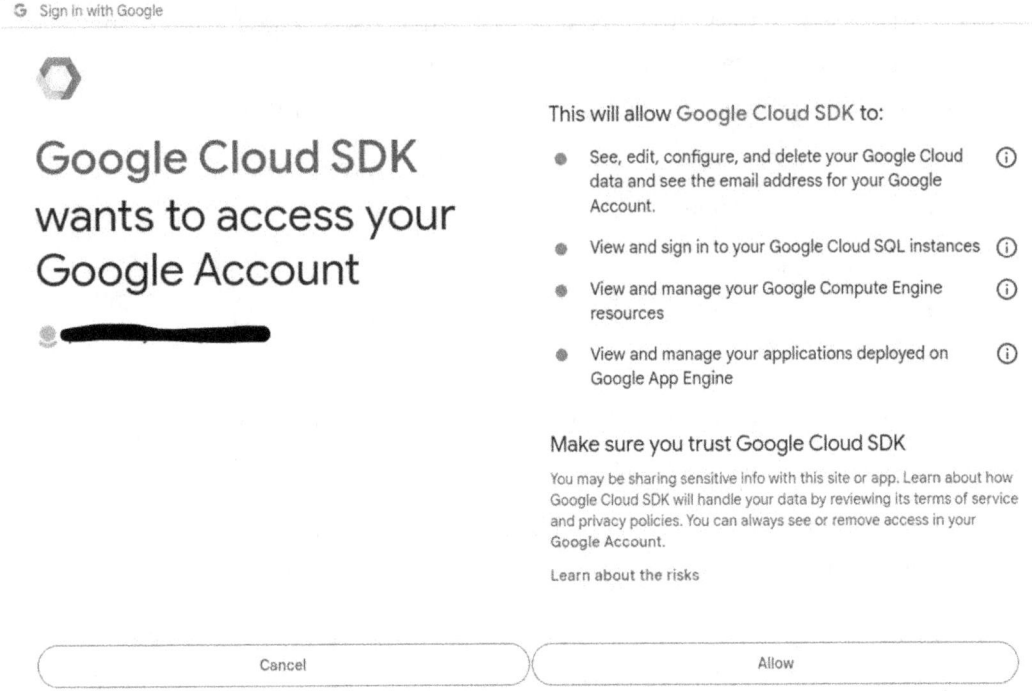

Figure 16-2. Google Cloud SDK authentication page

If you are working on MS Windows operating system, then open a PowerShell tab and enter the following:

```
(New-Object Net.WebClient).DownloadFile("https://dl.google.com/dl/
cloudsdk/channels/rapid/GoogleCloudSDKInstaller.exe", "$env:Temp\
GoogleCloudSDKInstaller.exe")
& $env:Temp\GoogleCloudSDKInstaller.exe
```

Follow the prompts that the Google installer provides. Open Google Cloud CLI and enter the following to get started with GCP CLI:

```
gcloud init
```

You are all set to utilize Google Cloud from your command line.

CHAPTER 16   ENGINEERING DATA PIPELINES USING GOOGLE CLOUD PLATFORM

Google Compute Engine

Google offers virtual machines on demand to meet the needs of various enterprise applications. You can stand up a database server, build an application, host a website, and perform various other tasks using the Compute Engine virtual machines. Google offers various types of compute engines:

- General-purpose engine, which is meant for everyday tasks

- Storage optimized, where the workload is mostly regarding storage and consumes less compute

- Compute optimized, where the performance is prioritized with multiple cores and higher clock performance

- Memory optimized, which offers more memory per core, suitable for memory-intensive workloads

- Accelerator optimized, good for GPU-intensive workloads, massively parallel CUDA workloads, and high-performance computing purposes

Let us get started by enabling the Compute Engine API. Navigate to the Google Cloud console and look for a virtual machine option or Google compute option; we need to enable the API service for consumption.

Here is how that may look:

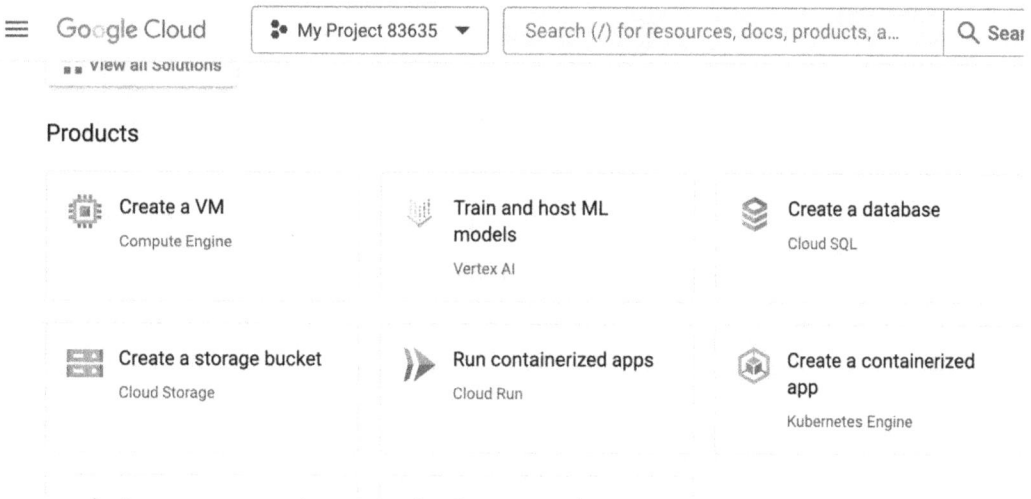

Figure 16-3. *Creating a virtual machine in Google Cloud*

537

CHAPTER 16 ENGINEERING DATA PIPELINES USING GOOGLE CLOUD PLATFORM

Click Create a VM, which will lead us to the following page, where you would be provided the option to enable the Google Compute Engine API.

Click ENABLE, to enable the compute API.

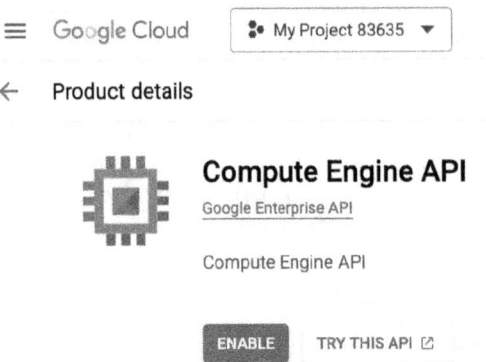

Figure 16-4. Enabling the Compute Engine API

Once enabled, you would be able to manage the API. It would appear like this:

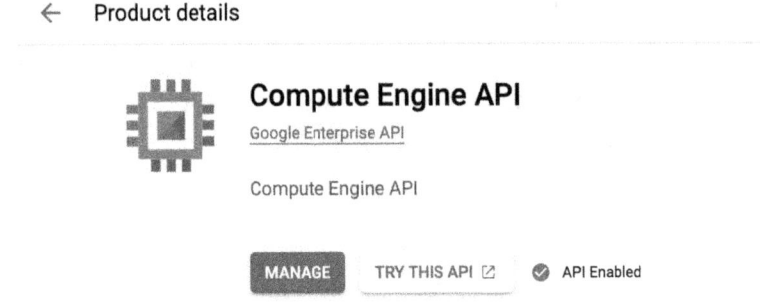

Figure 16-5. Compute Engine API enabled

With the Compute Engine API enabled, you may proceed to create a new virtual machine.

Now, we are going to create a Linux virtual machine. We will begin by creating a new compute instance, selecting the appropriate Linux distribution and the version, and enabling access by allowing incoming HTTP traffic.

CHAPTER 16   ENGINEERING DATA PIPELINES USING GOOGLE CLOUD PLATFORM

First, let us visit the menu on the left, locate the Compute Engine option, and click VM Instances and create a new instance. You would be provided with a page, where you can specify the name, location, and type of instance to begin with. You also have the choice of either specifying a pre-packed machine type with a specific number of cores and memory, or you can customize your configuration as well.

Based on the choices you make, Google would present you costs that may be incurred every month. Here is how this may look for you:

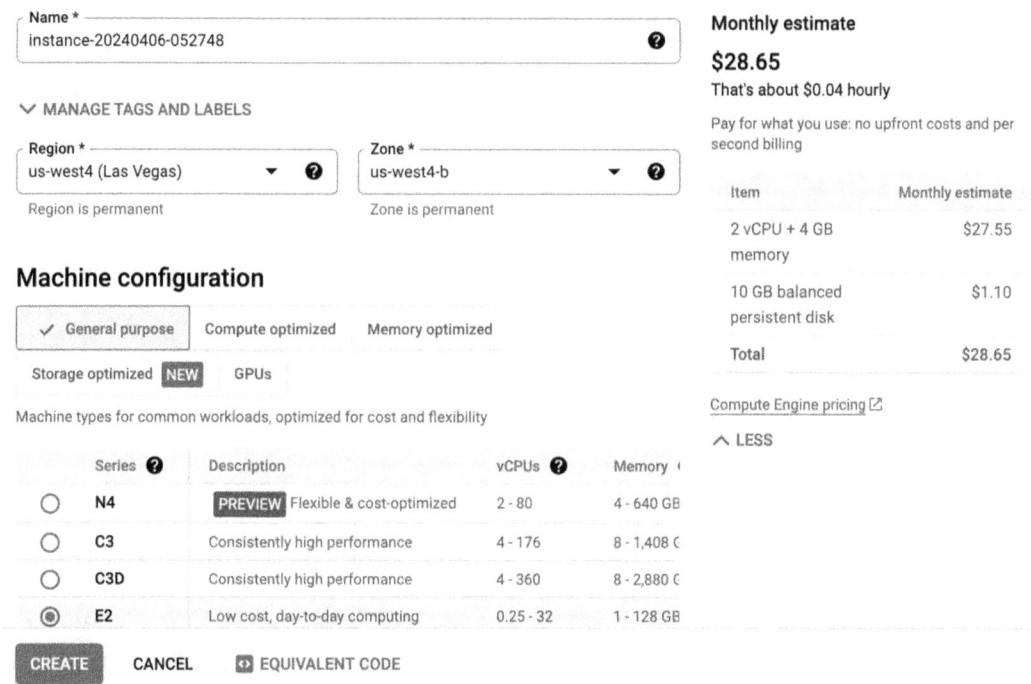

Figure 16-6. Provisioning a new virtual machine in Google Cloud

CHAPTER 16   ENGINEERING DATA PIPELINES USING GOOGLE CLOUD PLATFORM

You also have the option of specifying the access control, firewall rules, and other security configurations within the same page. Once you have chosen your desired configuration, you may click Create. You may also choose the most minimum configuration and go with the bare essentials. Here is how the dashboard may look:

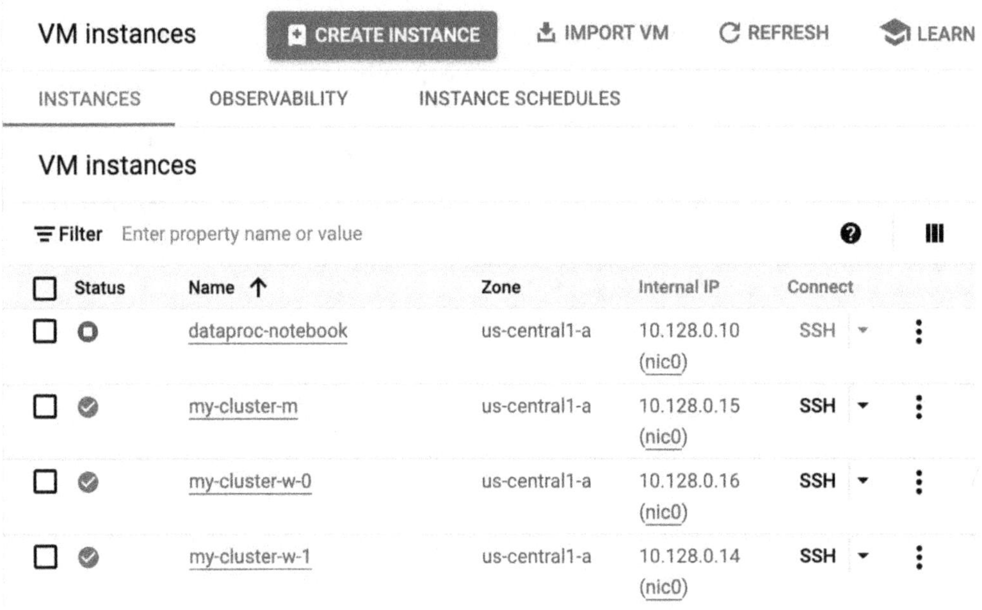

Status	Name	Zone	Internal IP	Connect
○	dataproc-notebook	us-central1-a	10.128.0.10 (nic0)	SSH
✓	my-cluster-m	us-central1-a	10.128.0.15 (nic0)	SSH
✓	my-cluster-w-0	us-central1-a	10.128.0.16 (nic0)	SSH
✓	my-cluster-w-1	us-central1-a	10.128.0.14 (nic0)	SSH

Figure 16-7. List of all VM instances in Google Cloud

Please wait for a few minutes for the instance to be created. Once it is created, Google will provide an SSH-in-browser session. Here is how that may look for you:

```
Welcome to Ubuntu 22.04.3 LTS (GNU/Linux 6.5.0-1013-gcp x86_64)

 * Documentation:  https://help.ubuntu.com
 * Management:     https://landscape.canonical.com
 * Support:        https://ubuntu.com/pro

  System information as of Tue Feb 13 09:02:18 UTC 2024

  System load:  0.0               Processes:             100
  Usage of /:   21.7% of 9.51GB   Users logged in:       1
  Memory usage: 7%                IPv4 address for ens4: 10.138.0.2
  Swap usage:   0%


Expanded Security Maintenance for Applications is not enabled.

6 updates can be applied immediately.
To see these additional updates run: apt list --upgradable

Enable ESM Apps to receive additional future security updates.
See https://ubuntu.com/esm or run: sudo pro status



The programs included with the Ubuntu system are free software;
the exact distribution terms for each program are described in the
individual files in /usr/share/doc/*/copyright.

Ubuntu comes with ABSOLUTELY NO WARRANTY, to the extent permitted by
applicable law.

dev_instance@instance-20240213-083922:~$
```

Figure 16-8. *Accessing your VM instance through SSH*

Cloud SQL

Cloud SQL is a fully managed relational database service offered by Google. Cloud SQL is completely managed by the provider, Google, meaning that it does not require system maintenance, administration, and operational support. Here is how this may look for you:

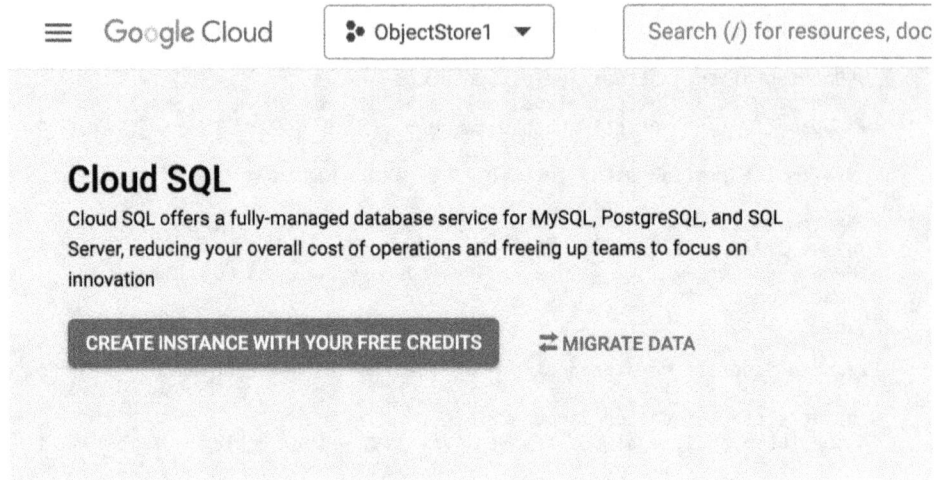

Figure 16-9. *Home page of Google Cloud SQL*

Google automatically performs backups, applies patches to databases, and scales up the storage capacity as required. The criteria and settings among other things need to be defined though.

Google provides the option of choosing the database driver for your database instance. The current options are Postgres, MySQL, and MS SQL Server. When you create a new instance of Cloud SQL with your choice of a database driver, Google will provision the instance in a single zone by default. Let us visit the Google console, select the Cloud SQL service, and create a new database.

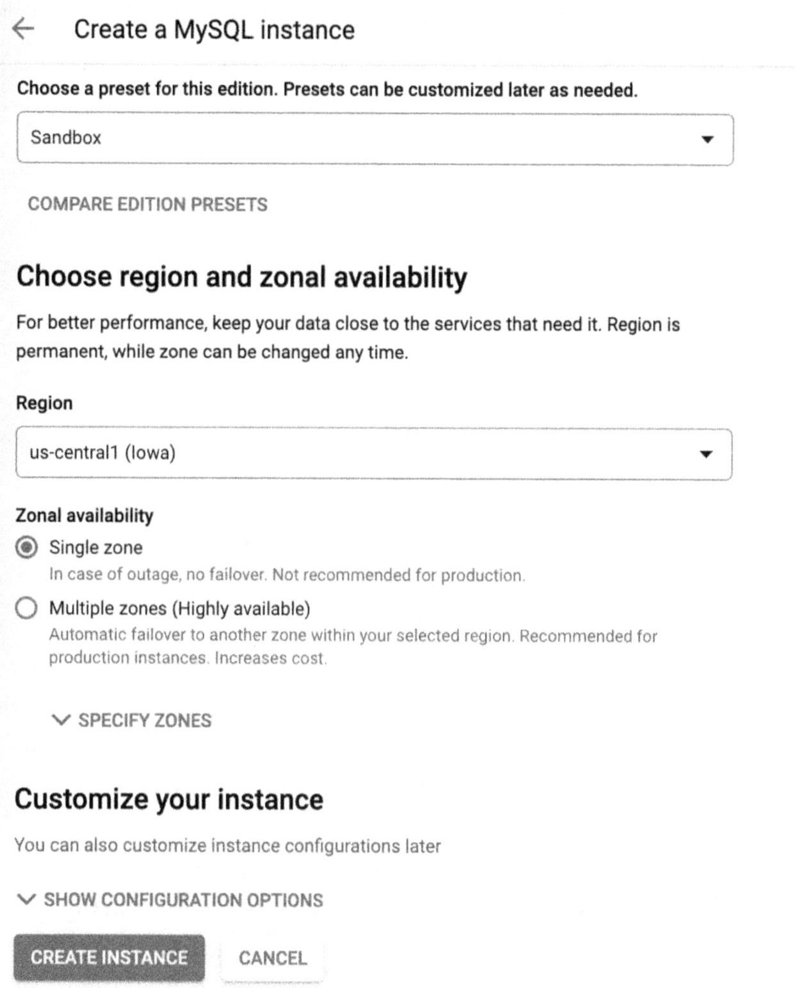

Create a MySQL instance

Choose a preset for this edition. Presets can be customized later as needed.

[Sandbox ▼]

COMPARE EDITION PRESETS

Choose region and zonal availability

For better performance, keep your data close to the services that need it. Region is permanent, while zone can be changed any time.

Region

[us-central1 (Iowa) ▼]

Zonal availability

◉ Single zone
In case of outage, no failover. Not recommended for production.

○ Multiple zones (Highly available)
Automatic failover to another zone within your selected region. Recommended for production instances. Increases cost.

∨ SPECIFY ZONES

Customize your instance

You can also customize instance configurations later

∨ SHOW CONFIGURATION OPTIONS

[**CREATE INSTANCE**] [CANCEL]

Figure 16-10. *Creating a new database instance in Google Cloud SQL*

As we discussed earlier, Cloud SQL will provide you an option to choose the database engine. Once you have chosen the database engine, proceed to provide the compute instance name, a password for the root user, a sandbox for the preset, a region close to your location, and a single zone, and choose to create the instance.

CHAPTER 16   ENGINEERING DATA PIPELINES USING GOOGLE CLOUD PLATFORM

Please wait a few minutes for the instance to be created. When it is created, here is how the screen would look:

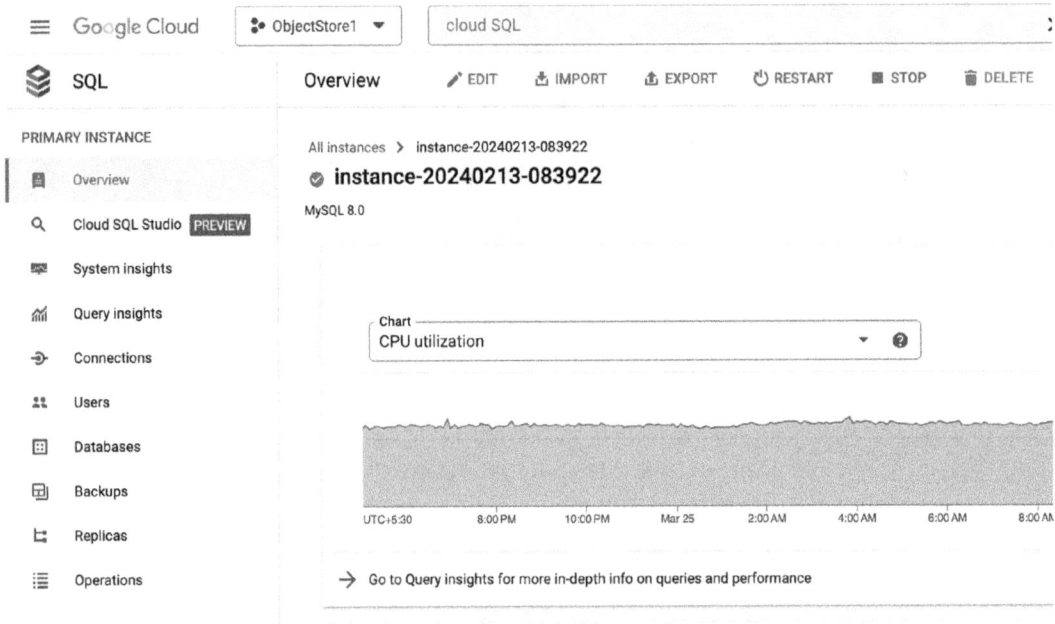

Figure 16-11. *A live database instance in Google Cloud SQL*

Once you have provisioned the Cloud SQL service, you can proceed to create a database and enable shell access to your Cloud SQL instance among other things. Let us create a database here. Go to your instance, choose Databases in the left menu, and click "Create a database." Provide a database name, leave the character set to UTF-8, and click Create.

Here is a simple application to load data programmatically from your local workstation to your newly created Cloud SQL database:

```
from google.cloud.sql.connector import Connector
import sqlalchemy
import csv
import configparser

config = configparser.ConfigParser()

path = '/Users/<your-folder-path-here>/credentials.cfg'
```

```python
config.read(path)

# initialize Connector object
connector = Connector()

# function to return the database connection
def getconn():
    conn= connector.connect(
        config.get('gcp','instance_connection_name'),
        "pymysql",
        user=config.get('gcp','database_user'),
        password=config.get('gcp','database_password'),
        db=config.get('gcp','database')
    )
    return conn

pool = sqlalchemy.create_engine(
    "mysql+pymysql://",
    creator=getconn,
)

filepath = "/Users/<your-folder-path-here>/gcp/polars_dataset.csv"

with pool.connect() as db_conn:

    db_conn.execute(' \
                    CREATE TABLE onedata( \
                        firstname varchar(200), \
                        lastname varchar(200), \
                        gender varchar(10), \
                        birthdate varchar(100), \
                        type varchar(20), \
                        state varchar(100), \
                        occupation varchar(100) \
                        ) \
                    ')
```

CHAPTER 16 ENGINEERING DATA PIPELINES USING GOOGLE CLOUD PLATFORM

```python
    with open(filepath, mode='r',encoding='utf-8') as csv_file:
        csv_reader = csv.DictReader(csv_file)

        for row in csv_reader:
            insert_query = 'INSERT INTO onedata ( \
                                    firstname, \
                                    lastname, \
                                    gender, \
                                    birthdate, \
                                    type, \
                                    state, \
                                    occupation \
                                ) \
                            VALUES (%s, %s, %s, %s, %s, %s, %s)'

            # Tuple of values from the CSV row
            data_tuple = (row['firstname'],
                          row['lastname'],
                          row['gender'],
                          row['birthdate'],
                          row['type'],
                          row['state'],
                          row['occupation']
                         )

            # Execute the query
            db_conn.execute(insert_query, data_tuple)

    db_conn.commit()

    result = db_conn.execute("SELECT * from onedata").fetchall()

    for row in result:
        print(row)

    #db_conn.execute("DROP TABLE onedata")
```

CHAPTER 16   ENGINEERING DATA PIPELINES USING GOOGLE CLOUD PLATFORM

In this preceding application, we connect to Google Cloud SQL, execute a DDL script creating a table, iteratively read a flat file and insert the rows into Cloud SQL, and commit the transaction. Here is how the data looks in Cloud SQL Studio:

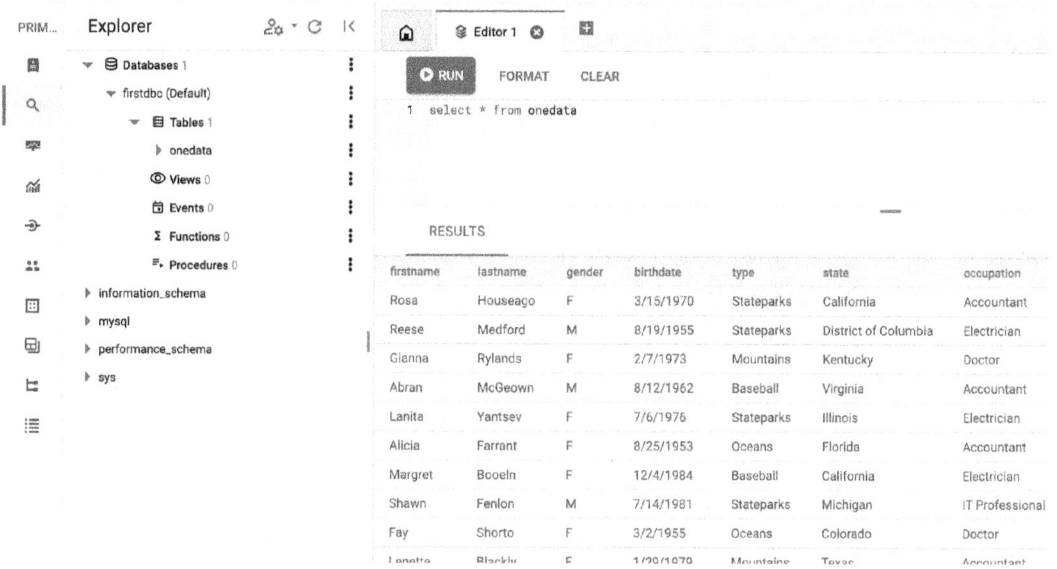

Figure 16-12. Creating and loading a table in Google Cloud SQL

Google Bigtable

Bigtable is a NoSQL offering of Google Cloud. This NoSQL data store is designed for high-velocity incoming data and low-latency database application. Bigtable is used in applications where data is required rapidly, applications like stock trading, real-time gaming, and streaming media platforms. Bigtable is a denormalized database; there is no support for joins between tables.

Note A low-latency database is a type of database management system that provides very fast response times, say in the lines of milliseconds.

Google Bigtable is a key-value store, with no support for secondary indexes (other than the key), and columns need not be in the same order. You can still store the data in tables, and Bigtable can have several tables within a given instance. In the absence of a schema, you can attempt to store data of similar structure (or schema) in the same table. Moreover, all operations are atomic at the row level. I believe that when an operation writes multiple rows to Bigtable and it fails, then it may be possible that some rows may have been written and some may not have been written at all.

When designing Bigtable row keys, it is important to elucidate how the data will be queried and how one would retrieve data. You can incorporate multiple identifiers within the row key that are separated by delimiters. Rows are sorted lexicographically by row keys in Bigtable, and so it may be beneficial to design a row key that starts off with a common value and completes with the most granular value.

Google Bigtable defines a table as a logical organization of values indexed by row key. Let us create a Bigtable instance. Navigate to the GCP console and type in "Bigtable." Once you are able to locate the service, choose to create an instance.

You can also install the Google Bigtable client for Python in your workstation. Here is how you can do that, as per the documentation (https://cloud.google.com/python/docs/reference/bigtable/latest):

```
pip install google-cloud-bigtable
```

You always create a Bigtable instance. An instance typically has one or more clusters, located in different time zones. A given cluster has one or more compute nodes. These compute nodes are responsible for managing your data and performing other tasks. To create an instance, you have to specify the storage, where you have the choice of hard disk or SSD.

Here is how this may look for you:

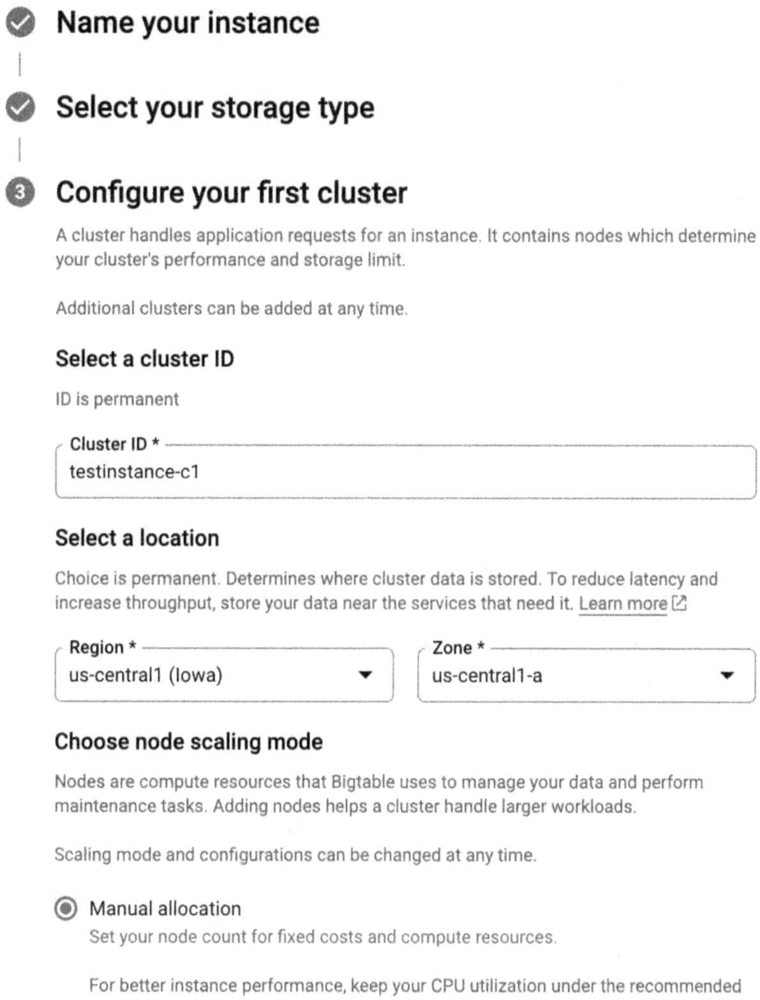

CHAPTER 16 ENGINEERING DATA PIPELINES USING GOOGLE CLOUD PLATFORM

✓ **Name your instance**

✓ **Select your storage type**

③ **Configure your first cluster**

A cluster handles application requests for an instance. It contains nodes which determine your cluster's performance and storage limit.

Additional clusters can be added at any time.

Select a cluster ID

ID is permanent

Cluster ID *
testinstance-c1

Select a location

Choice is permanent. Determines where cluster data is stored. To reduce latency and increase throughput, store your data near the services that need it. Learn more

Region *
us-central1 (Iowa)

Zone *
us-central1-a

Choose node scaling mode

Nodes are compute resources that Bigtable uses to manage your data and perform maintenance tasks. Adding nodes helps a cluster handle larger workloads.

Scaling mode and configurations can be changed at any time.

◉ Manual allocation
  Set your node count for fixed costs and compute resources.

  For better instance performance, keep your CPU utilization under the recommended

Figure 16-13. *Creating a new Google Bigtable instance*

Let us try to upload JSON documents into Google Bigtable.

Here is how the JSON document structure looks like:

```
{
    "id": 2,
    "first_name": "Rube",
    "last_name": "Mackett",
    "email": "rmackett1@japanpost.jp",
    "gender": "Male",
    "ip_address": "153.211.80.48"
}
```

CHAPTER 16 ENGINEERING DATA PIPELINES USING GOOGLE CLOUD PLATFORM

Here is an example application of loading JSON to Google Bigtable:

```python
from google.cloud import bigtable
from google.cloud.bigtable import column_family
import json
import configparser

config = configparser.ConfigParser()

path = '/Users/pk/Documents/de/gcp/credentials.cfg'

# Configuration
project_id = config.get('bigtable','project_id')
instance_id = config.get('bigtable','instance_id')
table_id = config.get('bigtable','table_id')
json_file_path = "/Users/pk/Downloads/randomdata.json"

# Authenticate and connect to Bigtable
client = bigtable.Client(project=project_id, admin=True)
instance = client.instance(instance_id)
table = instance.table(table_id)

max_versions_rule = column_family.MaxVersionsGCRule(2)
column_family_id = "cf1"
column_families = {column_family_id: max_versions_rule}

if not table.exists():
    table.create(column_families=column_families)

column_family_id = "cf1"

with open(json_file_path, 'r') as f:
    records = json.load(f)

for record in records:
    row_key = "row-key-{}".format(record["id"]).encode()
    row = table.direct_row(row_key)
```

CHAPTER 16   ENGINEERING DATA PIPELINES USING GOOGLE CLOUD PLATFORM

```
    for key, value in record.items():
        row.set_cell(column_family_id,
                     key,
                     str(value).encode(),
                     timestamp=None)

    # commmit
    row.commit()

print("data loaded to Bigtable successfully.")
```

You may observe the output at Bigtable Studio by running a simple query. Here is how that may look for you:

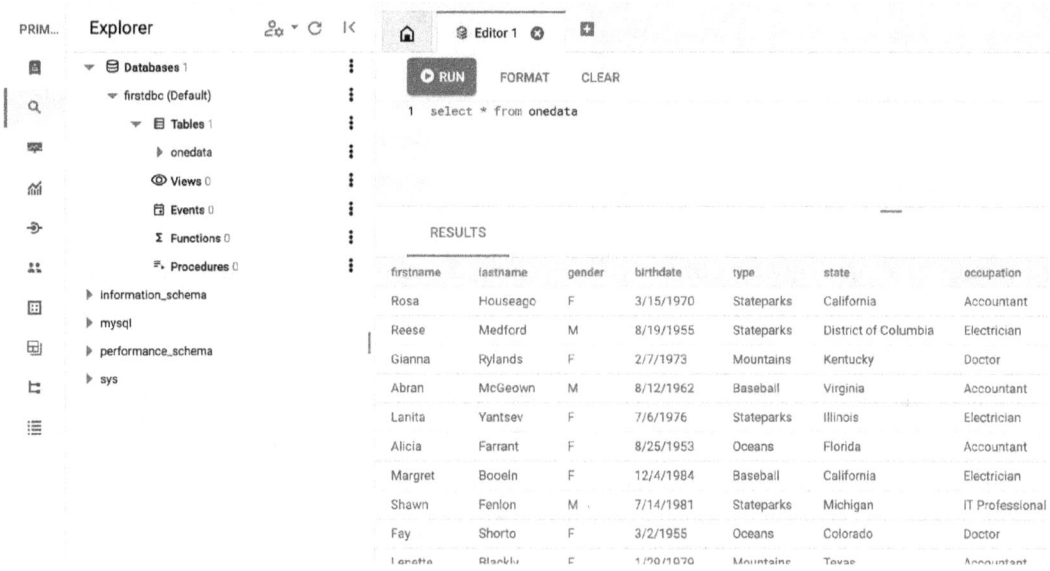

Figure 16-14. *Loading a JSON dataset in Google Bigtable*

Here is how a sample credential file may look like:

```
[bigtable]
    project_id = "value-here"
    instance_id = "value-here"
    table_id = "value-here"
```

551

Google BigQuery

Google BigQuery is a fully managed analytical data warehousing solution by Google. As it is a fully managed service, there is no need to set up, provision, or maintain the underlying infrastructure. The storage for BigQuery is automatically provisioned and also automatically scaled when there is more data coming in. BigQuery supports high-throughput streaming ingestion in addition to high-throughput reads. In Google BigQuery, the storage and compute are separated. Both the storage and compute can be scaled independently, on demand.

Google BigQuery is a columnar store, where data is automatically replicated across multiple availability zones to prevent data loss from machine failure or zone failure. In BigQuery, data is stored mostly in tabular format. Tabular data includes standard data, table clones, materialized views, and snapshots. The standard tabular data is a table data that has a schema, and each column has an assigned data type. The table clones are lightweight copies of standard tables, where the delta of the table clone and base table is stored (and so it is lightweight).

Table snapshots are point-in-time copies of tables, where you have the option to restore a given table from a given point in time. The last one is materialized views, where the results of a query are precomputed and cached and updated periodically. In Google BigQuery, the cached results of a query are stored in temporary tables. Google does not charge for cached query results.

Google BigQuery lets you manage the security and quality of data throughout its lifecycle through access control, data stewardship, and data quality modules. You can manage and control the access of BigQuery users and the level of access they have with respect to tables, views, snapshots, and so on. There are also provisions to control and manage access to certain rows and columns, enabling fine-grained access to sensitive data. You have the choice of obtaining the detailed user profiles and system activity by obtaining audit logs.

In addition, you can safeguard personally identifiable information in the BigQuery tables and views, by applying data masking to the PII fields. This will safeguard the datasets against accidental data disclosure. BigQuery automatically encrypts all data, both during the transit and at rest. You also have the option to check the data quality by running statistical metrics (mean, median, unique values, etc.) on your dataset. You can also validate your data against predefined rules and check data quality and troubleshoot data issues.

To get started using Google BigQuery, we need to locate the service within Google Cloud and enable it. Here is how that process may look:

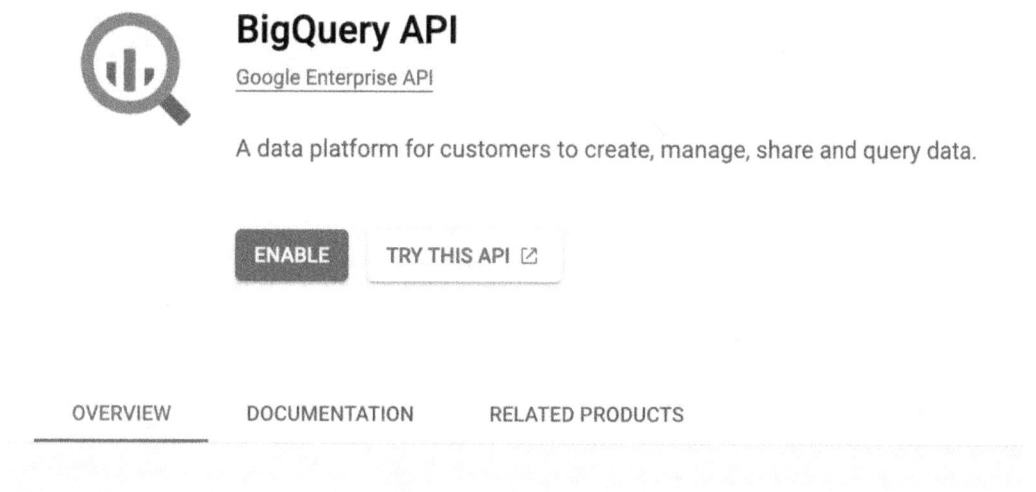

Figure 16-15. *Enabling the Google BigQuery API*

Once you click ENABLE, you may wait a few more minutes till you get automatically redirected to Google BigQuery. There is no need to provision any infrastructure, compute, or even storage. When you start using BigQuery, depending upon your usage, you will be billed for storage and compute.

Here is how that may look for you:

CHAPTER 16 ENGINEERING DATA PIPELINES USING GOOGLE CLOUD PLATFORM

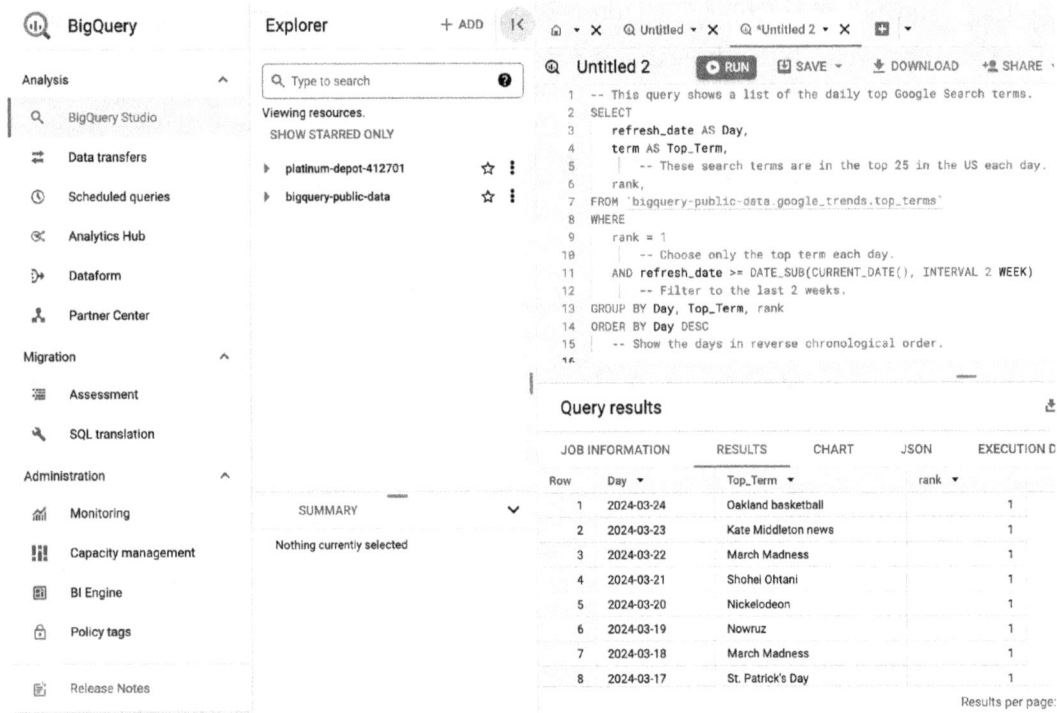

Figure 16-16. Main console of Google BigQuery

Google Dataproc

Google Dataproc is a fully managed service that provides you a Hadoop cluster or a Spark cluster for you to run big data processing in a distributed environment to take advantage of big data tools. The Dataproc service also integrates well with various other Google Cloud services like BigQuery and others, enabling you to set up a complete data platform as well.

To get started with Google Dataproc, enable the API from the cloud console:

CHAPTER 16 ENGINEERING DATA PIPELINES USING GOOGLE CLOUD PLATFORM

Product details

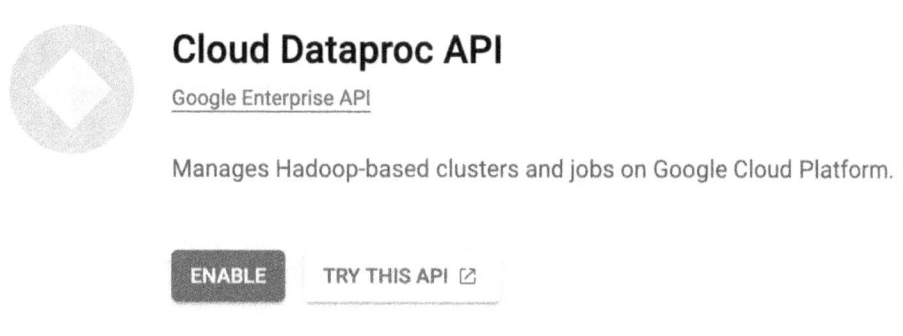

Figure 16-17. Enabling the Cloud Dataproc API

Google Dataproc's Spark cluster is a serverless model, which means you pay only for your consumption. The Hadoop cluster is also relatively less expensive. Google Dataproc takes very little time to set up a new cluster. It also has an autoscaling feature, which would scale up or down depending upon the memory usage.

There are three types of clusters you can set up using Google Dataproc. They are single-node cluster, standard cluster, and high-availability cluster. The single-node cluster consists of one master and no worker nodes. The standard cluster consists of one master and "n" workers, whereas the high-availability cluster consists of three master nodes and "n" worker nodes. For a single-node cluster, the autoscaling feature is not available.

Let us create a simple cluster in Google Dataproc with Compute Engine. To get started, let us click Create a cluster and choose the Compute Engine option:

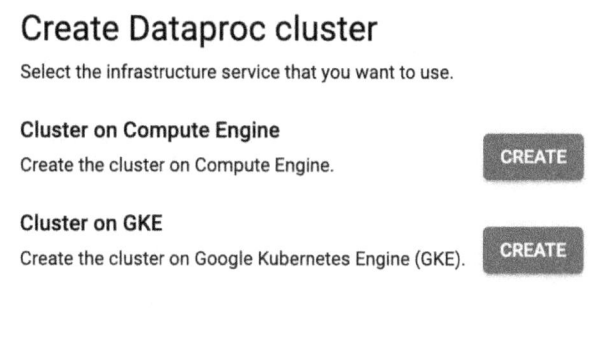

Figure 16-18. Selecting the infrastructure for the Dataproc cluster

555

CHAPTER 16　ENGINEERING DATA PIPELINES USING GOOGLE CLOUD PLATFORM

In our illustration, we will create a single-node cluster with 1 master and 0 workers. Let us provide a cluster name, region, and type of cluster and choose a disk image for the compute engine. Here is how that looks:

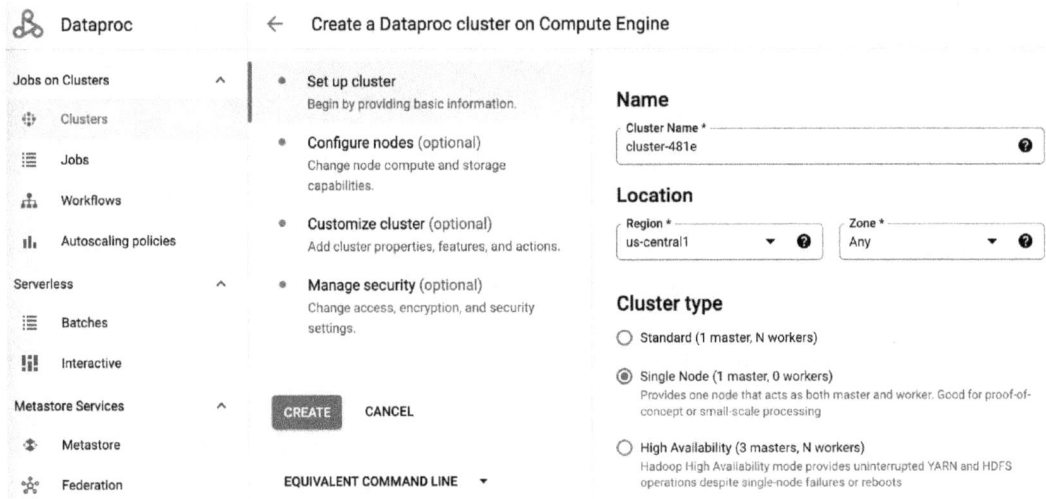

Figure 16-19. Creating a new Dataproc cluster

As you scroll down, make sure check the Enable component gateway checkbox and also the Jupyter Notebook option. You may wish to choose other options depending upon your needs. Here is how this may look:

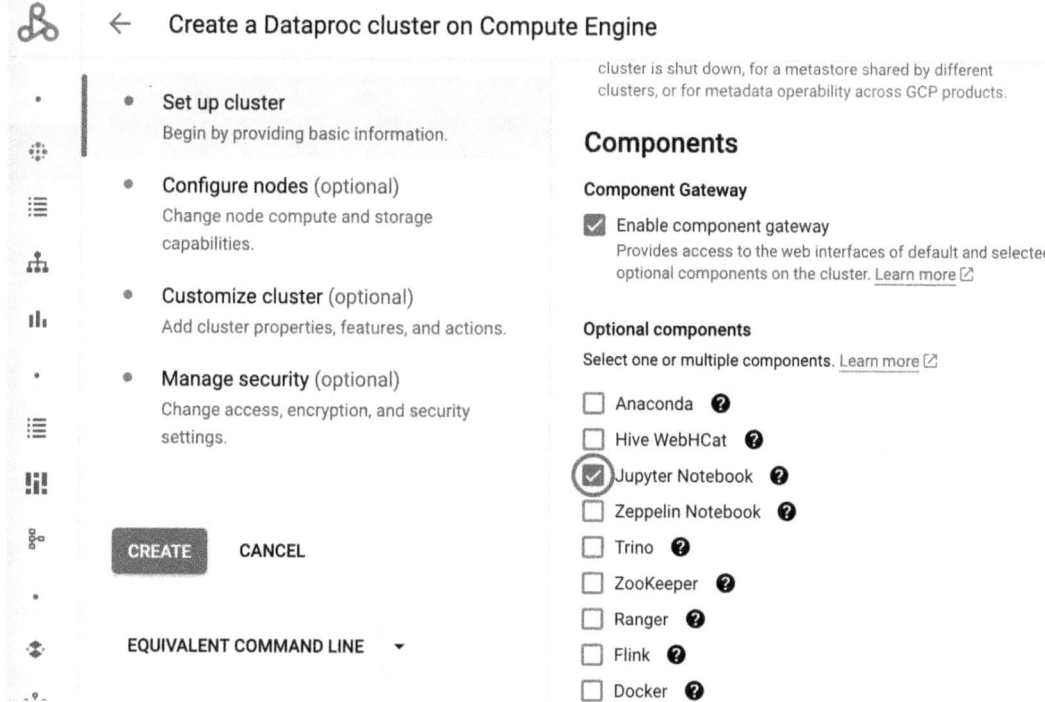

Figure 16-20. *Specifying the components for the new Dataproc cluster*

Now, click Configure nodes, where you have the option to specify the type of compute you wish to use. If you are creating a cluster for a development environment, then you may be able to work comfortably with a less powerful compute.

In that case, you may wish to choose a general-purpose compute and select the "N1 series" option if you see it. You may also select similar options for the worker node as well.

CHAPTER 16   ENGINEERING DATA PIPELINES USING GOOGLE CLOUD PLATFORM

Here is how that may look for you:

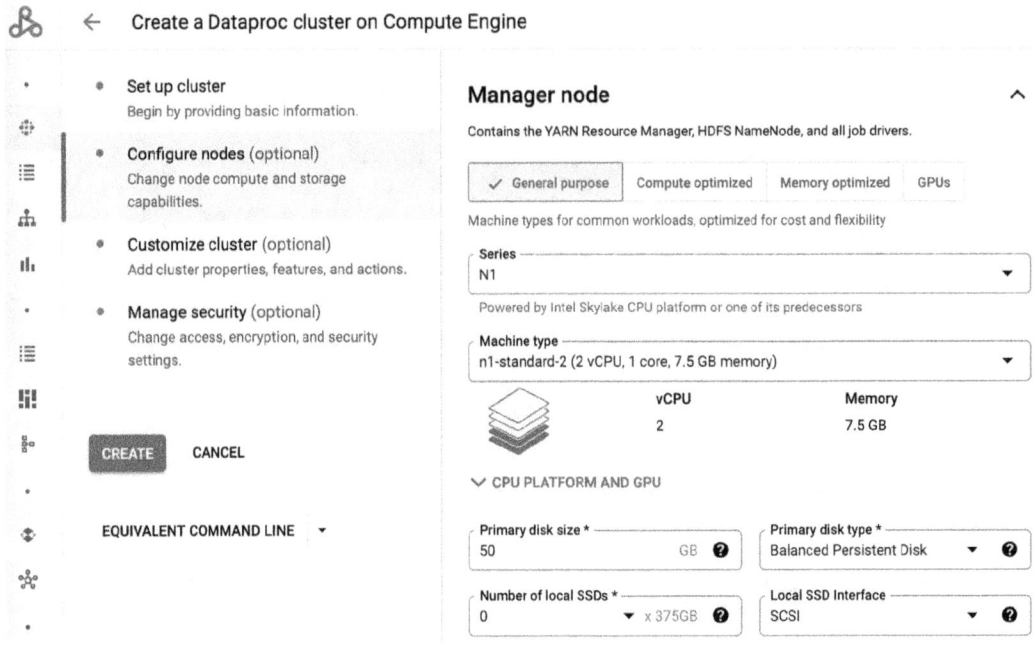

Figure 16-21. *Configure the node for the Dataproc cluster*

You can also create a cluster from your Google Cloud command line interface. Here is the command for that:

```
gcloud dataproc clusters create my-cluster \
--enable-component-gateway\
--region us-central1\
--no-address\
--master-machine-type n1-standard-2\
--master-boot-disk-type pd-balanced\
--master-boot-disk-size 50\
--num-workers 2\
--worker-machine-type n1-standard-2\
--worker-boot-disk-type pd-balanced\
--worker-boot-disk-size 50\
--image-version 2.2-debian12\
--optional-components JUPYTER\
--project {your-project-id}
```

When the cluster is created, you may be able to see it in the dashboard. Here is how this may look for you:

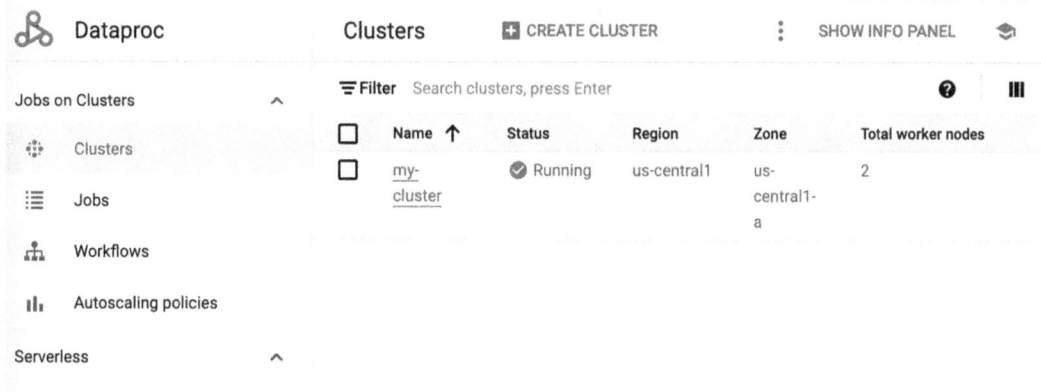

Figure 16-22. Dashboard of active clusters in Dataproc

Let us create a Spark data pipeline and use Cloud Dataproc to execute the data pipeline. We will use Vertex AI Workbench service to create a Jupyter notebook.

Google Vertex AI Workbench

Google Vertex AI Workbench service is a fully managed Jupyter Notebook–based development environment. Google Vertex AI also integrates with GitHub. Let us navigate to the GCP console and enable the Vertex AI API and Notebook API, respectively. Then navigate to Google Vertex AI Workbench and create a notebook. You will see the option of creating a PySpark instance on the Dataproc cluster. Here is how that may look:

CHAPTER 16 ENGINEERING DATA PIPELINES USING GOOGLE CLOUD PLATFORM

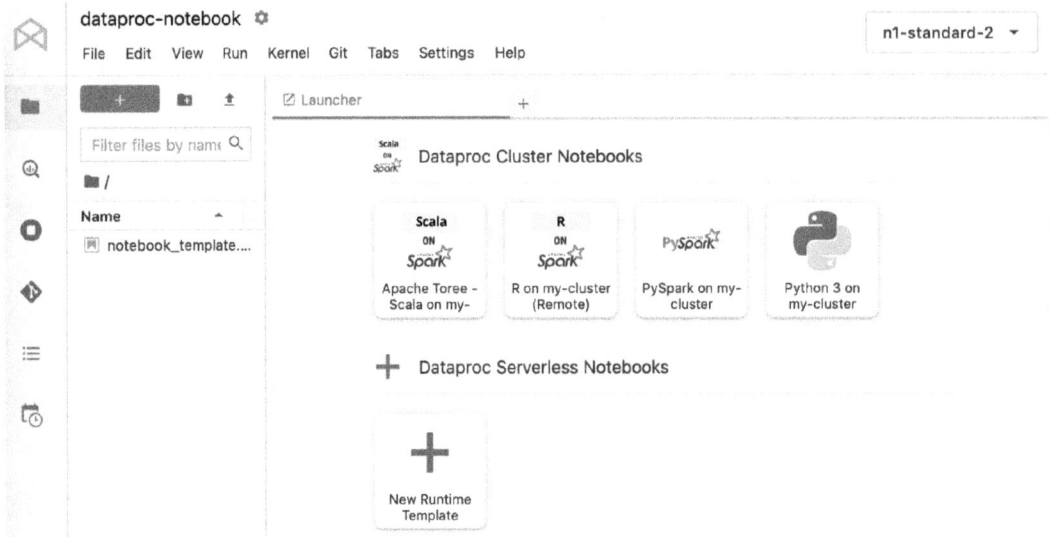

Figure 16-23. *Dashboard of Google Vertex AI Workbench*

Vertex AI Workbench integrates well with Cloud Storage and BigQuery services. It comes with various machine learning libraries preinstalled and supports both PyTorch and TensorFlow frameworks. You have the option of using CPU-only instances or GPU-only instances.

Here is a sample code for you that you can run on Google Cloud Storage data:

```
from pyspark.sql import SparkSession

bucket = "your-bucket"
filepath = "gs://your-bucket/polars_dataset.csv"

spark = SparkSession.builder\
    .appName("PySpark-my-cluster")\
    .config("spark.jars", "gs://spark-lib/bigquery/spark-bigquery-latest.jar")\
    .getOrCreate()

df = spark.read.csv(filepath, header=True, inferSchema=True)
df.show(3)

occ_report = df.crosstab(col1="state",col2="occupation")
occ_report.show(5)
```

CHAPTER 16 ENGINEERING DATA PIPELINES USING GOOGLE CLOUD PLATFORM

Note The "spark.jars" file is used with Spark to read and write data with Google BigQuery. You may obtain the appropriate Spark jars file at Google Dataproc service. Visit the dashboard and navigate to the "Submit a job" page within Dataproc. Provide the appropriate information and click Submit to obtain the jars file.

Here is how this code can be executed on the Jupyter notebook:

```
[9]: from pyspark.sql import SparkSession

[10]: bucket = "onto-bigquery"
      filepath = "gs://onto-bigquery/polars_dataset.csv"

[11]: spark = SparkSession.builder\
          .appName("PySpark-my-cluster")\
          .config("spark.jars", "gs://spark-lib/bigquery/spark-bigquery-latest.jar")\
          .getOrCreate()

[12]: df = spark.read.csv(filepath, header=True, inferSchema=True)
      df.show(3)
+---------+--------+------+---------+----------+--------------------+-----------+
|firstname|lastname|gender|birthdate|      type|               state| occupation|
+---------+--------+------+---------+----------+--------------------+-----------+
|     Rosa|Houseago|     F|3/15/1970|Stateparks|          California| Accountant|
|    Reese| Medford|     M|8/19/1955|Stateparks|District of Columbia|Electrician|
|   Gianna| Rylands|     F| 2/7/1973| Mountains|            Kentucky|     Doctor|
+---------+--------+------+---------+----------+--------------------+-----------+
only showing top 3 rows

[14]: occ_report = df.crosstab(col1="state",col2="occupation")
      occ_report.show(5)
[Stage 36:>                                                        (0 + 1) / 1]
+--------------------+----------+------+-----------+---------------+------+
|    state_occupation|Accountant|Doctor|Electrician|IT Professional|Lawyer|
+--------------------+----------+------+-----------+---------------+------+
|           Minnesota|         1|     2|          3|              4|     1|
|            Nebraska|         2|     3|          2|              1|     2|
|            Oklahoma|         3|     4|          6|              5|     4|
|District of Columbia|         7|     6|          8|              5|    11|
|            Missouri|         1|     6|          2|              5|     5|
```

Figure 16-24. Running PySpark code on Google Vertex AI Workbench

Google Vertex AI

Google's Vertex AI is a fully managed machine learning platform that lets you train and deploy machine learning models. Once you train a model, you can deploy it on Vertex, enabling the model consumers to receive real-time intelligence and insights aimed at generating value to the business. This machine learning platform does have so many features and is highly comparable to other leading machine learning platforms offered by AWS and Microsoft.

Let us build a custom machine learning model from scratch, and this time we will use the AutoML feature. AutoML stands for automated machine learning and is the process of automating machine learning models, including preprocessing, feature engineering, model selection, and tuning the hyperparameters.

Google Vertex AI offers two development environments for building machine learning models; they are Vertex AI Workbench and Colab notebooks. We have already seen Vertex AI Workbench in the previous section and utilized the notebook to build machine learning models. These workbench environments are provided through virtual machines, and they are customizable with support for GPU as well.

Google Colab notebooks are a managed notebook environment that lets you collaborate with others, while Google manages the underlying infrastructure for you. You have the option of customizing the underlying compute and configuring runtimes. The upload size for a tabular file cannot exceed 20 megabytes in Google Colab, whereas in the case of Vertex AI Workbench, you can work with a tabular dataset with size up to 100 megabytes.

Let us look at building a machine learning model and host the model as an endpoint using Google Vertex AI. We are using the iris flowers dataset for a multi-class classification problem. Here is the machine learning code for your reference. I encourage you to use this code as a starter and experiment with other classifiers:

```
import pandas as pd
import pandas as pd
from xgboost import XGBClassifier
from sklearn.model_selection import train_test_split
from sklearn.metrics import accuracy_score
from sklearn.preprocessing import LabelEncoder
import pickle
```

```python
training_dataset = pd.read_csv('iris.csv')
dataset = training_dataset.values
# split data into X and y
X = dataset[:,0:4]
Y = dataset[:,4]

label_encoder = LabelEncoder()
label_encoder = label_encoder.fit(Y)
label_encoded_y = label_encoder.transform(Y)
seed = 7
test_size = 0.33
X_train, X_test, y_train, y_test = train_test_split(
                                    X,
                                    label_encoded_y,
                                    test_size=test_size,
                                    random_state=seed
                                   )
xgmodel = XGBClassifier()

xgmodel.fit(
     X_train,
     y_train
   )

print(xgmodel)

# make predictions for test data
predictions = xgmodel.predict(X_test)

# evaluate predictions
accuracy = accuracy_score(
             y_test,
             predictions
           )
print("Accuracy: %.2f%%" % (accuracy * 100.0))

filename = "model.pkl"
pickle.dump(xgmodel, open(filename, 'wb'))
```

Note Pickle files are binary files, used to serialize Python objects like dictionaries, lists, etc. The process of converting a Python object to a byte stream is called pickling or serialization. The inverse also holds. You can deserialize or unpickle a pickled file. They are considered secure as pickle files are Python specific. Pickling is commonly used in machine learning modeling in the Python ecosystem.

Once we have the model available in pickle file format, we can now add this model in Google Vertex AI model registry. To get started, visit Vertex AI from the Google console, navigate to deployment, and click Model Registry.

Here is how that may look for you:

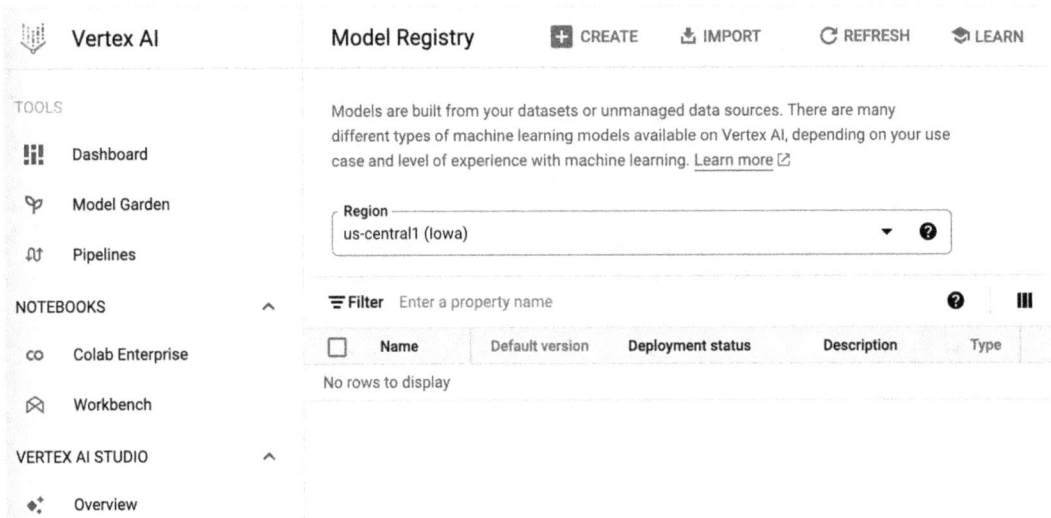

Figure 16-25. Model Registry page of Google Vertex AI

You can either create a new machine learning model interactively or import the pickle file you have generated. In this case, we will import our model as a pickle file. You have the option of choosing the region for your model.

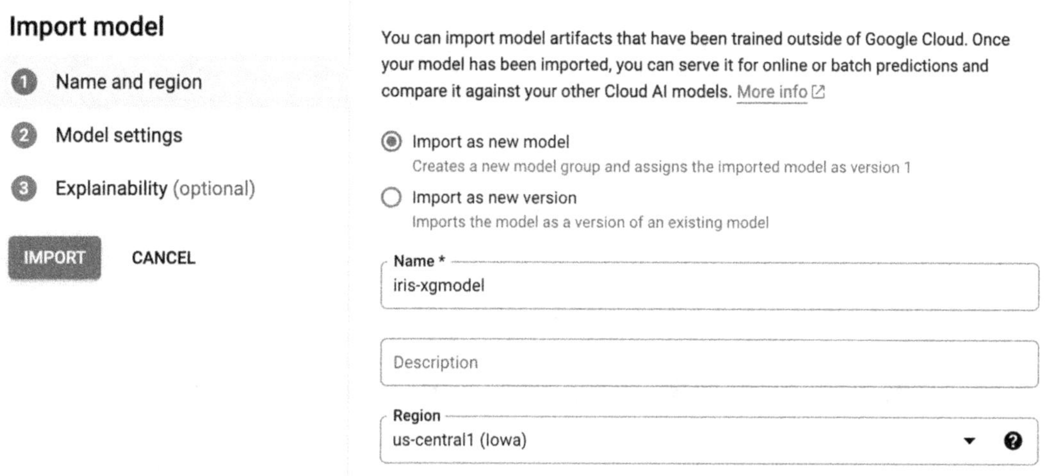

Figure 16-26. Importing the model, naming, and specifying the region

You may specify a name, description, and the region for this model. You can choose to import as a new model. If by any chance you have retrained this same model (with the iris dataset) and generated another model pickle file, then you may import the same as a newer version of the existing model. We have trained our model using the XGBoost model framework. If you have used scikit-learn or TensorFlow as a model framework, then choose the appropriate model framework and model framework version from the drop-down box. And for the model artifact location, provide the cloud storage folder location, either by browsing visually or just by entering the URI.

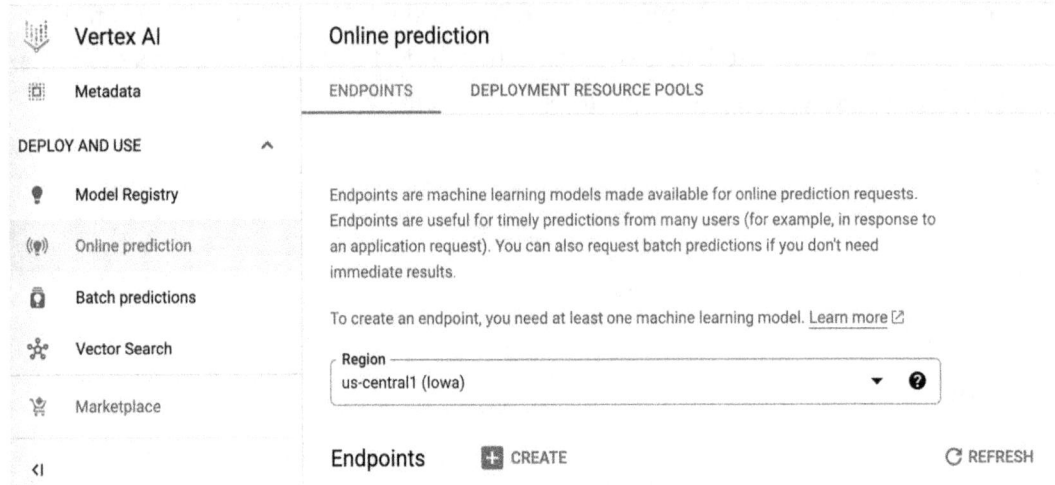

Figure 16-27. Further settings for importing the model

For the model artifact location, please specify only the folder and not the exact model path. The column would auto-populate the "gs://" prefix, and so be mindful of the cloud storage path.

Once you have deployed the machine learning model pickle file in the model registry, then it is time to create an endpoint and expose the same for consumption. Let us navigate to the "DEPLOY AND USE" tab within Vertex AI and click Online prediction. Here is how that may look:

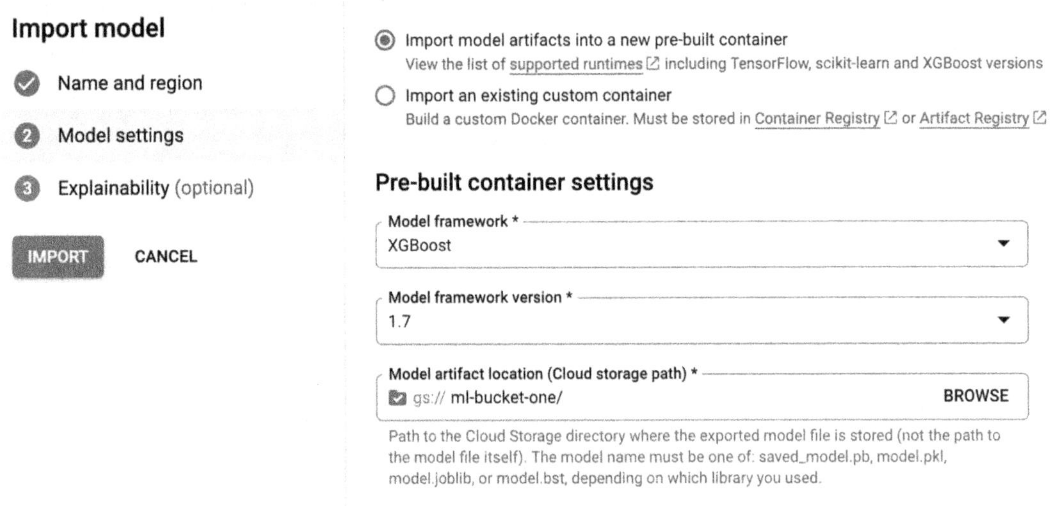

Figure 16-28. Creating an endpoint for our ML model

Chapter 16   Engineering Data Pipelines Using Google Cloud Platform

The concept of online prediction is only when you are looking to consume a machine learning model by submitting a request to an HTTP endpoint. Let us deploy our iris machine learning model as an endpoint. You may choose a region where you wish to host this endpoint and click Create new endpoint.

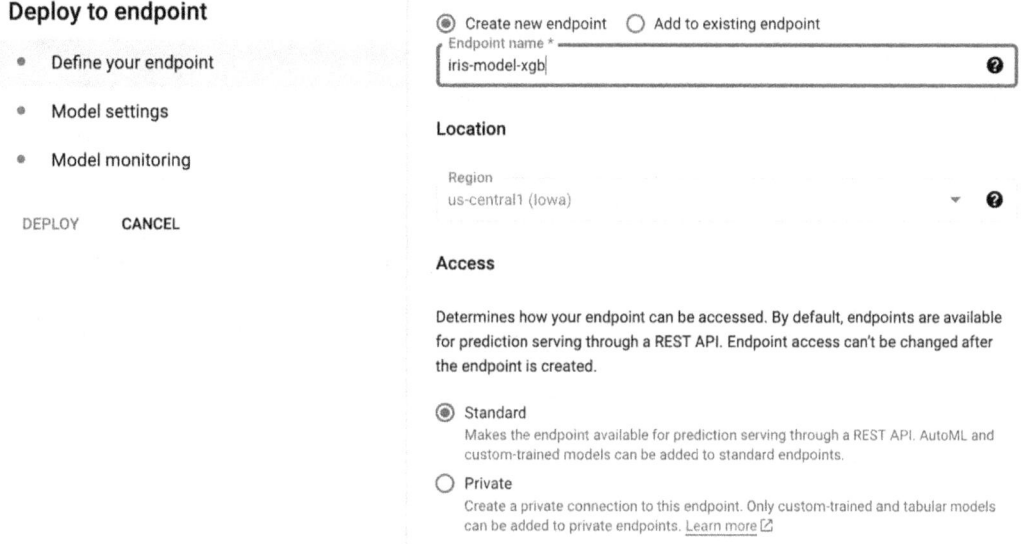

Figure 16-29. *Defining the endpoint and specifying the region*

Once you click to create an endpoint, provide the name for the endpoint and make a note of that name separately. To access the endpoint, we will choose the standard option here.

The traffic split is the model is basically specifying how much of the incoming traffic should be routed to this endpoint. For our illustrative purposes, we can leave the value as is. This parameter can be changed at a later point, if we are hosting multiple endpoints for a machine learning model.

Furthermore, you can specify the number of compute nodes for this one endpoint. Also, you can specify the maximum number of compute nodes you would like to have should there be higher traffic. You can also specify custom machines with specific compute and physical memory; there are also options to choose from specialized machines like CPU intensive, memory intensive, etc.

CHAPTER 16 ENGINEERING DATA PIPELINES USING GOOGLE CLOUD PLATFORM

Here is how that may look:

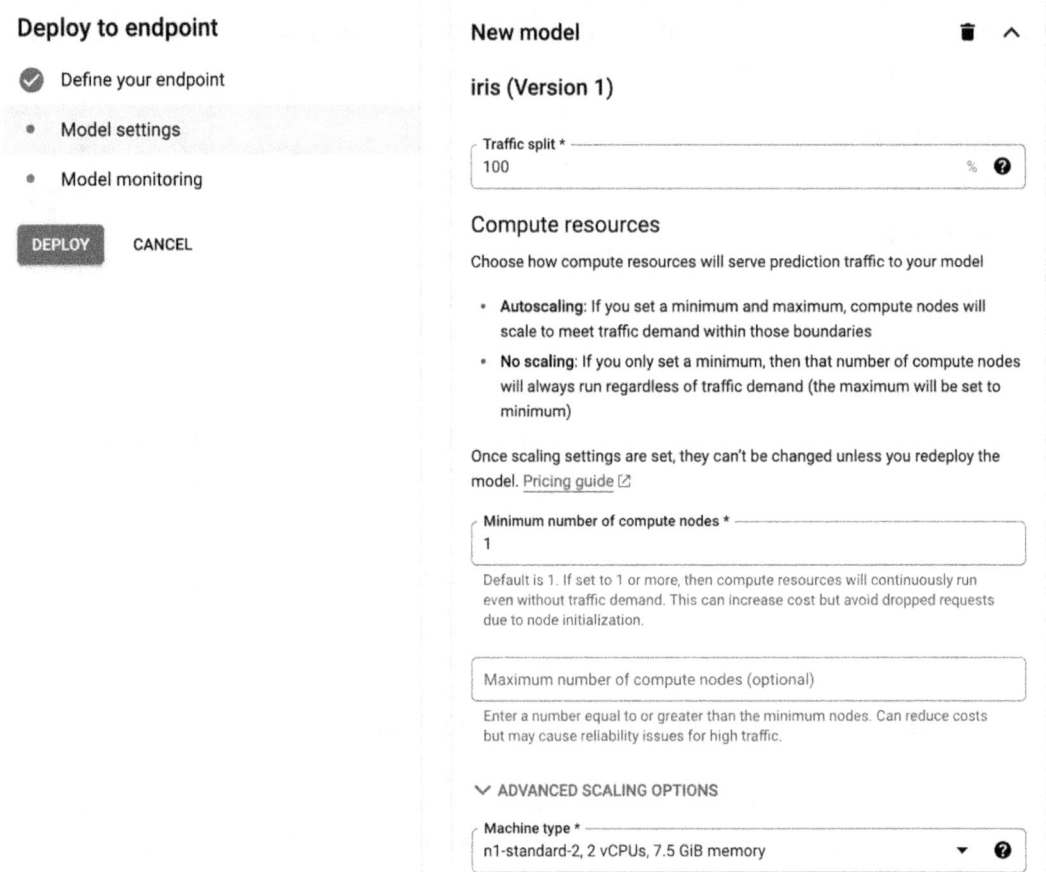

Figure 16-30. Model settings for deploying to an endpoint

For the remaining options, you may leave as is and choose to deploy the model. The process may take several minutes. Once the model is deployed, you may be able to see the entry in your deployment dashboard.

Here is how that may look for you:

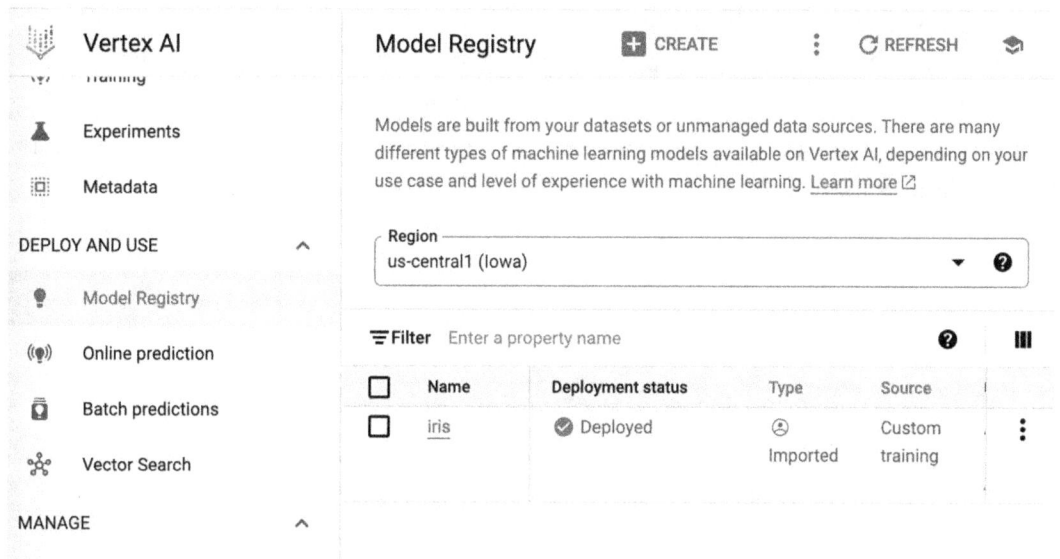

Figure 16-31. *A deployed model that can be seen in the Google Vertex AI dashboard*

You can test the endpoint in various forms. Postman client would be simplest. Here is an illustration of using the Google Cloud command line interface:

```
gcloud ai endpoints predict 2954719796345700352 \
--region us-central1 \
--json-request request.json
```

The request.json file would contain the sample data.

Conclusion

So far, we were able to look at Google Cloud Platform with a focus on data engineering and machine learning projects. We explored a wide range of services from Cloud Storage to BigQuery. GCP offers flexible compute options through the Compute Engine API and managed database services like Bigtable and Cloud SQL. We looked at Dataproc and BigQuery for running distributed computing workloads and also Vertex AI Workbench for developing machine learning models. This was followed by Google Vertex AI, where we looked at building a custom machine learning model, deploying the same to model registry, and exposing it as an endpoint for real-time predictions. As GCP evolves and continues to add more services, they continue to offer managed services that reduce the operational overhead for data and ML engineers, enabling them to build data solutions that are stable and reliable.

CHAPTER 17

Engineering Data Pipelines Using Microsoft Azure

Introduction

Azure cloud computing is a fully managed computing service by Microsoft. The Azure concept was introduced early in 2008 at a conference, and Microsoft launched the Windows Azure service in 2010, which was rebranded as Microsoft Azure. Currently it provides a comprehensive suite of computing, database, analytics, and other services through various deployment modes. Microsoft has data centers all across the globe and serves various industries. In this chapter, we will look at some of Azure's key components and services, with a focus on data engineering and machine learning.

By the end of this chapter, you will learn

- How to set up a Microsoft Azure account and Azure's resource hierarchy
- Microsoft's object storage called Azure Blob Storage and its concepts and setup
- The relational database service called Azure SQL and its concepts and installation
- Cosmos DB, a NoSQL database, and its architecture and concepts

- A fully managed data warehousing environment
- Serverless functions and data integration services
- Azure Machine Learning Studio and its concepts

As you embark on your journey exploring the Microsoft Azure platform, please may I request you to be mindful of managing the cloud resources, especially when you are done using them. In the case of traditional infrastructure, you may switch off a computer when you are done. However, in the case of cloud infrastructure, where you often procure a plan that is based on a pay-as-you-go model, you will be billed for your resources even if they are not being used. Signing off Azure does not mean powering off the services you have procured. This is particularly important for virtual machines, databases, and any other compute-intensive resources. Please, always, make it a habit to review your active resources and shut down these specific resources. If you are no longer planning on using a service, you may delete the same. And so, you are only paying for the services you actively use and keep your costs under control. I also would like to thank Azure for providing free credits for someone who is a new user and having actively supported new learners in bringing their cost down.

Introduction to Azure

Azure is Microsoft's cloud computing platform offering several of the services that are comparable to Google's and Amazon's cloud service offerings. Azure's offerings are particularly attractive to organizations who are heavily invested in Microsoft-based technologies.

Microsoft has had a long-standing reputation in the data management field. Microsoft is one of the early adopters of building ETL tools, which were instrumental in facilitating data movements between various sources. To get started with Microsoft Azure, you need to visit Microsoft's website to sign up for a Microsoft account, before attempting to obtain access to Azure. Once you have a Microsoft account, you may wish to sign up for free Azure services. Like AWS and Google Cloud, you will enter various information to Microsoft to obtain access to Azure services.

Let us get ourselves familiar with the resource hierarchy of Microsoft Azure. It helps to configure and govern the provisioning of services and workload by a given account, especially an organizational account. The highest level is called the management group. The management group governs utilization of various services by a given account.

CHAPTER 17 ENGINEERING DATA PIPELINES USING MICROSOFT AZURE

Management groups enable an organization to apply various policies across all users within a management group and apply access control to one or more Azure service offerings. From the root management group, you can create one or more management groups that may correspond to a business group or a certain type of billing model. Simply search for the term "management group" from your Azure dashboard to get into that section, and you may wish to create one for your organization. You can name your management group as your organization or your department within the organization or a specific project name.

If you visit a service offering page more often, then it may be convenient to pin that page to your dashboard. To perform that, simply click the pin-like icon next to the title of your management group. Here is how that looks:

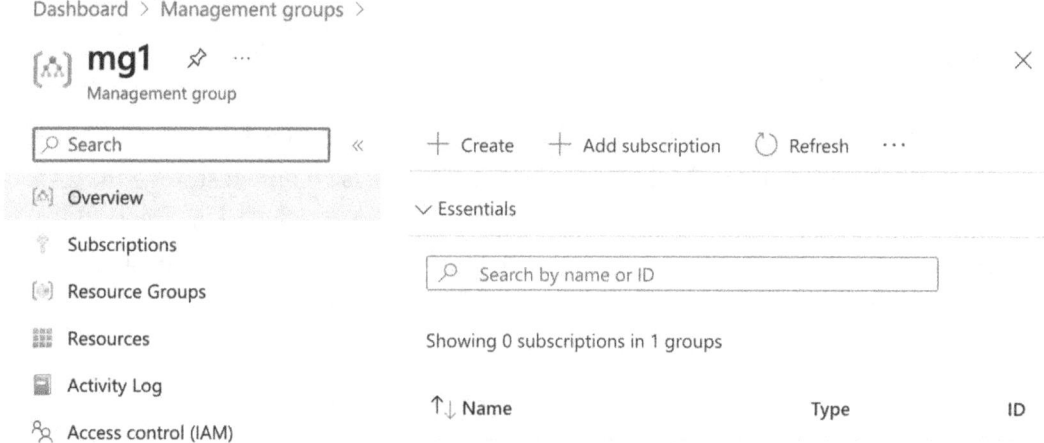

Figure 17-1. *Azure management group dashboard*

Within a given management group, you can have one or more subscriptions. You can see the option of subscriptions in the Management groups section. A subscription is a scheme that enables one or more users to access one or more Azure service offerings. For instance, you can have two subscriptions within a management group for a specific team, namely, development and production. You may allow more resources on the production subscription and enable certain constraints on the development subscription. Another example could be creating a subscription for every product or service that the organization offers so that one may be able to calculate the costs.

Within any given subscription, we have a concept called resource group. This is a project-based grouping of resources or one or more Azure services. For instance, if you are looking to set up a data warehousing project and need databases, data warehouses, data integration services, secrets management, data modeling tool services, and logging and monitoring services, then you can create one resource group that can contain these services. This can be beneficial in the context of governance and setting up access policies.

Finally, the lowest abstraction is called the resources in Azure. These are various services that Azure offers. It can be a database or a tool that you would pay to use. Resources are anything that consist of networking, storage, or compute entities. For instance, a serverless data integration service that uses compute to move data from source to sink is an example of a resource.

From a data products and services point of view, let us explore the storage systems, database services, and data ingestion and integration services among others.

Azure Blob Storage

As described in earlier sections, object storage is a versatile storage system and has been a successful offering of cloud vendors. Azure Blob Storage is an object storage that is a comparable solution to S3 offered by AWS and Bigtable by Google Cloud. Azure Blob Storage is a highly scalable and secured storage platform that can store both structured and unstructured data. Azure Blob Storage provides policy- and role-based access control and data lifecycle management.

Note "Blob" is a database term that expands to binary, large objects. Blob can store any type of data into binary objects. It is perfect for images, audio, video, or PDF content. These blob data are not indexable though. There is also clob, which expands to character large objects, that stores large amounts of text data like documents, large code snippets, etc.

To create a new blob storage, choose Storage from the main dashboard and create a new storage. Provide the name of the instance and choose a closely located region. In addition, choose standard performance and low-cost replication. You may be able to see other options like choosing customer-managed encryption keys among other critically important options. Microsoft will take a moment to run validation to review your choices before enabling the "Create" button. Once you have created it, wait a few moments for deployment of the resource in your account. Once deployed, here is how that looks:

CHAPTER 17 ENGINEERING DATA PIPELINES USING MICROSOFT AZURE

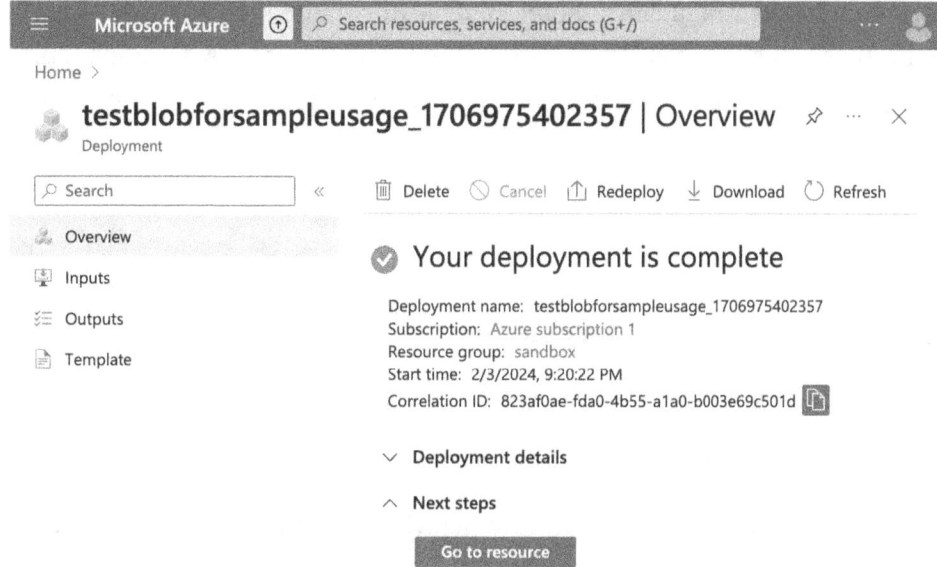

Figure 17-2. *Deployment of a new Azure blob storage*

You now have a storage account configured for your use. Let us now create a container by clicking "Upload" and choosing "Create new."

Here is how that looks:

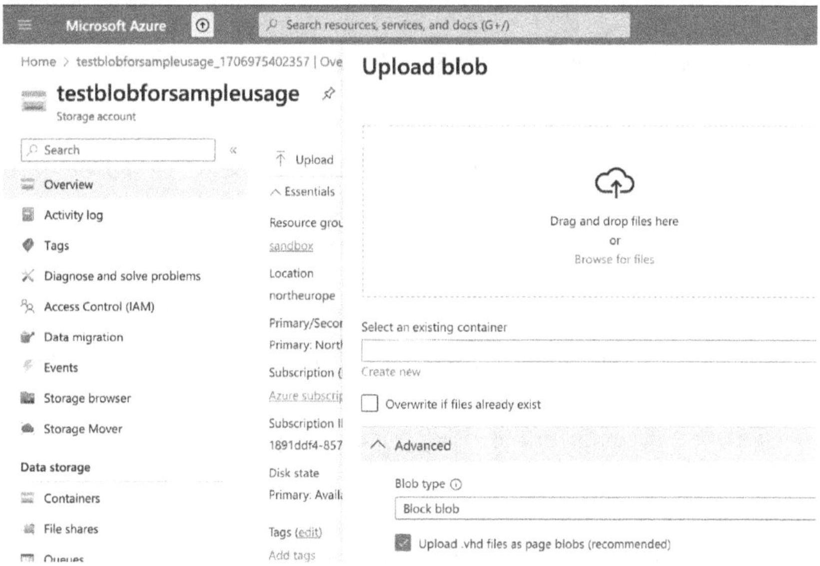

Figure 17-3. *Uploading data in Azure Blob Storage*

575

CHAPTER 17   ENGINEERING DATA PIPELINES USING MICROSOFT AZURE

Let us now programmatically upload a flat file into Azure Blob Storage using Python. To begin with, we need to install the software development kit for Azure Blob Storage for Python programming language. Open a terminal and start a virtual environment session or conda environment session:

```
python3 -m venv azure
```

```
source azure/bin/activate
```

If you are on MS Windows, then please use

```
azure/Scripts/activate
```

```
pip install azure-storage-blob azure-identity
```

In order for us to programmatically interact with the cloud storage service, we need the connection string, container name, and local file path inputs, respectively. We will set the connection path to the file that needs to be uploaded and define the container name and connection string, respectively. Then we will initialize the blob client and the container client, respectively. A container is a logical grouping of objects in the blob storage service. You can enforce container-level access control as well. Once we initially use the blob client, we can upload our file from there on.

Here is how this all may look:

```
from azure.storage.blob import BlobServiceClient
import os

# connection string, container name and path to the file to be uploaded

connection_string = 'DefaultEndpointsProtocol=https;\
                     AccountName=testblobforsampleusage;\
                     AccountKey=YjsfMmriMCRCn5Po/DdzCQAiidxPsfgfsdgs45S
                     FGS$%SGDFFADSFGSR$#$AGDFGHDFGHcpjKd6AINtVWMiO9+AS
                     tc9nKVA==;\
                     EndpointSuffix=core.windows.net'

container_name = "firstcontainer"
local_file_path = "/Users/username/hello.csv"

# Initialize blob service and create containers clients
```

```
blob_service_client = BlobServiceClient.from_connection_
string(connection_string)

container_client = blob_service_client.get_container_client(container_name)

# I am just using the file name from the local file path here
blob_name = os.path.basename(local_file_path)

# get the blob client to access the upload method
blob_client = container_client.get_blob_client(blob_name)

# upload blob
with open(local_file_path, "rb") as data:
    blob_client.upload_blob(data)
```

Similar to AWS and Google Cloud, Microsoft Azure offers various storage classes where you can organize data in a specific tier based on the frequency of accessing that data. Currently Azure provides hot tier, cool tier, cold tier, and archive tier for storage. The hot tier is the storage class that is designed for most frequent access (reading or writing). The hot tier is the most expensive of all. The cool tier is for data that is infrequently accessed; however, the data should be stored for a minimum of 30 days. The cold tier is similar to the idea of the cool tier except the data should be stored for a minimum of 90 days. The minimum storage period is 180 days for the archive period, which is also known as long-term backup.

Azure SQL

Azure SQL is a managed database service that can be availed in three different methods. Recall our discussion on Infrastructure as a Service, Platform as a Service, and Software as a Service.

You can deploy Azure SQL on an Azure virtual machine. Azure provides you images of preinstalled SQL Server on operating systems. You can pick the ones you prefer, though keep in mind you will be responsible for operating system upgrades and patches. You have the option to choose memory optimized or storage optimized for a virtual machine. The service is named "SQL database," and this type of setup is the Infrastructure as a Service.

You can also avail an Azure managed database instance, where the fully managed Azure SQL database service automates most of the software updates, security updates, and patches. Here we just choose the appropriate database version and not worry about the underlying operating system. You can still configure the underlying compute (and have the option to scale up or down the number of virtual cores currently allocated to the managed instance). This is referred to as a Platform as a Service model.

The other option is availing a service called "SQL Server" that operates as a SaaS (Software as a Service). In this option, you do not need to worry about the underlying infrastructure. Microsoft will manage both the application and the underlying operating system. You have the option to go to the serverless model where you pay only for what you consume. Azure bills you based on the number of seconds you have utilized SQL Server. And you can choose various tiers for the underlying infrastructure, and you have the flexibility of changing to scale up or down the compute as well.

Let us create an Azure SQL database. In the process of creating a new SQL database, we will also create an Azure SQL logical server. The Azure SQL logical server is like a parent resource for the SQL database. With the Azure SQL logical server, you can enable role-based access control, logins, threat protection, and firewalls to name a few. It also provides an endpoint for accessing databases.

Let us search for "SQL database" and choose to create a new database instance. You will have to mention the subscription and resource group details. You will also have to enter a name for the database and choose a server. The server option is for the Azure SQL logical server. If no server currently exists, you may create a new one. Here is how this may look:

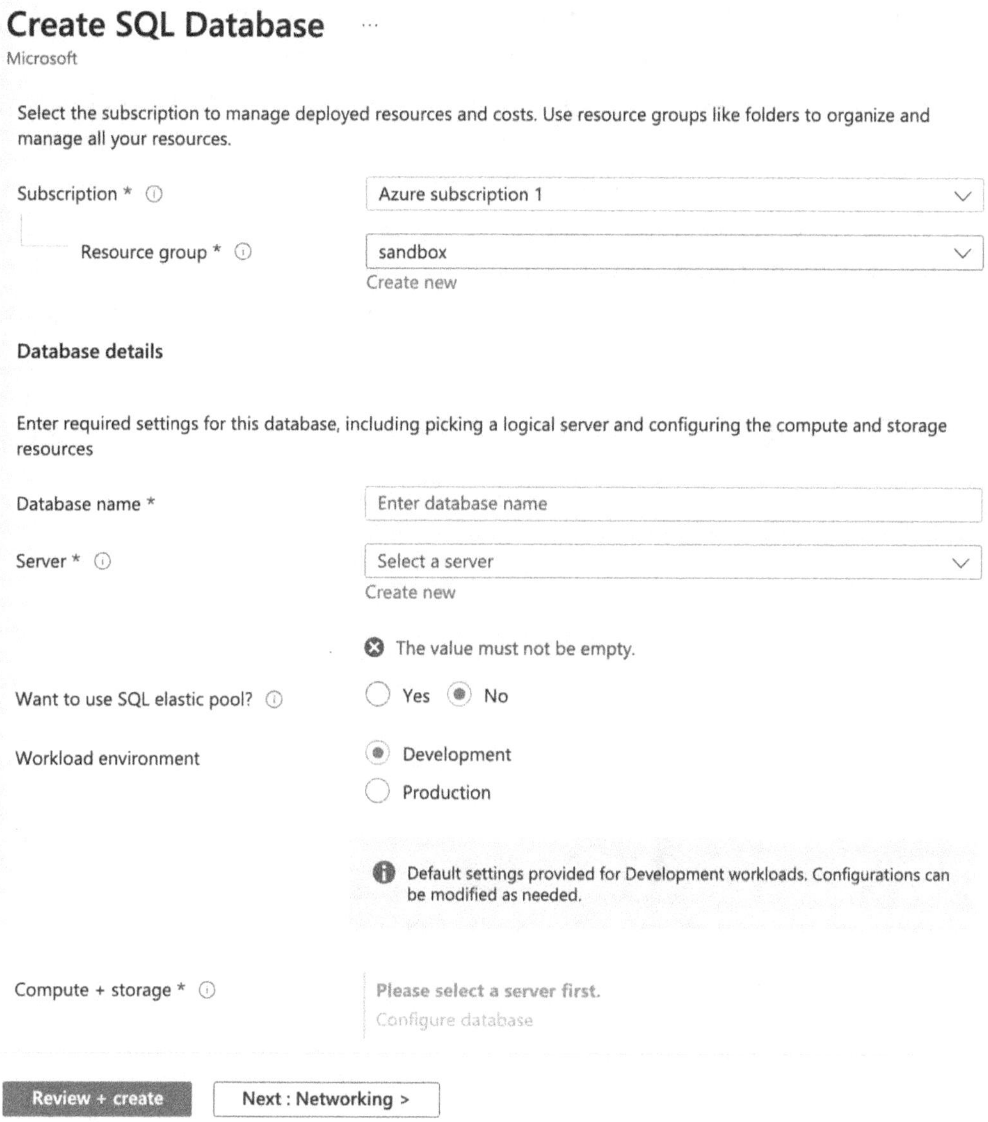

CHAPTER 17 ENGINEERING DATA PIPELINES USING MICROSOFT AZURE

Figure 17-4. Creating a new SQL database

You can also choose whether to use an elastic SQL pool. An elastic SQL pool is basically the underlying compute, storage, and IO for the database that can be shared with other database instances as well. If you create "n" single database instances, then you would have "n" underlying compute and other resources. If you use an elastic SQL pool, then you can have less than "n" compute and storage resources and "share" them among "n" database instances.

In addition to the option of a SQL elastic pool, we also have the choice of specifying whether this instance we are creating is for development or production purposes. The SQL elastic pools are billed by eDTUs. An eDTU is a unit of currency that represents a certain amount of compute, storage, and IO resources. You may wish to buy a certain number of eDTUs in advance and set the limit for the databases on how many such units they can consume in a given amount of time.

Before we create a new SQL database, we may have to create a new Azure SQL logical server. Let us click Create SQL database server, which would provide the page for creating a new logical server. Here is how that may look:

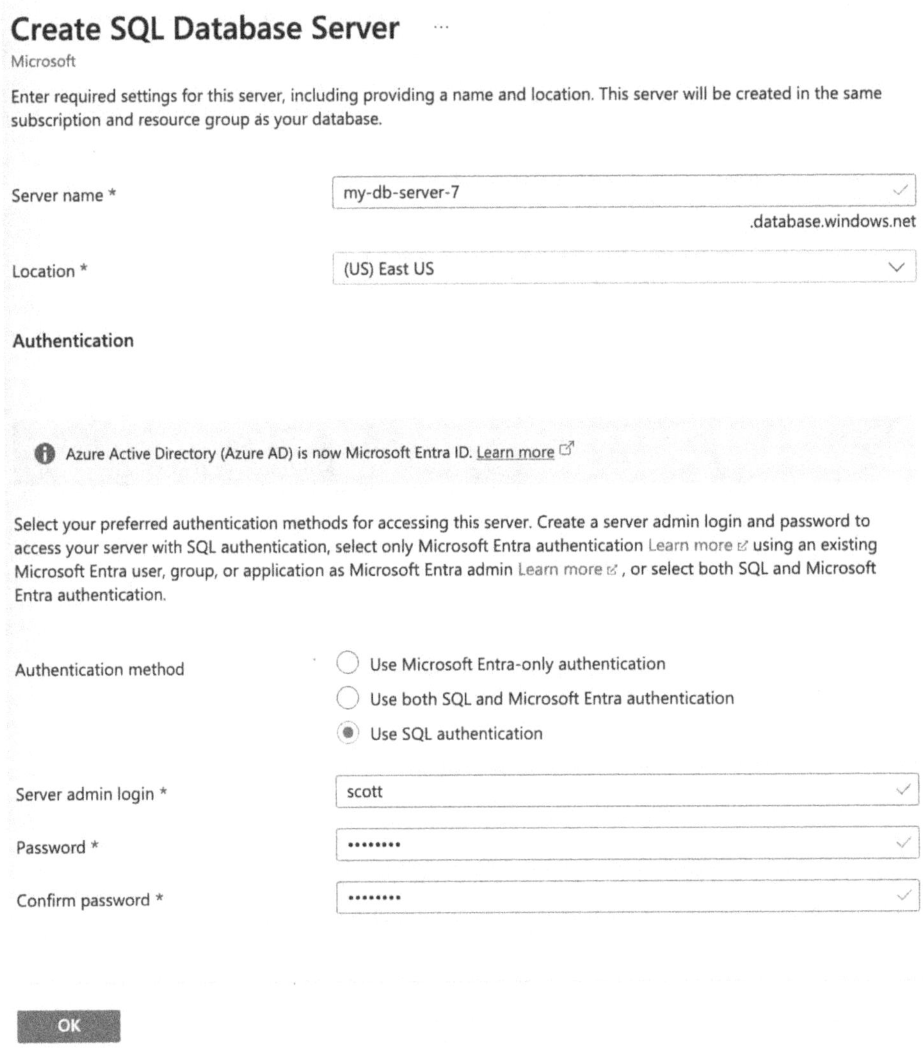

Figure 17-5. Creating a SQL database server in Azure

CHAPTER 17 ENGINEERING DATA PIPELINES USING MICROSOFT AZURE

You may specify a name and a location for the logical server; also, you have the option of specifying the authentication method. Once the information is supplied, you may click OK to create a new logical server for your SQL database. For our instances, we are choosing to create a single instance of a SQL database. Hence, we are choosing not to use an elastic SQL pool.

Once the logical server is ready, you may have the option of choosing the compute and storage for the SQL database. Here is where you can choose the underlying hardware for your SQL database; you may wish to explore various hardware that Azure offers along with their pricing, which will be computed and provided to you instantly. In addition, you have the option of choosing the compute tier, whether you want pre-allocated compute resources or you wish to choose serverless, where you will be billed only on the resources you have consumed.

Here is how this may look:

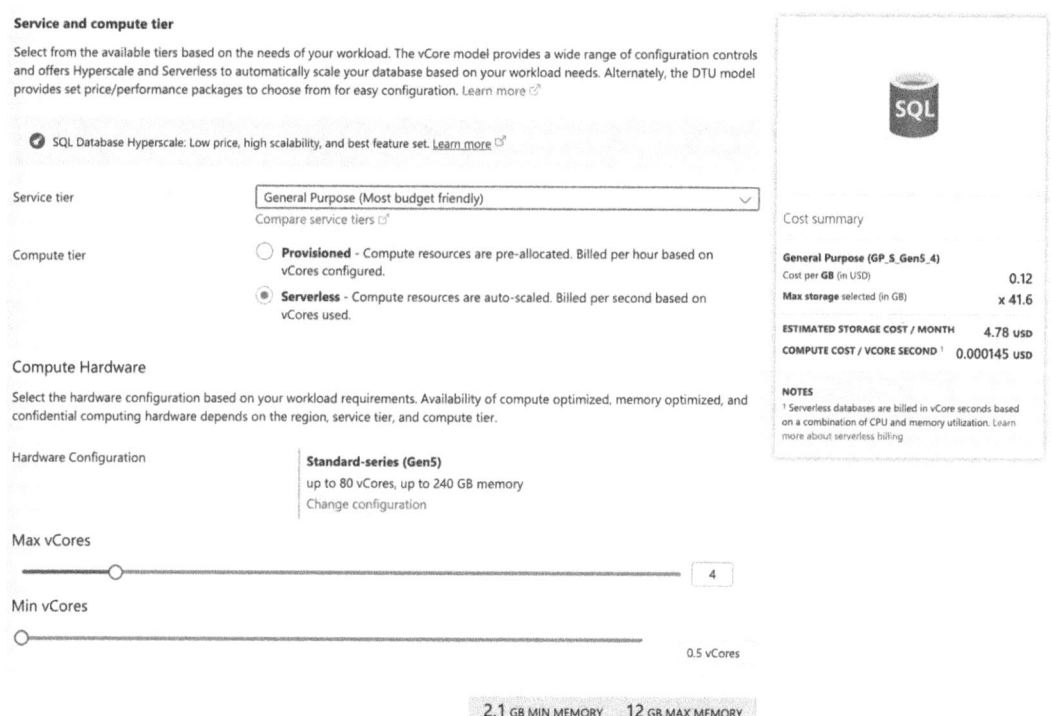

Figure 17-6. Allocation of hardware resources for our database server

581

CHAPTER 17   ENGINEERING DATA PIPELINES USING MICROSOFT AZURE

When it comes to choosing backup storage redundancy, you may wish to choose either locally redundant backup storage or the zone-redundant option for your development purposes. You may leave other options as is and choose to create the database.

Here is a sample Python application that would connect to a database, create a table, and insert rows to that table. If you do not have the "pyodbc" package already, please issue the following command in your terminal:

```
pip install -u pyodbc
import pyodbc
from dotenv import load_dotenv
import os

server = os.getenv("server")
database = os.getenv("database")
username = os.getenv("username")
password = os.getenv("password")
driver = os.getenv("driver")

try:

    cnxn = pyodbc.connect('DRIVER=' + driver +
                    ';SERVER=' + server +
                    ';DATABASE=' + database +
                    ';UID=' + username +
                    ';PWD=' + password)

    cursor = cnxn.cursor()

    print('Connection established')

    try:

        query1 = "CREATE TABLE SampleTable \
                (Id INT PRIMARY KEY, \
                Name NVARCHAR(100));"

        cursor.execute(query1)
        cnxn.commit()
```

```
        print("Table created")

        try:

            query2 = "INSERT INTO \
                    SampleTable (Id, Name) \
                    VALUES (1, 'John'); \
                  INSERT INTO \
                    SampleTable (Id, Name) \
                    VALUES (2, 'Jane'); \
                  INSERT INTO \
                    SampleTable (Id, Name) \
                    VALUES (3, 'Alice');"

            cursor.execute(query2)
            cnxn.commit()

            print("Rows inserted")

        except:

            print("unable to insert rows into the table")

            cursor.close()
            cnxn.close()
    except:

        print("unable to create the table")

except:

    print('Cannot connect to SQL server')
```

Here is how the contents of the environment file may look like:

```
server = "your-azure-database-server-instance"
database = "your-database"
username = "your-username"
password = "password"
driver = "driver"
```

Azure Cosmos DB

Azure Cosmos DB is a managed NoSQL database service provided by Microsoft. It is a versatile service offering that can provide both SQL and NoSQL database development including relational data models, document stores, graph databases, and vector databases as well. Cosmos DB offers multiple APIs, each one simulating a different database engine. These include relational SQL, MongoDB for document databases, Cassandra DB for column-oriented databases, Gremlin for graph databases, and Table for key–value databases. In a way, Cosmos DB has a multimodel database service, where you can use appropriate APIs for your database application needs. In the case of database migration, you may choose the appropriate API corresponding to your current relational or NoSQL database that you intend to migrate.

You can avail Cosmos DB in two methods, namely, provisioned throughput and serverless models. In provisioned throughput, as the name suggests, you provision how much throughput you expect the database to deliver while also defining the performance levels you wish to obtain for your database application. The unit of measurement here is called a request unit. The other method of availing Cosmos DB is through serverless.

The serverless model is such that it only charges you based on how many requests that it has received and processed. As far as billing is concerned, provisioned throughput will compute the bill based on how many request units have been provisioned and not consumed, whereas, in the case of serverless, billing is computed based on how many number of request units consumed over an hour by performing database operations.

Note A request unit is a way of measuring how you have consumed the service. Every operation that you perform with Cosmos DB will consume a certain amount of compute, storage, and physical memory. The request unit is calculated based on such consumption. For instance, a query that is sent to Cosmos DB will cost a certain number of request units to retrieve the results. The exact number of request units required to obtain the results may depend on the size of the table, query compilation time, and number of rows, to name a few parameters.

CHAPTER 17   ENGINEERING DATA PIPELINES USING MICROSOFT AZURE

Let us look at creating a Cosmos DB instance with the MongoDB API. Once you make an API selection, it remains the same for that instance. Should you choose to work with another API, you may wish to create another instance with that respective API. In our case, we choose the MongoDB API. If you are familiar with MongoDB, then the usage will be very similar to Cosmos DB. Here is where you may choose the API within Cosmos DB:

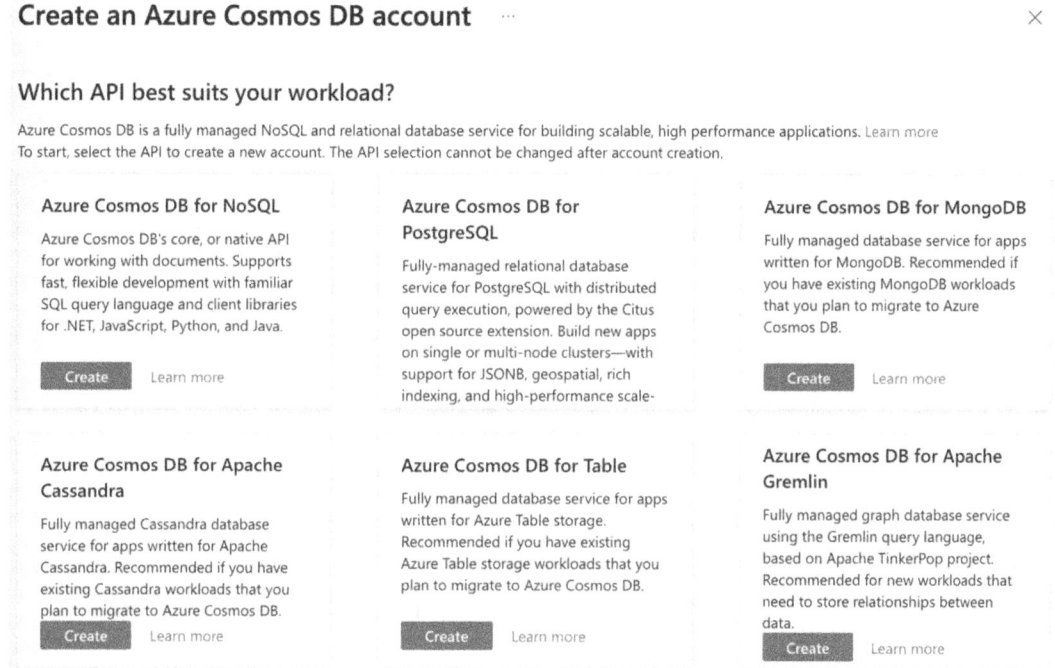

Figure 17-7. *Choosing the right API for your Azure Cosmos DB instance*

Next, you will be prompted to choose the resource for your instance. There are two categories: one is based on request units, while the other is based on virtual cores. Both the options are fully managed by Azure and scalable as the workload increases. Let us choose the virtual core option, where we have the option to choose the underlying hardware.

Here is how this might look:

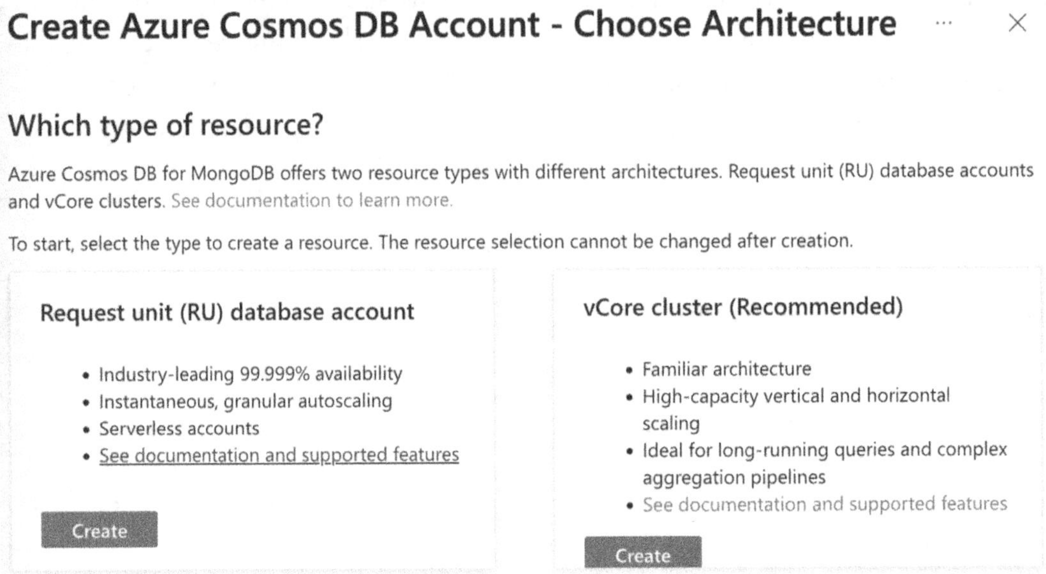

Create Azure Cosmos DB Account - Choose Architecture

Which type of resource?

Azure Cosmos DB for MongoDB offers two resource types with different architectures. Request unit (RU) database accounts and vCore clusters. See documentation to learn more.

To start, select the type to create a resource. The resource selection cannot be changed after creation.

Request unit (RU) database account
- Industry-leading 99.999% availability
- Instantaneous, granular autoscaling
- Serverless accounts
- See documentation and supported features

Create

vCore cluster (Recommended)
- Familiar architecture
- High-capacity vertical and horizontal scaling
- Ideal for long-running queries and complex aggregation pipelines
- See documentation and supported features

Create

Figure 17-8. Choosing the hardware resource for Cosmos DB

Once you choose the virtual core option, you will be directed to configure your Cosmos DB. Here, you will have to specify the subscription or billing account and appropriate resource group. A resource group is something that shares similar policies and permissions. In addition, choose an administrator username and password and name and location of the cluster, among other details.

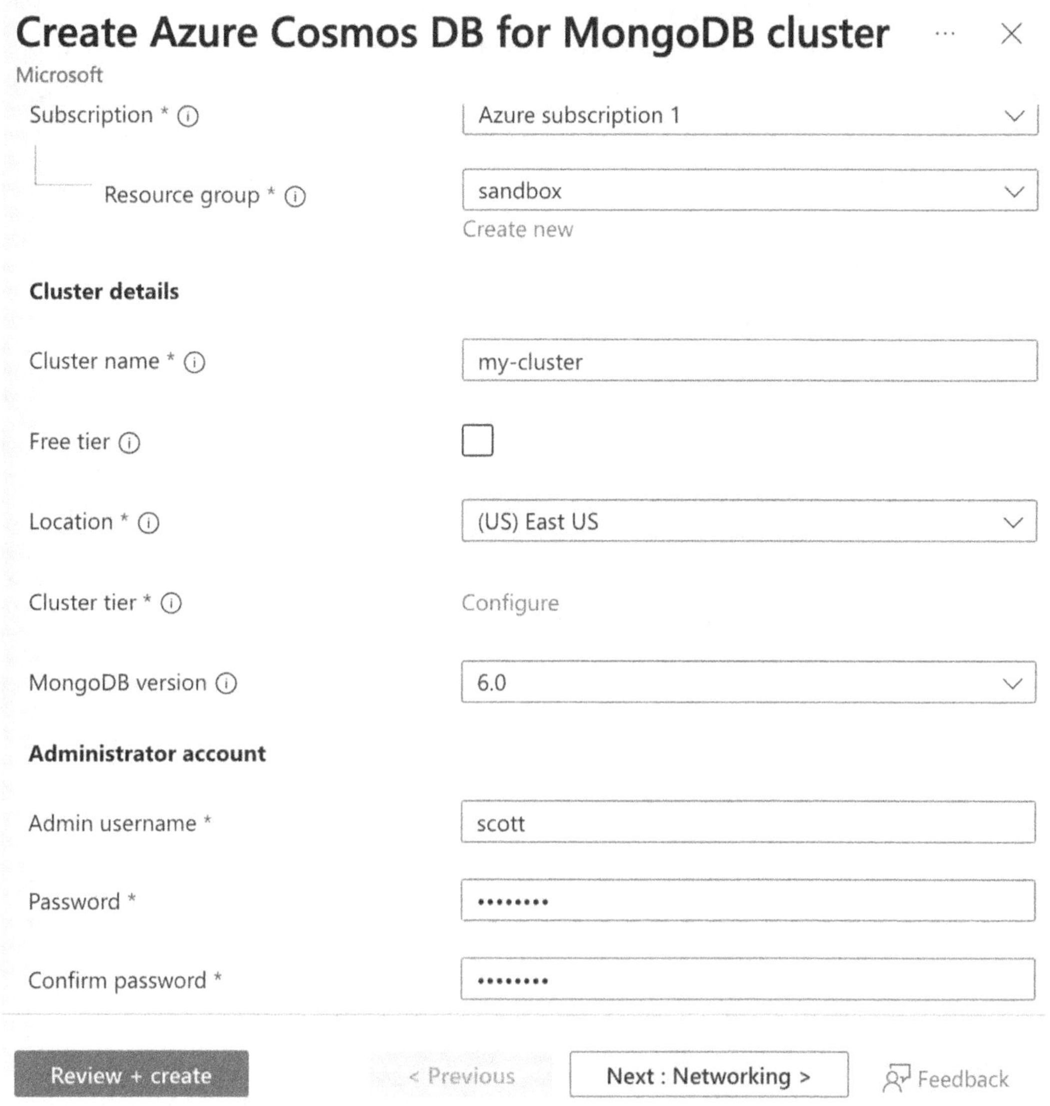

CHAPTER 17   ENGINEERING DATA PIPELINES USING MICROSOFT AZURE

Create Azure Cosmos DB for MongoDB cluster

Microsoft

Subscription * — Azure subscription 1

Resource group * — sandbox
Create new

Cluster details

Cluster name * — my-cluster

Free tier — ☐

Location * — (US) East US

Cluster tier * — Configure

MongoDB version — 6.0

Administrator account

Admin username * — scott

Password * — ········

Confirm password * — ········

[Review + create]   < Previous   [Next : Networking >]   Feedback

Figure 17-9. Creating a new Cosmos DB instance for a MongoDB cluster

You have the option of choosing the free tier for the underlying cluster. Let us look at configuring our underlying cluster for this Cosmos DB instance. Under configuration, you will have various development- and production-level hardware instances, where you can customize your compute, storage, and physical memory. In addition, you also have the option to choose high availability and database shards as well.

587

Database sharding in MongoDB is a method where you distribute the data between multiple worker nodes within a given cluster. You can also create replications of your data, and so you will gain speed in read/write operations. For instance, let us say a node that contains a specific shard may process "n" operations per second; by introducing "m" shards containing the same data, you may be able to process "m × n" operations per second.

Here is how cluster configuration may look:

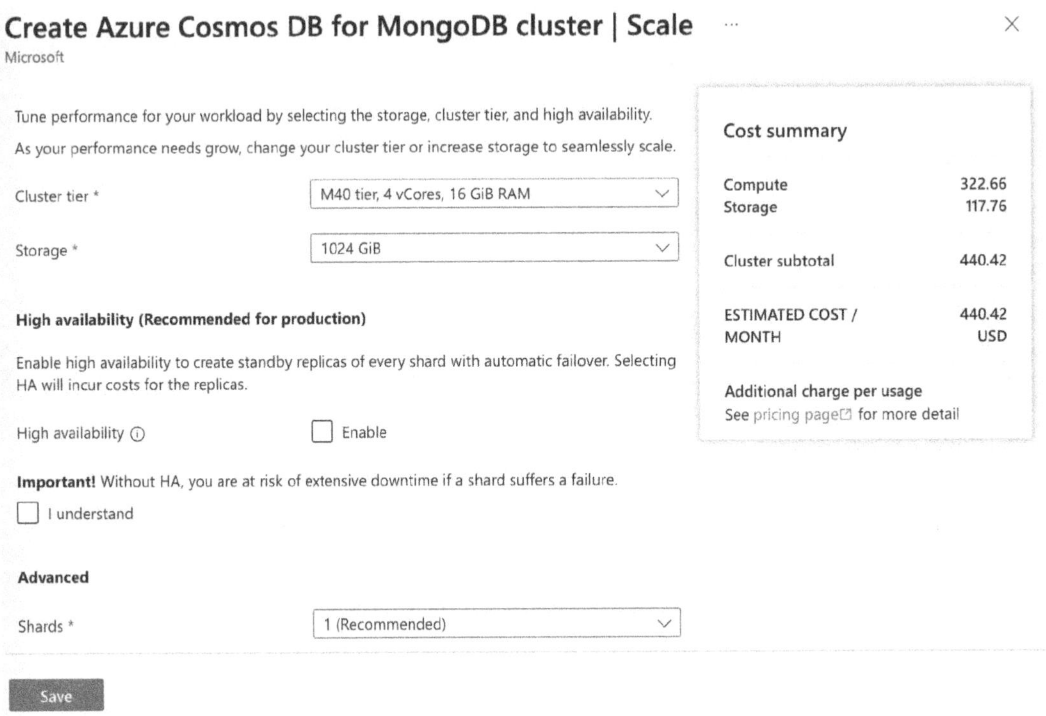

Figure 17-10. Hardware configuration for Cosmos DB for a MongoDB cluster

You also have the option of specifying the firewall rules and connectivity methods within the networking option. Here is how that may look for you:

CHAPTER 17   ENGINEERING DATA PIPELINES USING MICROSOFT AZURE

Network connectivity

Public access assigns public IP addresses to your cluster and permits connections only from authorized public IPs. Private access, which assigns private IP addresses to the shards in your cluster, is coming soon.

Encrypted connections

This cluster enforces encrypted connections using Transport Layer Security (TLS). For information on TLS version and certificates, refer to connecting with TLS/SSL. Learn more

Connectivity method *
- ● Public access (allowed IP addresses)
- ○ Private access

  ⓘ Connections from the IP addresses configured in the Firewall rules below will have access to this cluster. By default, no public IP addresses are allowed. Learn more

Firewall rules

☐ Allow public access from Azure services and resources within Azure to this cluster

\+ Add current client IP address (122.178.91.33)   + Add 0.0.0.0 - 255.255.255.255

Rule Name	Start IP address	End IP address	
ClientIPAddress_2024-4-2-12-47-...	122.178.91.33	122.178.91.33	🗑

Figure 17-11. *Firewall rules specification for Cosmos DB*

You may leave the networking settings and other settings as is, and click Create. It might take several minutes for Azure to provision a new cluster. Let us look at connecting and uploading a JSON document to Cosmos DB:

```
from azure.cosmos import CosmosClient
import config
import json

url = config.settings['host']
key = config.settings['master_key']
database_name = config.settings['database_id']
container_name = config.settings['container_id']
path_to_json = "/Users/pk/Documents/mock_data.json"
```

```python
def connect():
    try:
        client = CosmosClient(url,credential=key)
        print("Established connection with the client")

        try:
            database = client.get_database_client(database_name)
            print("Database obtained")

            try:
                container = database.get_container_client(container_name)
                print("Container obtained")

            except:
                print(f"unable to select the container {container_name}")

        except:
            print(f"unable to select the database, {database_name}")

    except:
        print("Unable to connect to the database")

    return container

def upsert(container):
    with open(path_to_json, "rb") as file:
        json_doc = json.load(file)

    for js in json_doc:
        upload_json_doc - container.upsert_item(js)
        print(f"id uploaded is : {js['id']}")

def primaryfunc():
    c_name = connect()
    upsert(c_name)

primaryfunc()
```

Azure Synapse Analytics

So far we have looked at object storage, relational database solutions, and NoSQL database solutions. Azure Synapse Analytics is a modern storage solution offered by Microsoft that enables storage of relational, unstructured, and semistructured data into a single storage solution. Once the data has been in Synapse Analytics, it is then easy to use various computing layers like Hadoop, Spark, or Databricks to query and generate analytical insights from the data. Besides, the Synapse data lake provides options to keep the data in Parquet format, so that data takes less space.

You can avail Synapse Analytics into forms, namely, dedicated SQL pool and serverless SQL pool. A dedicated SQL pool is formerly known as Azure SQL Data Warehouse. A dedicated SQL pool is a collection of compute resources that are provisioned during the creation of the instance. Similar to Azure SQL and Cosmos DB, the size of these instances is measured by a unit called data warehouse unit. A data warehouse unit represents a certain amount of compute, memory, and IO resources.

Serverless SQL pools, as the name goes, are serverless. There is no underlying infrastructure that requires to be configured, and no cluster maintenance is required. You are charged only for the data processing that you run, and so there are no charges for the underlying infrastructure. You can get up and running querying your data in a matter of a few minutes. This service creates a querying layer on top of your data lake, and you can run T-SQL queries on top of your data.

From an architectural point of view, compute is separated from storage in Synapse Analytics. You have the choice of scaling the compute independently of the data. Internally, Synapse SQL uses a node-based architecture that consists of a control node and compute nodes. The T-SQL code that is submitted is routed to control nodes. The control node uses a distributed query engine to optimize the queries for parallel processing and shares tasks for compute nodes to execute the queries in parallel.

Let us create an Azure Synapse Analytics instance. We start by navigating to the Azure console and searching for Synapse Analytics. Once the Synapse Analytics service page is rendered for us, we would choose to create a new Synapse workspace. We start by specifying the subscription group, resource group, and managed resource group. A managed resource group is where supplemental resources that Azure creates for the Synapse workspace are found.

Here is how that looks:

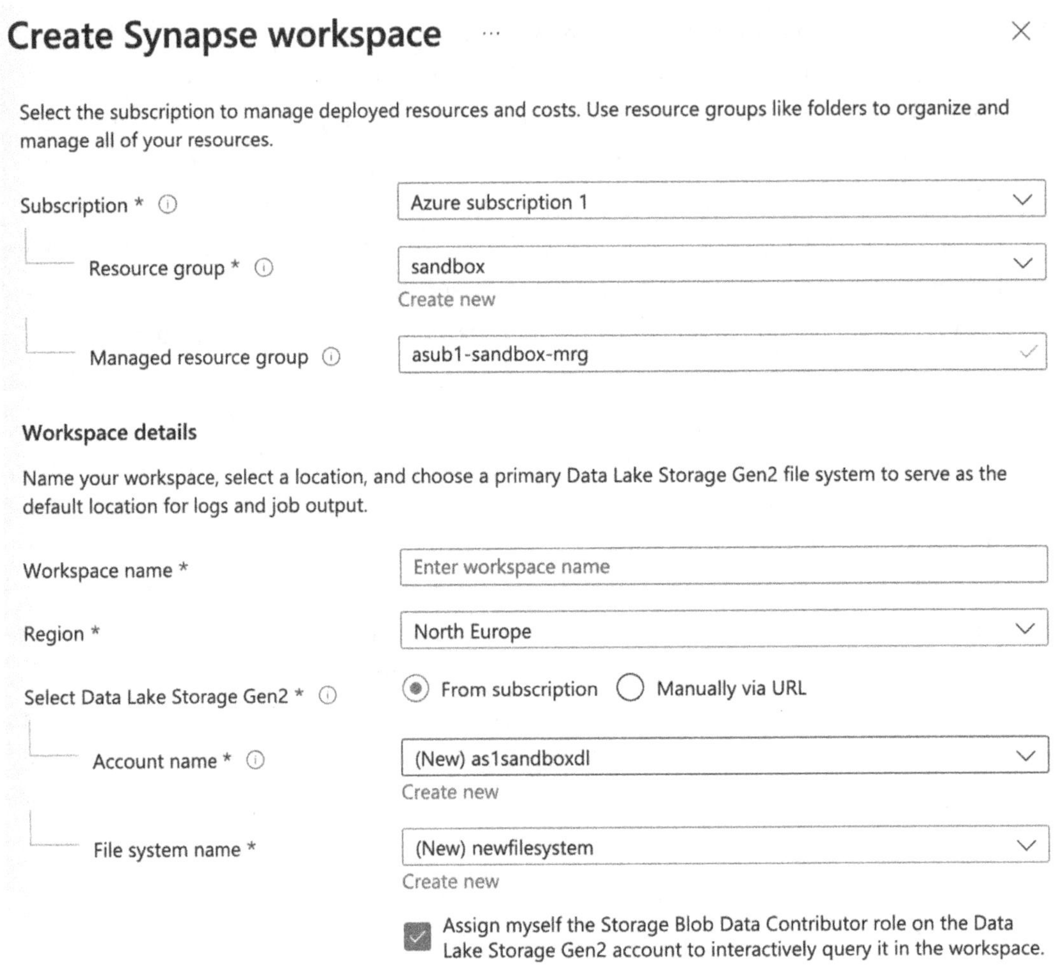

Create Synapse workspace

Select the subscription to manage deployed resources and costs. Use resource groups like folders to organize and manage all of your resources.

Subscription *	Azure subscription 1
Resource group *	sandbox
	Create new
Managed resource group	asub1-sandbox-mrg

Workspace details

Name your workspace, select a location, and choose a primary Data Lake Storage Gen2 file system to serve as the default location for logs and job output.

Workspace name *	Enter workspace name
Region *	North Europe
Select Data Lake Storage Gen2 *	◉ From subscription ○ Manually via URL
Account name *	(New) as1sandboxdl
	Create new
File system name *	(New) newfilesystem
	Create new

☑ Assign myself the Storage Blob Data Contributor role on the Data Lake Storage Gen2 account to interactively query it in the workspace.

Figure 17-12. Creation of an Azure Synapse workspace

We also have to define the name and location of the workspace; in addition, we have to either specify the current data lake or create a new data lake, part of configuring the workspace. Once we have entered the basics, let us specify a local user in security settings.

CHAPTER 17   ENGINEERING DATA PIPELINES USING MICROSOFT AZURE

Once Synapse Analytics is up and running, you can view it in your dashboard. Here is how that may look:

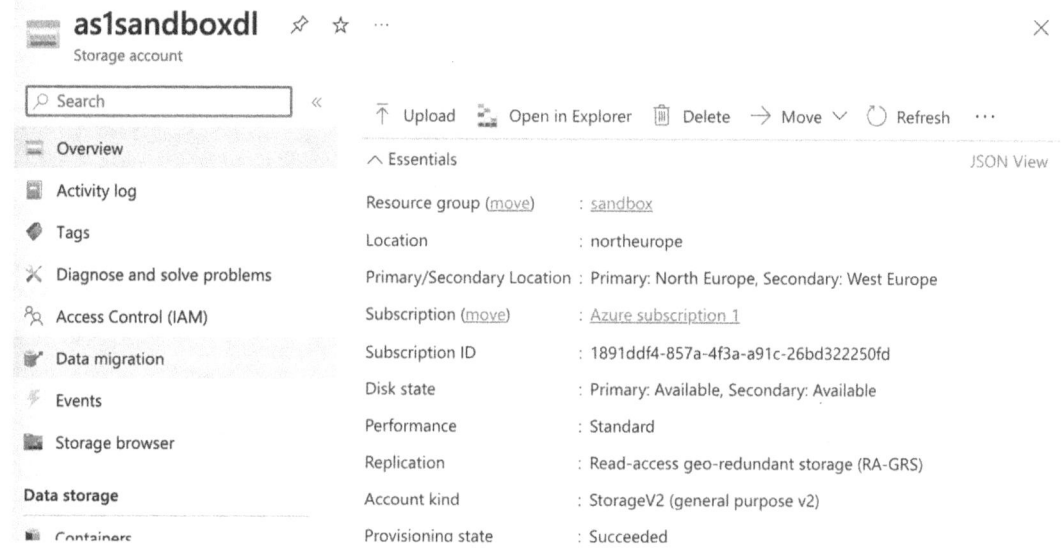

Figure 17-13. Dashboard of the Azure Synapse workspace

Azure Data Factory

Azure Data Factory is a fully managed data integration service. The service is offered serverless as you do not have to pay for the underlying hardware. You can either design data integration pipelines visually or programmatically. It has several built-in connectors that enable smoother integration; it also integrates well with git so you can push code changes where applicable. Azure Data Factory lets you create custom event triggers that automate execution of certain actions when a certain event is triggered.

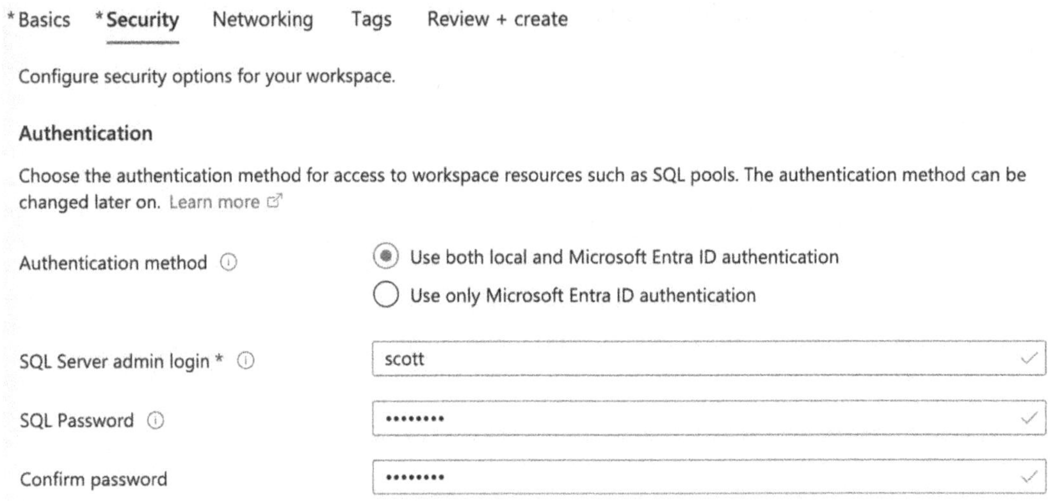

CHAPTER 17   ENGINEERING DATA PIPELINES USING MICROSOFT AZURE

*Basics   *Security   Networking   Tags   Review + create

Configure security options for your workspace.

Authentication

Choose the authentication method for access to workspace resources such as SQL pools. The authentication method can be changed later on. Learn more

Authentication method
- (●) Use both local and Microsoft Entra ID authentication
- ( ) Use only Microsoft Entra ID authentication

SQL Server admin login *   scott

SQL Password   ••••••••

Confirm password   ••••••••

Figure 17-14. Azure authentication

Using Azure Data Factory, it is possible to compress the data during loading from one system to another, thereby consuming less bandwidth. To get started with Azure Data Factory, visit the link `https://adf.azure.com/en/` or find the same link by searching for Azure Data Factory in the console. A new tab will open where you can specify a name for your new data factory and associate your subscription account as well.

Here is how that may look:

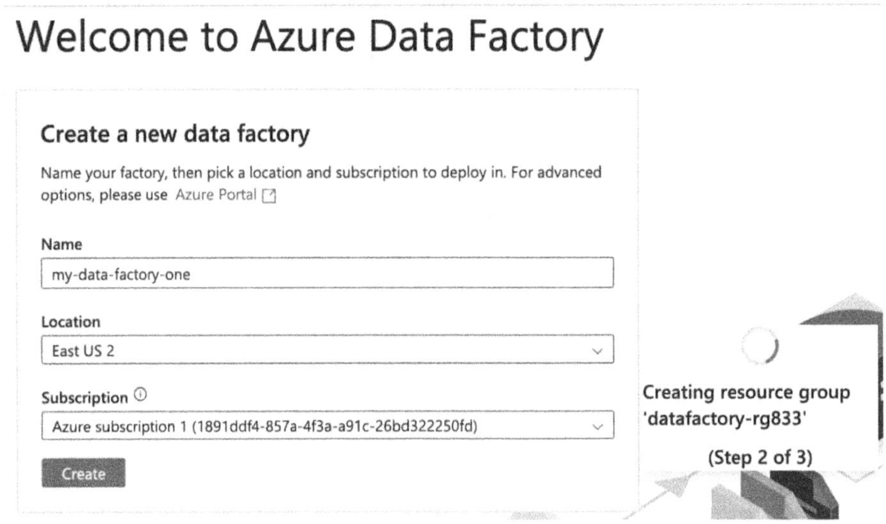

Welcome to Azure Data Factory

Create a new data factory

Name your factory, then pick a location and subscription to deploy in. For advanced options, please use Azure Portal

Name
my-data-factory-one

Location
East US 2

Subscription
Azure subscription 1 (1891ddf4-857a-4f3a-a91c-26bd322250fd)

[Create]

Creating resource group 'datafactory-rg833'

(Step 2 of 3)

Figure 17-15. Creating an Azure data factory page

CHAPTER 17   ENGINEERING DATA PIPELINES USING MICROSOFT AZURE

When you submit the information, you may wait a few minutes for Azure to prepare the Azure Data Factory service for you. Here is how it may look when it is completed:

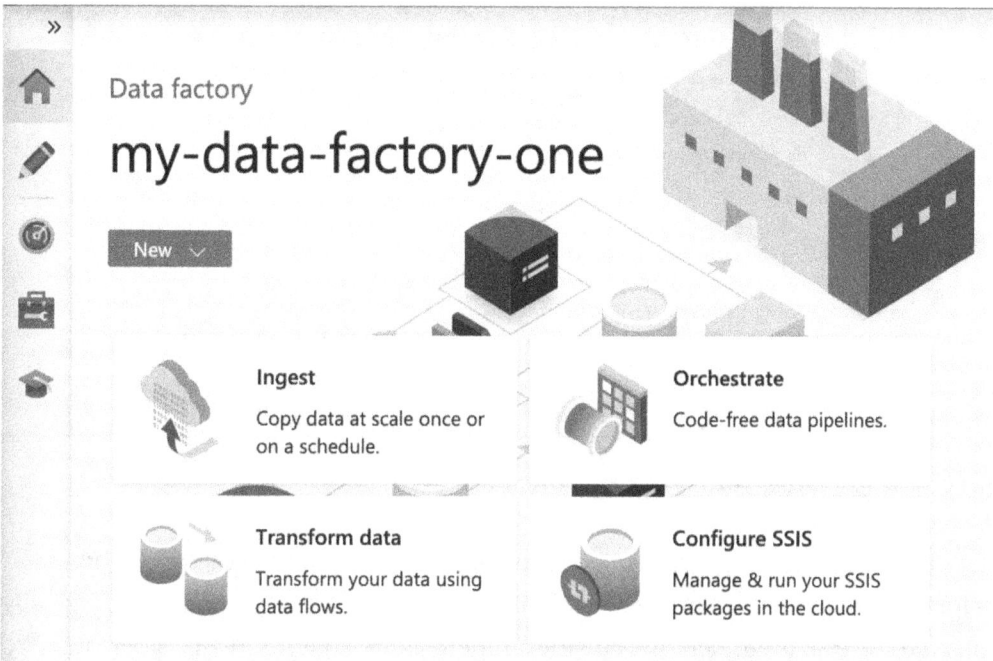

Figure 17-16. *Home page of Azure Data Factory*

On the left side, you will see five options, where Home is the main page, Author is your data integration IDE, Monitor is where you can orchestrate and observe the job runs, Manage is where you can control the git integration and other settings, and Learning Center is where you can obtain documentation and learning support for Azure Data Factory.

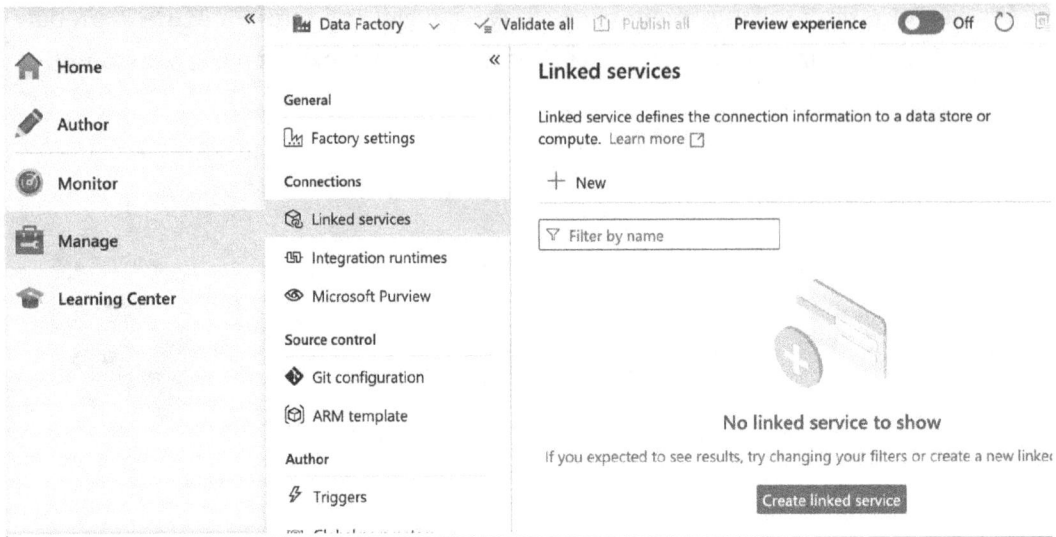

Figure 17-17. *Azure Data Factory console*

In designing a data integration job using Azure Data Factory, you can author a data integration job, internally called a pipeline, which is a group of activities logically arranged to be executed. Activities inside a pipeline are what perform a single task. You have the option of performing various activities that would connect source and sink data systems, transport data, and also perform various transformations.

You have to define a data system connection in order to connect to a data system. To define the data system connection information, navigate to the Manage page and click "Linked Services" under the Connection tab.

Let us try to create or add Azure Blob Storage as a new connection in Linked Services. You can click a new connection, search for Azure Blob on the data store, and click Continue. You may enter a name and description for the connection, followed by choosing an Azure subscription under Account selection method; select your appropriate subscription and container name.

You can test the connection and if it returns successful then you have successfully added Azure Blob Storage in your Azure Data Factory. Here is how that may look for you:

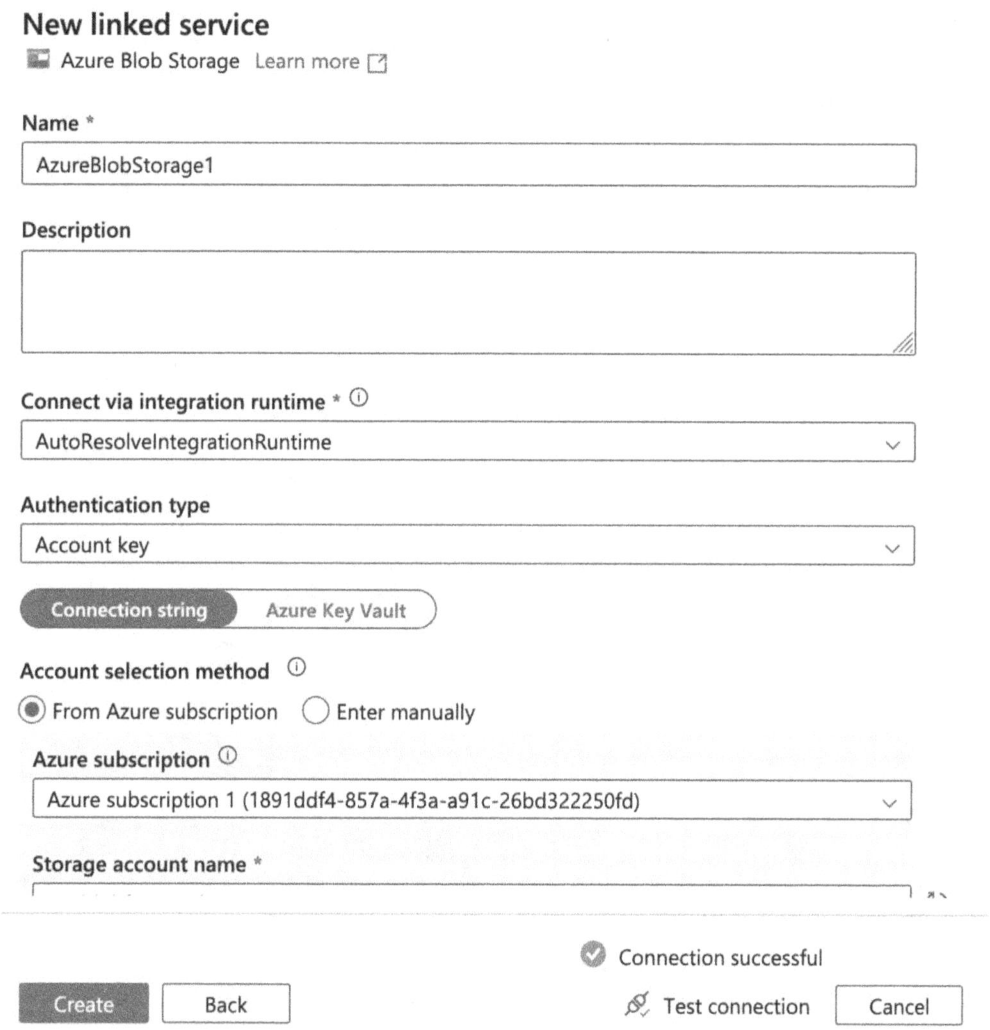

CHAPTER 17 ENGINEERING DATA PIPELINES USING MICROSOFT AZURE

Figure 17-18. Creating a new linked service in Azure Data Factory

To get acquainted with Azure Data Factory, let us look at copying data from Azure Blob Storage to an Azure SQL database. We will obtain a comma-delimited data file from a source and upload it to Azure Blob Storage. We would then use Azure Data Factory to initiate a copying pipeline from Blob Storage to an Azure SQL database. To begin, navigate to the main page and click the Ingest option and choose Built-in copy task.

CHAPTER 17 ENGINEERING DATA PIPELINES USING MICROSOFT AZURE

Here is how that may look:

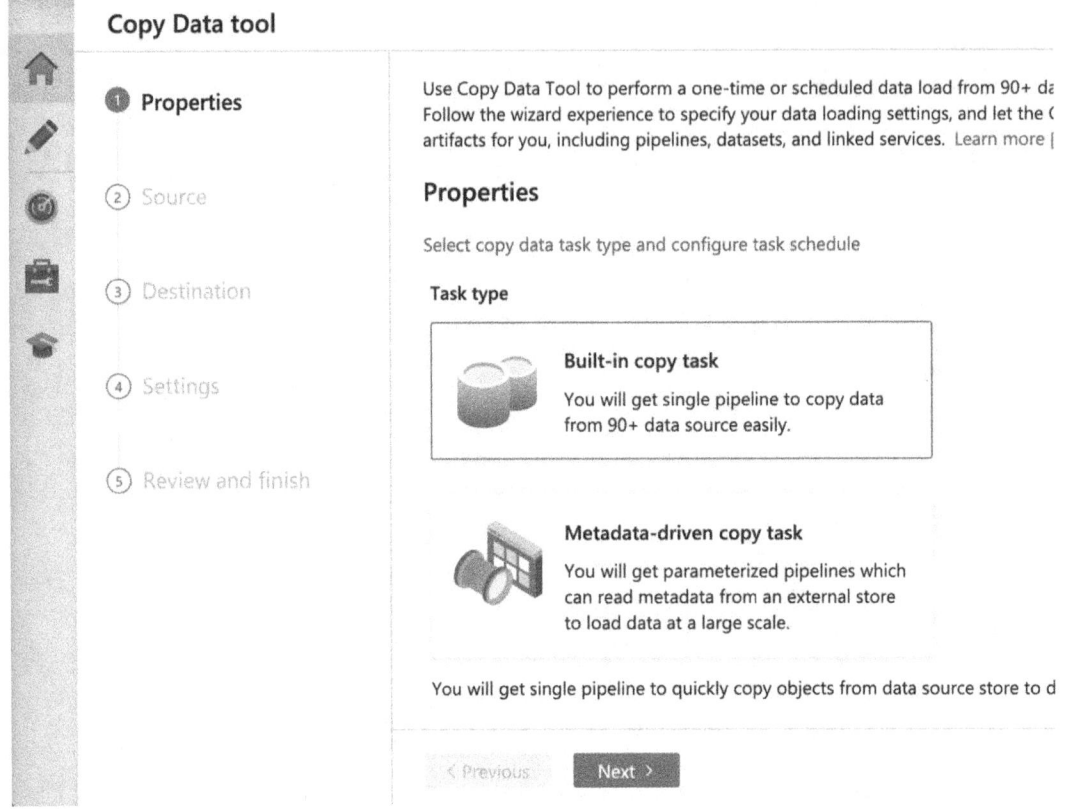

Figure 17-19. Azure Data Factory's Copy Data tool

We will specify the source data system in our next page, which is Azure Blob Storage. From the Source type drop-down menu, choose Azure Blob Storage, and from the Connection drop-down, choose your blob storage. This should be available if you have utilized the Linked Services option earlier. If not, the procedures are the same.

Once your blob storage is linked, you can browse the object storage to choose the appropriate flat file or a folder that contains the flat file. Choose the "Recursive" option if you plan to upload multiple flat files from one folder. The system would identify the type of flat file that is being selected and populate file format settings by itself.

CHAPTER 17 ENGINEERING DATA PIPELINES USING MICROSOFT AZURE

Here is how this may look:

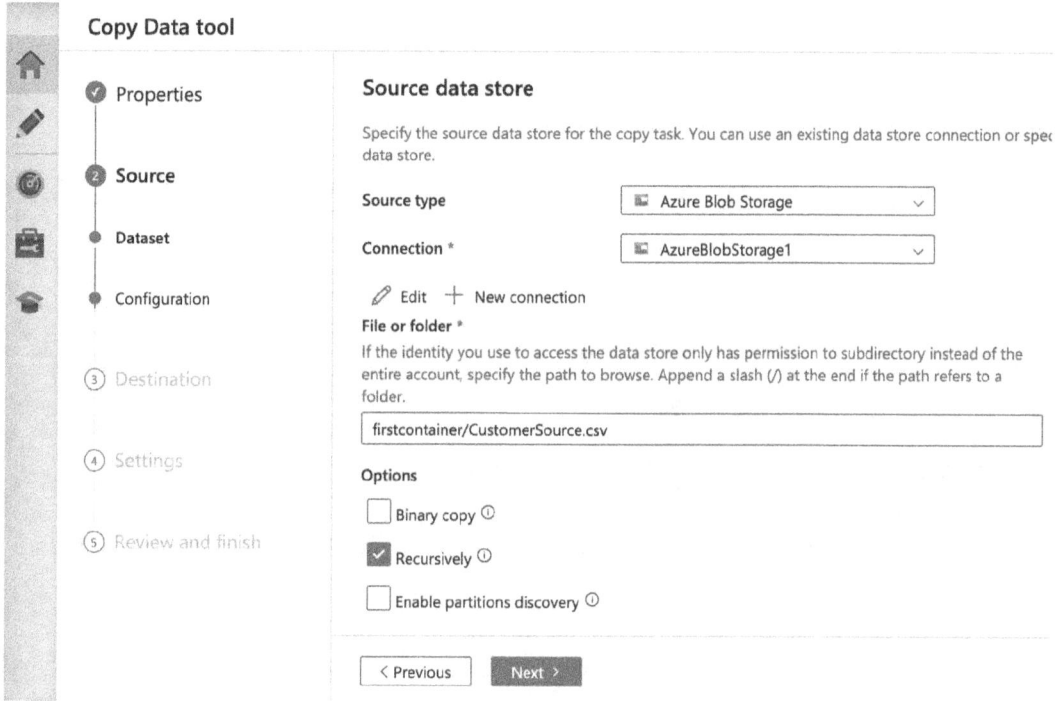

Figure 17-20. *Specifying the source data store in Azure Data Factory*

Now, let us specify the destination data system. You may choose an Azure SQL database and specify the connection, for which you may wish to create a new linked service, or choose the one that may have been already created. Since this is a relational database, we do have the option of creating a table schema on the target database and populating the flat file directly to the database. If not, we can choose to automatically create a data table and populate this data table as well. Here is how that may look:

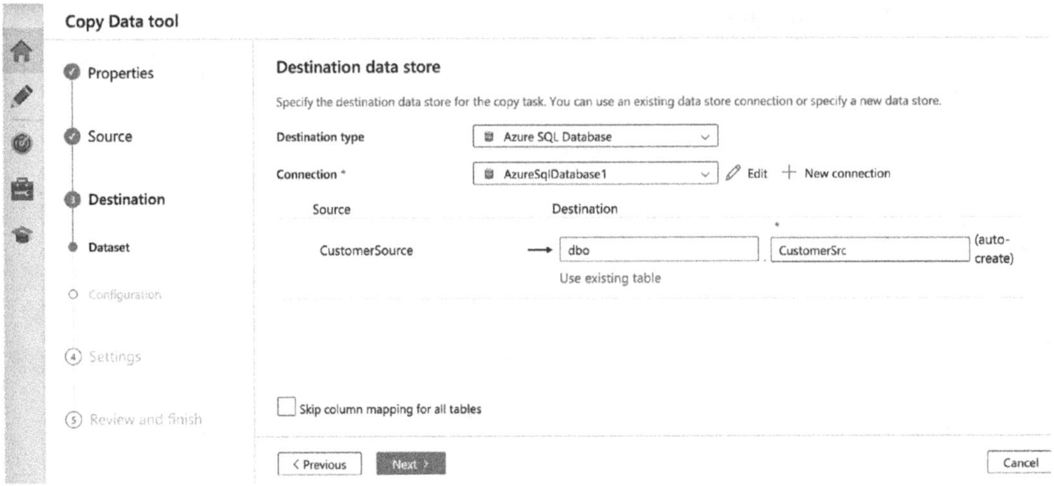

CHAPTER 17 ENGINEERING DATA PIPELINES USING MICROSOFT AZURE

Figure 17-21. Specifying the destination data store in Azure Data Factory

Once this step is completed, let us name this data pipeline under Settings. Here is how this may look for you:

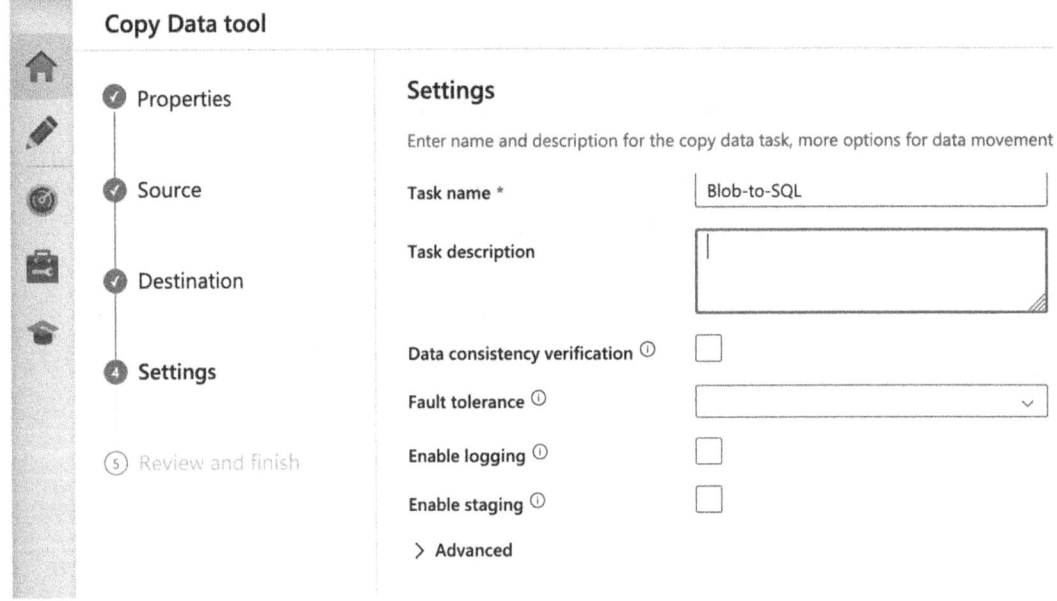

Figure 17-22. Adding the name and description of this data pipeline

Once all these steps are completed successfully, you will see a review screen that lists all the parameters of this data pipeline. If there are any errors or you may find something that needs to be changed, you have the option of editing those respective fields. If everything looks good, then you may click Next. Azure Data Factory will run the validation for this entire data pipeline. Once the validation process finishes successfully, you may get the following screen:

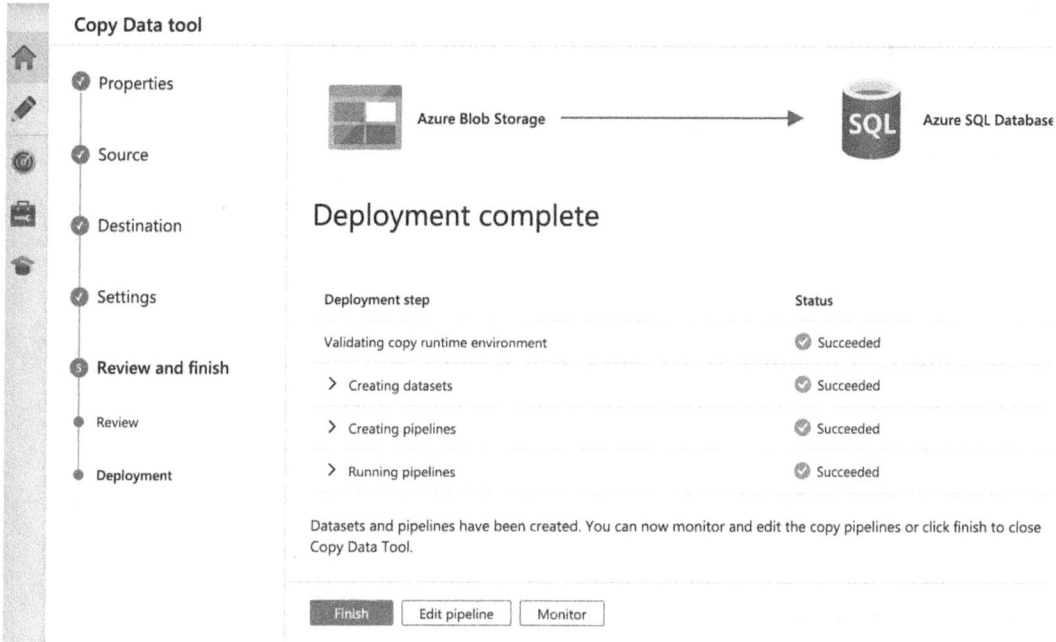

Figure 17-23. Validation and execution of the data pipeline

You may choose to click Finish, which will take you to the main page. On the dashboard of Azure Data Factory, click the Monitor page, which will list the current data pipelines that are in your repository. You can click "Rerun" for running the data pipeline. When the job completes successfully, you may get something like the following:

CHAPTER 17   ENGINEERING DATA PIPELINES USING MICROSOFT AZURE

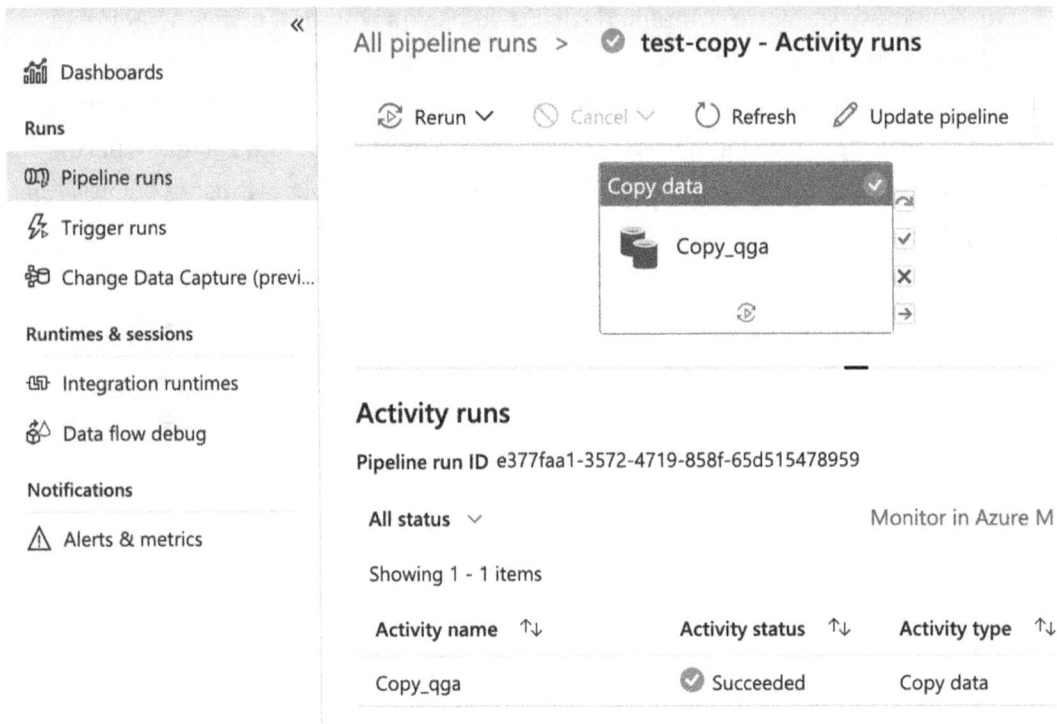

Figure 17-24. Monitoring and rerunning of data pipelines in Azure Data Factory

Azure Functions

Azure Functions is a serverless compute option that enables executing functions that perform certain tasks without having to worry about the underlying hardware. Azure automatically provides updated servers for these functions to execute and also scale up if required. These compute resources are allocated dynamically and autoscaled during runtime. You would get charged only for the code you execute (duration of the execution and compute resources used). You can use Azure Functions to build APIs, respond to an event happening on another service, etc.

The highlight of Azure Functions is the use of triggers and bindings. Using triggers and bindings, you can create functions that interact with other services. A trigger is basically an event or a condition that initiates the execution of a function. You can only have one trigger per function in Azure Functions. For instance, you can have an Azure function that has an HTTP trigger that receives a request and triggers the function to provide a response; you can have a service trigger that activates an Azure function when an event happens—new file uploaded in Blob Storage or even a simple timer that sets the function execution on a specific schedule.

Bindings in Azure Functions provide a method to connect to various data sources; thereby, you can interact (write and read data) with services. You can have more than one binding in a given Azure function. There are two types of bindings that Azure Functions supports, namely, input and output binding. Input bindings may include reading from data storage systems like Blob Storage, databases, file systems, etc., and output bindings could be sending data to another service. Azure Functions actually supports multiple input and output bindings in a given function.

To create a new Azure function, navigate to the Azure dashboard, and search for "Function app." Here is how that may look for you:

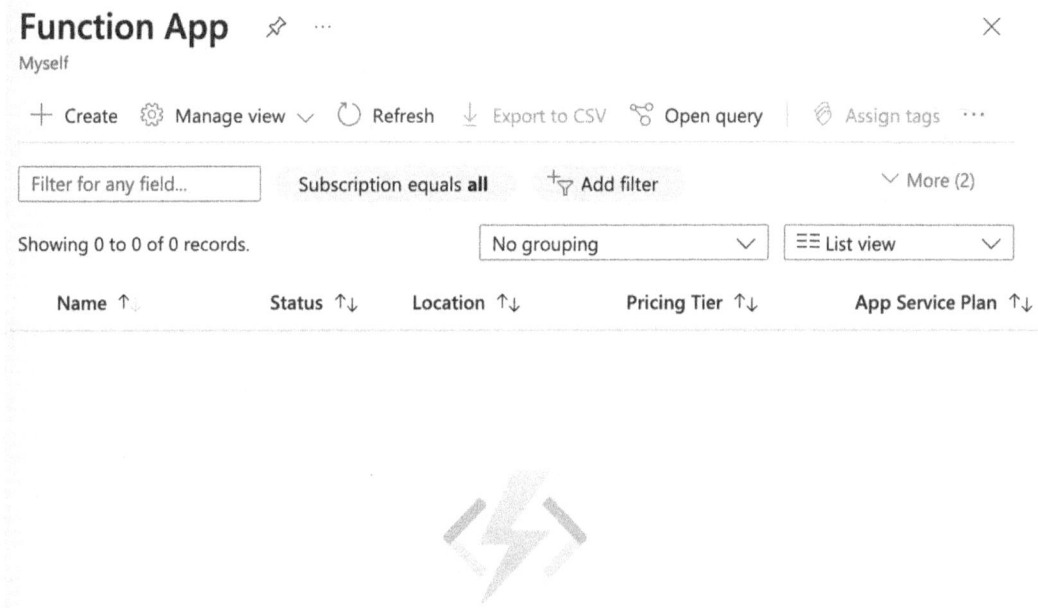

Figure 17-25. Function app—Azure Functions

Once you are in Azure Functions, create a new function app. You have to enter the subscription details; you would also define the instance details, including the name of the function app, the runtime stack, the environment, and the operating system.

You can also define how you choose to host the app. The cost-effective option is serverless; however, you have the option to reuse the compute from other Azure services as well.

Here is how that may look for you:

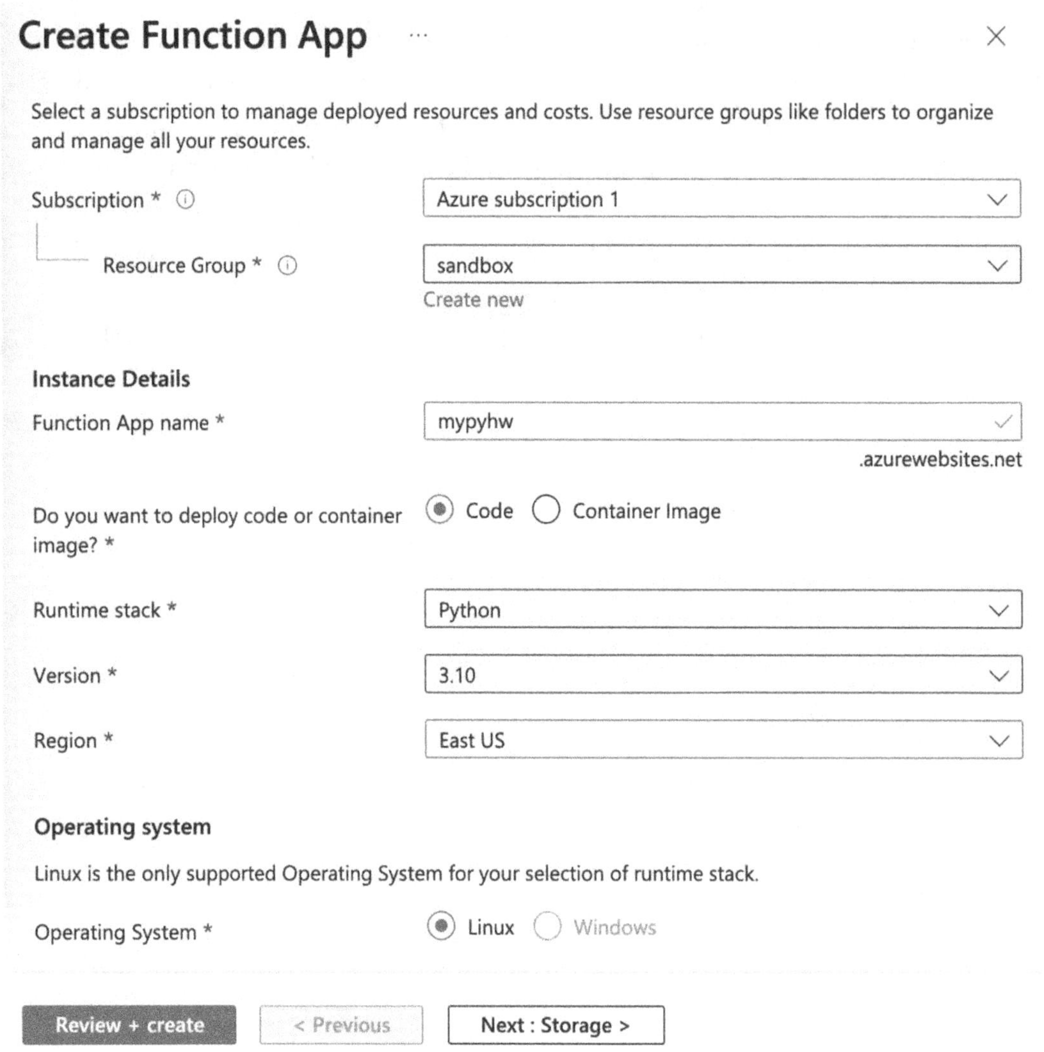

Figure 17-26. *Creating a new function app in Azure Functions*

Once you have created the function app instance, you can create your Azure functions within the same. Let us explore how to develop Azure functions locally by using the core tools package. Depending upon the operating system, you may wish to refer to these instructions to install the core tools.

Here is the link: https://learn.microsoft.com/en-us/azure/azure-functions/functions-run-local?tabs=macos%2Cisolated-process%2Cnode-v4%2Cpython-v2%2Chttp-trigger%2Ccontainer-apps&pivots=programming-language-python

To check if the installation is complete, you may open a terminal and type in "func," and it will render output relating to Azure functions. To create a new Azure function locally, enter

```
func init myfunction --worker-runtime python --model V2
```

It would set up the entire project structure with appropriate JSON elements. Here is how that may look for you:

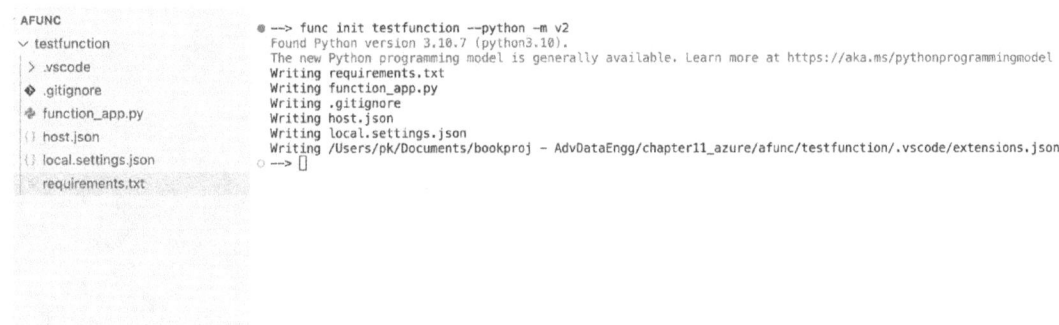

Figure 17-27. *Testing the installation of core tools in Azure Functions*

The folder structure on the left is created by Azure Functions core tools utility. You may navigate to the function_app.py and proceed to write your Azure function. Here is a sample Azure function for you:

```python
import azure.functions as func
import datetime
import json
import logging

app = func.FunctionApp()

@app.function_name('helloworld')
@app.route(route="route66")
def helloworld(requests: func.HttpRequest) -> func.HttpResponse:
    logging.info('received request for helloworld function')

    user = req.params.get('username')
    if user:
        message = f"Hello, {user}"
```

```
    else:
        message = "welcome to Azure functions"
    return func.HttpResponse(
        message,
        status_code=200
    )
```

Once you have the Azure function ready, you may run the function using "func start" from the command line. Here is how that looks:

```
○ (azure) pk@mac myfunction % func start
Found Python version 3.11.7 (python3).

Azure Functions Core Tools
Core Tools Version:       4.0.5611 Commit hash: N/A +591b8aec842e333a87ea9e23ba390bb5effe0655 (64-bit)
Function Runtime Version: 4.31.1.22191

[2024-04-04T15:44:39.366Z] Worker process started and initialized.

Functions:

        helloworld: http://localhost:7071/api/route95

        testfunction: http://localhost:7071/api/route90

For detailed output, run func with --verbose flag.
```

Figure 17-28. Executing the function in Azure Functions

When you navigate to the appropriate URLs, you may view the output of the function. If you wish to publish your code as an Azure function (currently, the function is developed and deployed locally), then you can utilize the following command from your terminal:

`func azure functionapp publish <your-azure-function-app>`

Azure Machine Learning

Azure Machine Learning is a fully managed machine learning service that helps develop code, train a machine learning model, execute, deploy, and manage the machine learning model that is deployed. Azure Machine Learning service offers serverless compute. The major services that Azure Machine Learning provides are Azure ML Studio and Python SDK for machine learning. With Azure ML Studio, you can write your own code using Jupyter Notebook and execute the same using Azure ML services. There is also a low-code option that Azure provides, which enables you to drag and drop ML data assets and components to design, train, and deploy machine learning models.

To get started with Azure Machine Learning, navigate to the management console and search for "Azure machine learning." You will find Azure Machine Learning in the marketplace option. Select the service and create a new machine learning workspace. Provide appropriate details for your subscription and workspace.

Here is how this may look for you:

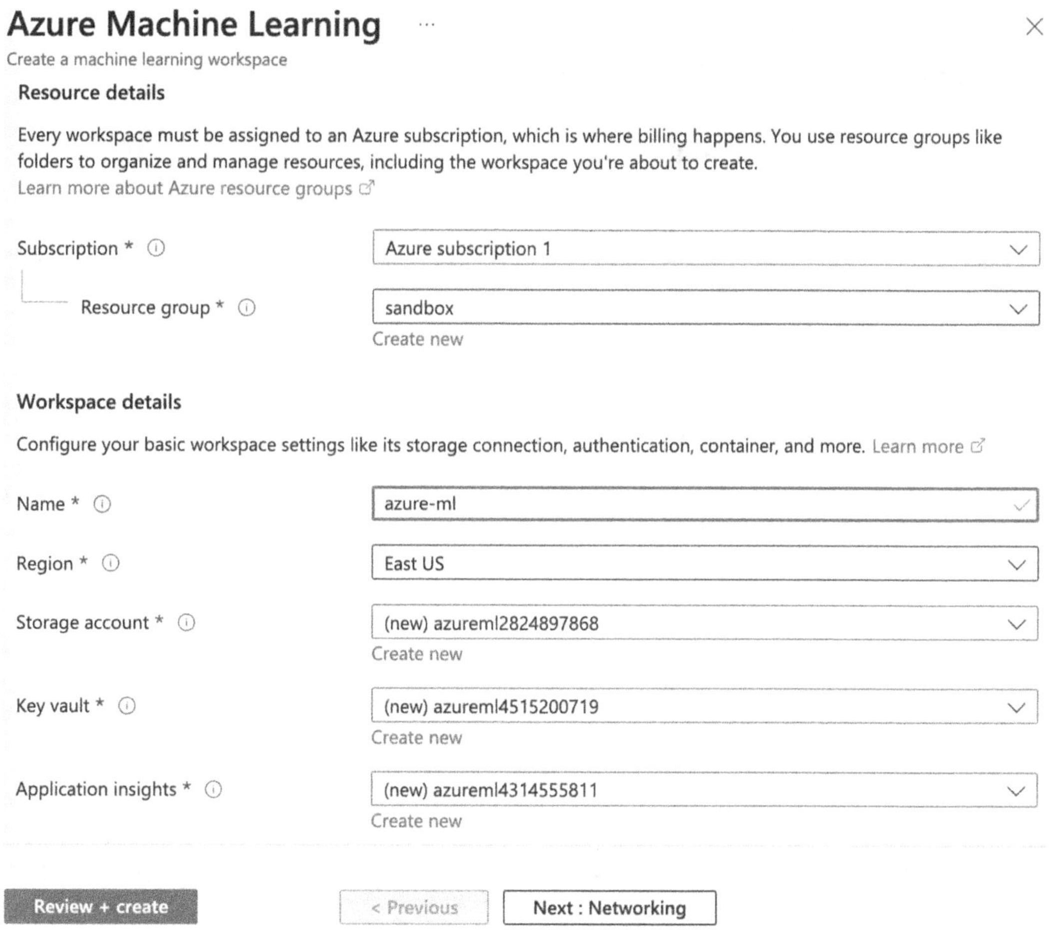

Figure 17-29. Creating a machine learning workspace in Azure Machine Learning

CHAPTER 17   ENGINEERING DATA PIPELINES USING MICROSOFT AZURE

The workspace option is to enable a centralized view of all resources that Azure Machine Learning provides. When you provision a new Azure Machine Learning workspace, Azure would automatically create a set of services dedicated to Azure machine learning. These are storage account, key vault, container registry, and insights.

The storage account is a newly created storage space for Azure machine learning data assets, logs, and reports (if any). There is also an option to create blobs. The key vault is a secrets management resource that is used to securely store sensitive information that is utilized in the machine learning workspace. Insights would provide operational insights and diagnostic information during monitoring of machine learning projects. The container registry handles Docker containers for machine learning models, development environments, etc.

Once the service is created and available, here is how this may look for you:

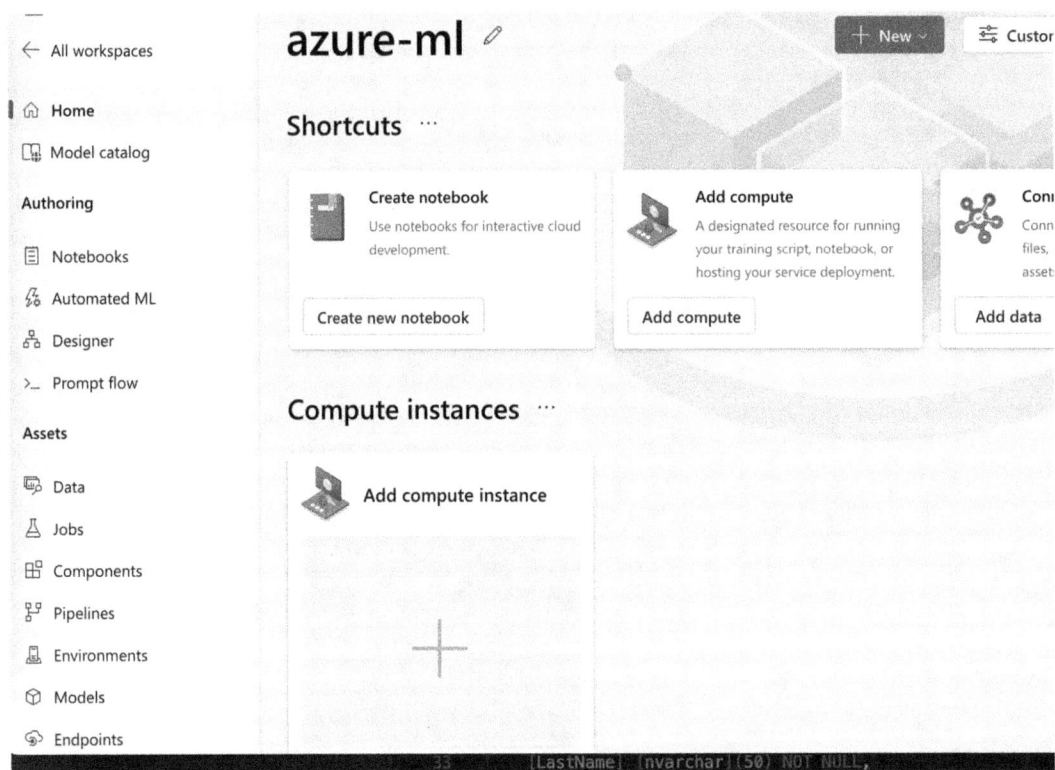

Figure 17-30. *Home page of Azure ML service*

Azure ML Data Assets

Formerly known as datasets in Azure ML Studio, data assets are like a repository within Azure Machine Learning, where you can create a reference to a dataset that is located elsewhere. You can create dataset references from Azure data stores, public URLs, and Azure open datasets. As the data continues to remain in its location, there are no storage costs incurred from your end. These datasets within data assets are lazily evaluated, which helps in efficient use of compute. There are two types of datasets that you can store in a data asset, namely, FileDataSet and TabularDataSet.

FileDataSets are references for one or more datasets that are located elsewhere. The source datasets can be in any format, and FileDataSets create references that can be used in machine learning workflows. FileDataSets are a good choice if the dataset you are referring to has already been cleaned and is ready to be used in machine learning pipelines.

TabularDataSet is the other type of datasets, wherein it parses the source dataset (one or more datasets) in its native format and converts it into tabular format. This tabular format dataset can then be imported into a Jupyter notebook or into a Spark environment. TabularDataSet can parse JSON files, Parquet files, results of a SQL query, and simply comma-delimited files as well.

Here is how you can create either of these dataset types from Azure ML Studio. You may click the "Data" option under "Assets," within Azure ML Studio.

CHAPTER 17 ENGINEERING DATA PIPELINES USING MICROSOFT AZURE

Here is how this may look:

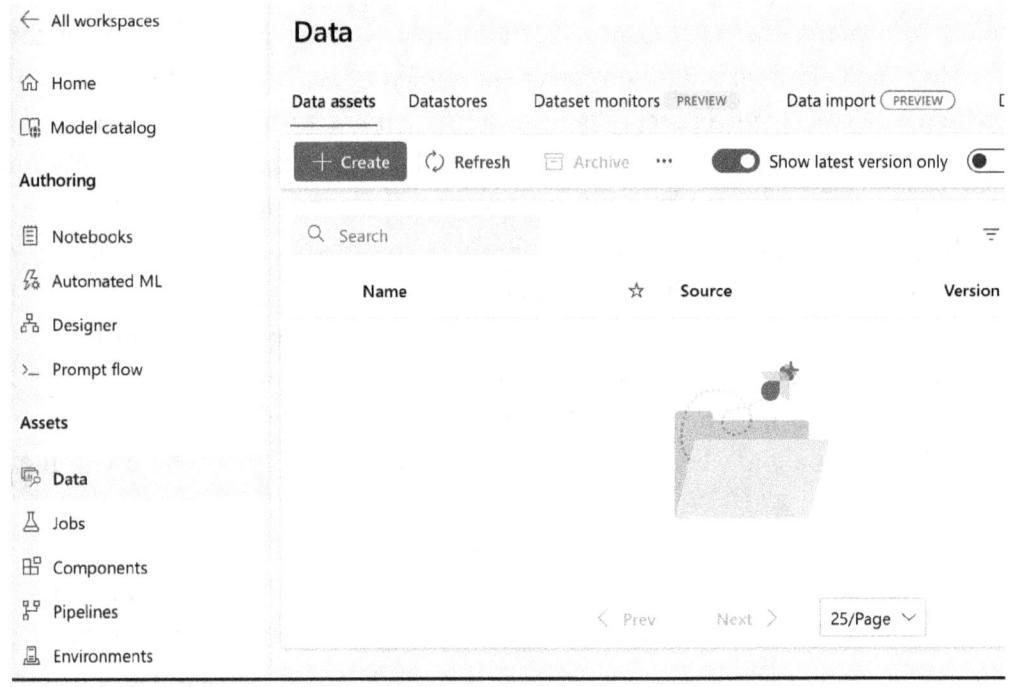

Figure 17-31. *Data assets in Azure ML Studio*

Once you are on this page, click Create. You can specify the name and description of the dataset and choose either File or Tabular from the "Type" drop-down.

Here is how this may look:

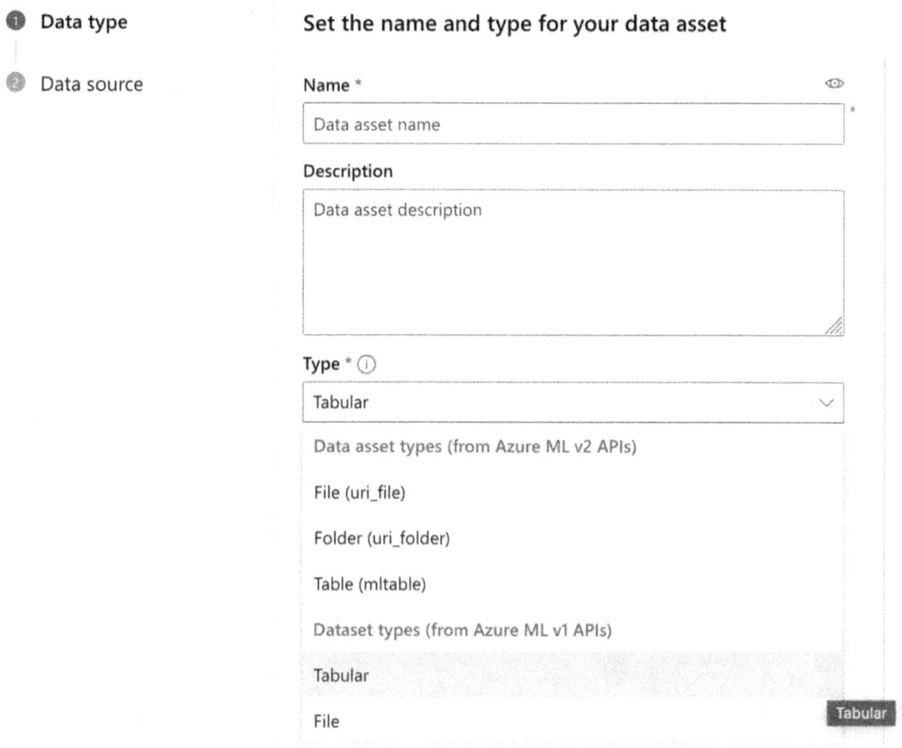

Figure 17-32. Specifying the data asset in Azure ML Studio

Azure ML Job

Azure machine learning is basically a set of machine learning tasks that runs on the underlying cluster. You can build a ML training job, which may consist of a set of tasks that trains a machine learning model; you can also build a sweep job, which may consists of a set of tasks that executes a machine learning model with multiple hyperparameters, and select the best model that best minimizes the cost function. These jobs also generate log entries and Azure ML Studio can track the metrics of each ML job.

To get started with developing machine learning models, you need a machine learning workspace and a compute instance. In previous sections, we have already created a workspace. To create a new compute instance, you may either choose the "Compute" option under "Manage" or simply click "Add compute instance" on the dashboard.

Here is how they may look for you:

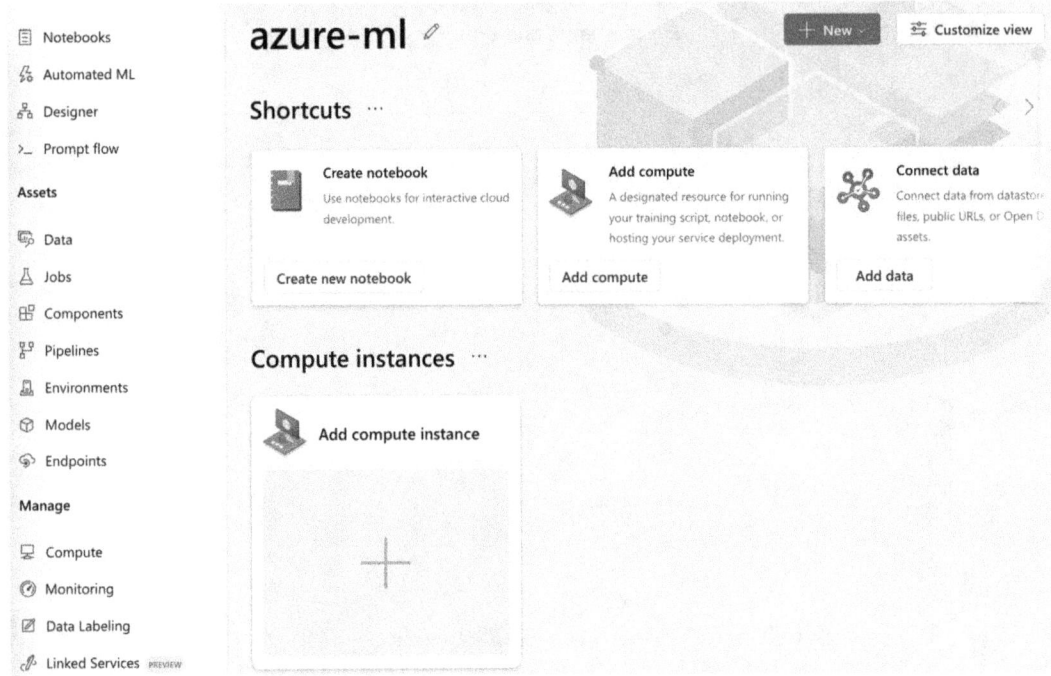

Figure 17-33. Add a new compute instance in Azure ML Studio

You can specify a name for the compute instance and choose whether you need a CPU-based instance or GPU-based instance. The choice on the type of instance would be based upon the nature of the machine learning task you are intending to perform.

GPUs are always associated with parallelism and high-performance computing, and so if your machine learning model leverages parallelism (through Dask or Ray), you can go for this choice. Certain applications like training a deep learning model with several layers or a compute vision task (image processing) are highly suited for GPUs. Certain machine learning applications like text modeling, large models, or building recommender engines where the underlying tasks benefit from parallelism can be trained by both CPU and GPU architectures. In the case of tabular data, CPU-based computation may be sufficient.

CHAPTER 17   ENGINEERING DATA PIPELINES USING MICROSOFT AZURE

Upon choosing the virtual machine, you may leave other options as is and proceed to create the compute instance. Here is how this may look for you:

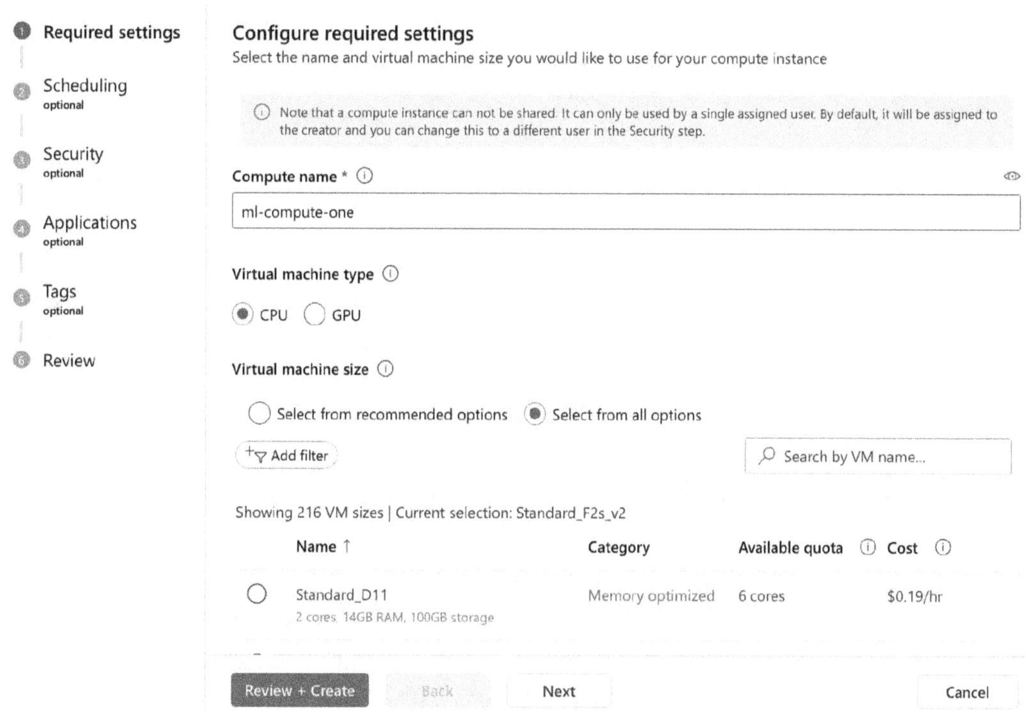

Figure 17-34.   Creating a new compute instance in Azure ML Studio

Once you have created the compute instance, you can start to utilize the compute for developing your machine learning model. You may choose the Jupyter Notebook option or the web-based VS Code editor as well. Here is how that may look for you:

CHAPTER 17   ENGINEERING DATA PIPELINES USING MICROSOFT AZURE

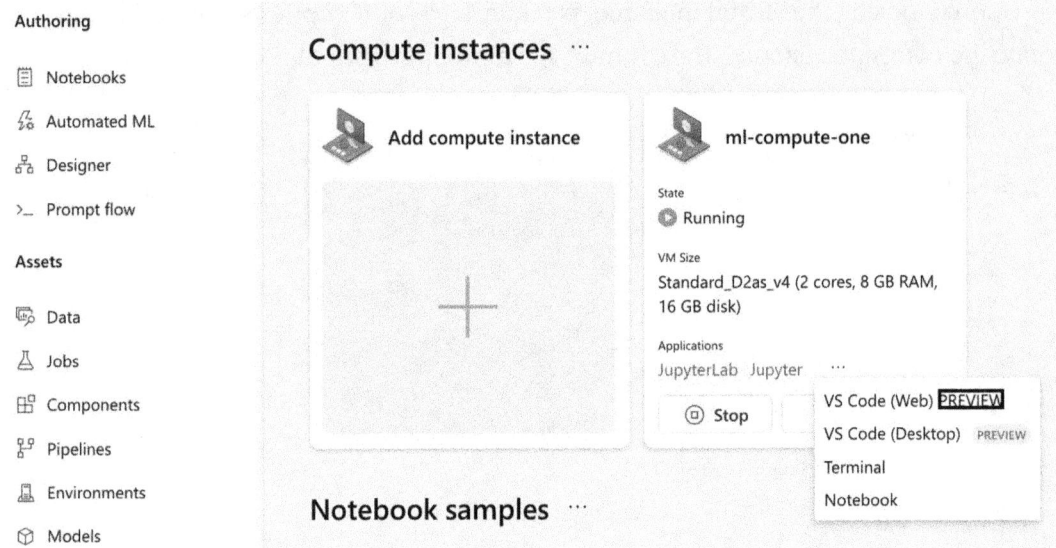

Figure 17-35. *Illustration of an active ML compute instance*

Let us build a classifier model that classifies whether a person is diabetic or not based on the given set of observations. Depending upon your convenience, you may either choose to work with a Jupyter notebook or access Visual Studio Code. The goal for us is to train the model using Azure ML Studio and deploy the model using Azure Functions.

Here is the model code for you:

```
import pandas as pd
import numpy as np
import pickle
from sklearn.ensemble import RandomForestRegressor
from sklearn.model_selection import train_test_split

df = pd.read_csv("diabetes.tsv",sep="\t")
#print("Top records of the dataset")
#print(df.head())
#print("Analytics on dataset")
#print(df.describe())

df = df.drop_duplicates()
df = df.fillna(0)
```

```python
# obtaining the least correlation data point with Y
eliminate = np.argmin(df.corr()["Y"].values)
# dropping the least important feature
final_df = df.drop(columns=df.columns[eliminate])

# parameters for random forest model
params = {
        "max_depth": 4,
        "n_estimators": 250,
        "min_samples_leaf": 4
        }
x = final_df.drop(["Y"], axis=1)
y = final_df["Y"]
X_train, X_test, y_train, y_test = train_test_split(x, y, test_size=0.2)

model = RandomForestRegressor(**params)

trained_model = model.fit(X_train, y_train)
pred = trained_model.predict(X_test)

file = 'model.pkl'
pickle.dump(model, open(file, 'wb'))
```

You can use this code as a starter, build your own features, and try out other classifiers as well. Once you have the pickle file exported, we can proceed to create a new Azure function to deploy the model. If you have followed along with the "Azure Functions" section earlier, then navigate to the terminal and create a new Azure function named diabetes.

Here is how that may look for you:

```
func init diabetes --worker-runtime python --model v2
```

We are going to deploy the model as a serverless API using Azure Functions. We will leverage the HTTP triggers. Now, change directory into the diabetes folder and navigate to the function_app.py file. This function has only been delayed locally. Let us deploy this function in the Azure function app.

Here is the code for deploying the model:

```
import azure.functions as func
import pickle
import json
import logging

app = func.FunctionApp()

classifier = ['Diabetic', 'Not-diabetic']

@app.function_name('pima-indian-diabetes-model')
@app.route(route="pidm", auth_level=func.AuthLevel.ANONYMOUS)
def main(requests: func.HttpRequest) -> func.HttpResponse:
    filepath = 'model.pkl'
    with open(filepath, 'rb') as file:
        model = pickle.load(file)

    data = requests.get_json()["data"]
    y_pred = model.predict(data)[0]
    pred_class = classifier[y_pred]

    logging.info(f"For the data points: {data}, the prediction is, {pred_class}")
    response = {"class": pred_class}
    return func.HttpResponse(json.dumps(response))
```

Conclusion

So far, we have explored a wide range of services from object storage to advanced analytics, to NoSQL solutions, to data integration using Azure Data Factory, to Azure Functions, and to Azure ML Studio. Azure Functions demonstrated how serverless computing can be utilized to execute code. From creating object storage solutions to deploying machine learning models, we looked at these services in detail along with code. While it's difficult to cover entire cloud computing offerings in a single chapter, I hope the ideas, services, and practical implementations we've explored will serve as a solid foundation for your cloud computing journey.

Index

A

Advanced data wrangling operations, CuDF
    apply() method, 154–156
    crosstab, 156, 157
    cut function, 157, 158
    factorize function, 158, 159
    group-by function, 152, 153
    transform function, 153, 154
    window function, 159
AGSI, *see* Asynchronous Server Gateway Interface (AGSI)
Airflow user, 390
Alpha, 188
Alternate hypothesis, 186, 187
Amazon, 335, 461, 469, 473, 489, 504
Amazon Athena, 504, 505
Amazon Glue, 506
Amazon RDS, 493–497
Amazon Redshift, 497–504
Amazon SageMaker, 519
    AWS SageMaker Studio, 519, 520
    create and configure, new Jupyter space, 521
    dashboard page, AWS Lambda, 526
    deploy API, 530
    home page, 520
    IAM role, 523, 524
    lambda function, 525
    linear methods, 522
    machine learning toolkits, 519
    POST method, 529
    REST API endpoint, 528
    SageMaker Domain, 519
    set environment variables, 527
Amazon's DynamoDB, 466
Amazon S3 Glacier, 489
Amazon S3 Standard, 489
Amazon Web Services (AWS)
    Amazon SageMaker (*see* Amazon SageMaker)
    data systems
        Amazon Athena, 504, 505
        Amazon Glue, 506
        Amazon RDS, 493–496
        Amazon Redshift, 497–502
    Lake Formation (*see* AWS Lake Formation)
Apache Airflow
    architecture
        configuration file, 391–393
        database, 390
        example, 393, 394
        executor, 390
        scheduler, 391
        web server, 390
    background process, 386
    DAGs (*see* Diagonal acyclic graph (DAGs))
    data storage and processing systems, 383
    definition, 389
    GUI, 387, 388
    home screen, 389

INDEX

Apache Airflow (*cont.*)
    initialization, 387
    initialize the backend database, 385, 386
    pip installation, 384, 385
    scheduler, 387
    set up, 384
    user credentials, text file, 388
    workflow, 389
    workflow management and orchestration platform, 384
Apache Arrow, 58, 94, 113, 136
Apache Cassandra, 466
Apache Hadoop, 59, 278
Apache Kafka, 277, 279
    architecture, 286, 322
        admin client, 286
        brokers, 282
        consumers, 283
        events, 281
        Kafka Connect, 284
        Kafka Streams, 284
        ksqlDB, 285, 286
        partitions, 282
        producers, 283
        replication, 282
        schema registry, 284
        topics, 282
    best practices, 321, 322
    callback function, 294
    config.py file, 292, 293, 295
    consumer.py, 295
    consumers, 283, 294–296
    dashboard, 290
    development environment, 292
    distributions, 279, 287
    elastic, 279
    fault-tolerant, 280
    functionalities, 279
    Kafka Connect, 321
    message brokers, 280
    messaging queue systems architecture, 281
    messaging systems architectures, 280
    print, 294
    produce method, 294
    producer.py file, 293
    producers, 296
    Protobuf serializer
        compiler, 303
        consumer/subscriber portion, 306
        guidelines, 301
        library, 305
        optional, 302
        ParseFromString() method, 307
        pb2.py file, 302
        producer script, 304, 305
        protoc, 302
        .proto file, 302
        Python file contents, 303, 304
        SerializeToString() method, 305
        student data, 305
        student model, 302, 303
    publisher-subscriber models, 322
    publisher-subscriber program, 292
    pub-sub messaging system, 281
    Python virtual environment, 292
    scalable, 279
    schema registry
        consumer configuration, 300
        consumer.py file, 299, 300
        JSONSerializer, 296, 298
        publisher.py, 296–298
        student data, 298, 301
    secure, 280
    servers and clients, 281

INDEX

setup/development
    API key/secrets, 291
    appropriate region, 289
    cluster dashboard, 290
    cluster type, 288
    config.py file, 291
    Confluent Kafka environment, 287
    topic, 290
uses, 279
Apache Spark, 455, 504
APIs, *see* Application programming interfaces (APIs)
Application programming interfaces (APIs), 324, 359
    development process, 327
    endpoints, 326, 327
    GraphQL, 325
    internal APIs, 325
    OpenWeather API, 324
    partner APIs, 325
    REST, 325, 328, 329
    SOAP, 325
    typical process, 326
    Webhooks, 325
apply() method, 154–156
Artifacts
    artifact key, 438
    data quality reports, 433
    debugging, 433
    documentation, 432
    link, 433, 434
    markdown, 434, 436
    persisted outputs, 432
    Prefect Cloud, 432
    Prefect Server, 432, 434
    share information, 432
    table, 436–438
    track and monitor, objects, 432

Async, 416
Asynchronous Server Gateway Interface (AGSI), 332
Automated machine learning (AutoML), 562
AutoML, *see* Automated machine learning (AutoML)
Avro, 59–60, 103, 144, 230, 284
AWS, *see* Amazon Web Services (AWS)
AWS Athena, 505, 515–517, 530
AWS console, 475
    access portal for respective user, 488
    assign users/groups, 487
    assign users to organizations, 487
    AWS organizational unit, 481
    configuration, new service control policy, 480
    create new organization, 477
    create organizational unit, 477, 478
    enable service control policies, 478, 479
    home page, 476
    IAM Identity Center, 482, 483, 486
    installation, AWS CLI, 488, 489
    organizational units, 475
    permission sets, 483
    predefined permission set, 484, 485
    service control policy, 478, 479
    sign up, new account, 474
AWS Glue, 469, 470, 502, 509, 515, 530
AWS Glue Data Catalog, 470, 502
AWS IAM Identity Center, 481–483, 488
AWS Lake Formation
    add permissions, IAM role, 511
    AWS Athena query editor, 515, 516
    create database, 508, 509
    create IAM role, 509, 510
    create new bucket, 506, 507

INDEX

AWS Lake Formation (*cont.*)
    create new crawler, 513, 514
    data lake service, 506
    grant permission, newly created role, 511–513
    new crawler initialization in AWS Glue, 515
    query data lake using AWS Athena, 516, 517
    query results, 518
    retrieval of results, AWS Athena, 517
    S3 buckets, 506
    trusted entity type, 510
AWS Lambda, 469, 524–526
AWS Organizations, 475, 477, 481, 486, 532
AWS RDS, 462, 493–496
AWS S3 (simple secure storage), 489
    AWS region, 491
    Bucket name, 491
    Bucket type, 491
    creation, new bucket, 491
    global bucket system, 490
    home page, 490
    stores data, 489
    uploading objects (files), 492, 493
AWS SageMaker Studio, 519, 520
Azure functions, 602, 614
    binding, 603
    core tools, 604, 605
    create, 605
    diabetes, 615
    execution, 606
    function app, 603, 604
    sample, 605, 606
    triggers, 602
    URLs, 606

B

Big data, 278, 454
    Apache Spark, 455
    Hadoop, 454
BigQuery, 367, 466, 552–554, 560
Bigtable, 466–468, 547–551, 574
Blob, 574, 596
Blocks, 418, 442–444
Bootstrapping, 258
Branching, 1, 16, 24, 408
BranchPythonOperator, 408, 410

C

Caching, 416, 460
cancel() method, 248
Capital expenditure, 452, 456
CD, *see* Continuous development (CD)
Central processing unit (CPU), 113, 133–136, 160, 225–228, 460, 560, 612
@check_input decorator, 193, 195
@check_output decorator, 193
Checkpoints, 199, 208, 216, 220
Checks, 183–186
Chi-square test, 257
Chunking, 232
Chunk loading, 60–61
Chunks, 231, 232, 236, 257, 266
CI, *see* Continuous integration (CI)
Clients, 234, 281, 324, 326, 332, 457
Client-server computing, 452
Client-server interaction, 452
client.submit function, 249
Cloud computing, 451, 456, 571
    advantages, 456
    architecture concepts
        caching, 460

INDEX

cloud computing vendors, 461
disaster recovery, 460
elasticity, 459
fault tolerance, 459, 460
high availability, 459
scalability, 458
deployment models
community cloud, 457
government cloud, 458
hybrid cloud, 457
multi-cloud, 458
private cloud, 457
public cloud, 457
GCP (*see* Google Cloud Platform (GCP))
networking concepts and terminologies (*see* Networking concepts)
security, 456
as web services, 473
Cloud computing services
CI, 469
Cloud storage, 463
compliance, 470
Compute, 463
containerization, 470
databases, 464
data catalog, 470
data governance, 470
data integration services, 469
data lakes, 468
data lifecycle management, 471
data protection, 470
data warehouses, 467, 468
identity and access management, 463
machine learning, 471
Not Only SQL (NoSQL) (*see* NoSQL database systems)

object storage, 463, 464
real-time/streaming processing service, 468
serverless functions, 469
Cloud computing vendors, 287, 453, 455–458, 461, 472
Cloud Functions, 469
Cloud infrastructure, 474, 532, 572
Cloud service models
categories, 461
infrastructure as a service, 462
platform as a service, 462
software as a service, 462
Cloud SQL, 541–544, 547, 569
Cloud storage, 463, 534, 560, 576
Cloud vendors, 413, 456, 459, 461, 464, 469–471
Colab notebooks, 140, 562
Column-oriented databases, 465–466, 584
Combining multiple CuDF objects
inner join, 150
left anti join, 151
left join, 148
left semi join, 151
outer join, 150
Combining multiple Pandas objects
cross join, 74
Cross join, 74
full outer joins, 72
inner join, 73
left join, 69
merge() method, 69
Pandas library, 69
right join, 72
Right join, 71
Combining multiple Polars objects
cross join, 119, 120
inner join, 117

621

INDEX

Combining multiple Polars objects (*cont.*)
    left join, 114–116
    outer join, 116
    semi join, 118, 119
Comma-separated values (CSV), 56, 102, 143, 171, 240, 250, 504, 518
Common table expression, 33–34, 39
Community cloud, 457–458
compare() method, 87–89
Compliance, 165, 457, 470–471, 482
Compute, 463
compute() function, 242, 244
compute() method, 231
Compute Unified Device Architecture (CUDA), 135–138, 537
Concurrency, 94, 229, 236
Concurrent processing, 226, 228, 229
ConcurrentTaskRunner, 447
Confluent's Kafka website, 287
Constrained types, 176, 183
Containerization, 470
Container registry, 470, 608
Content Delivery Network, 460
Context, 104–109
Continuous development (CD), 469
Continuous integration (CI), 469
CPU, *see* Central processing unit (CPU)
CPU-based data processing, 133
CPU environments, 161
CPU *vs.* GPU, 133, 134
create_link_artifact() method, 433
create_markdown_artifact() function, 434
Cron daemon process, 371, 372, 374
cronexample.sh, 375, 378
Cron job scheduler
    applications
        cron alternatives, 381
        database backup, 380
        data processing, 380
        email notification, 381
    concepts, 371, 372
    cron job usage, 375–380
    cron logging, 374, 375
    crontab file, 372–374
    in 1970s, 371
    shell script, 376, 377
    Unix operating system, 371
Cron logging, 362, 374, 375
crontab, 371, 372
crontab file, 372–374
crontab script, 379
Cross join, 69, 74–75, 119, 120
crosstab() function, 82–85
Cross tabulation, 82, 156, 157
Cross validation, 257, 266–270
Cryptographic signing and verification, 311
CSV, *see* Comma-separated values (CSV)
CUDA, *see* Compute Unified Device Architecture (CUDA)
CuDF
    advanced operations (*see* Advanced data wrangling operations, CuDF)
    basic operations
        column filtering, 145
        dataset, 144
        dataset sorting, 147, 148
        row filtering, 146
    combining multiple objects (*see* Combining multiple CuDF objects)
    cudf.pandas, 160, 161
    data formats, 143
    description, 136
    file IO operations
        CSV, 143

INDEX

JSON, 144
Parquet, 144
GPU-accelerated data manipulation library, 133
installation, 137–141
vs. Pandas, 136, 137
testing installation, 142, 143
cut function, 157, 158

D

DaaS, *see* Data as a service (DaaS)
DAGs, *see* Diagonal acyclic graph (DAGs)
Dask, 225
  architecture
    client, 234
    core library, 234
    initialization, Dask client, 235
    scheduler, 234
    task graphs, 236
    workers, 236
  banks, 230
  data structures
    dask array, 236–239
    dask bags, 239–241
    Dask data frames, 241–244
    dask delayed, 244–246
    dask futures, 246–248
  features
    chunking, 232
    Dask-CuDF, 233
    graph, 232
    lazy evaluation, 232
    partitioning, 232
    Picklingp, 233
    serialization, 232
    tasks, 231
  GPUs, 230
  healthcare, 230
  installation, 231
  national laboratories and organizations, 230
  optimize dask computations
    client.submit function, 249
    data locality, 250, 251
    priority work, 251
    Python garbage collection process, 250
    scheduler, 249, 250
    worker, 249
    work stealing, 251, 252
  Python library, 230
  set up, 230
  supports, 230
Dask-CuDF, 233
Dask data frames, 234, 241–244, 273
Dask distributed computing, 249, 251
Dask-ML
  data preprocessing techniques
    cross validation, 266, 267
    MinMaxScaler(), 264
    one hot encoding, 265
    RobustScaler(), 263
  hyperparameter tuning
    compute constraint, 269
    grid search, 269
    incremental search, 270
    memory constraint, 269
    random search, 270
  installation, 262
  ML libraries
    Keras, 262
    PyTorch, 261
    scikit-learn, 261
    TensorFlow, 262
    XGBoost, 261

INDEX

Dask-ML (*cont.*)
    setup, 262
    SimpleImputer, 272, 273
    statistical imputation, 272–274
DaskTaskRunner, 447
Data accuracy, 164–166
Data analysis, 41, 61, 93, 120, 160, 230, 370
Data as a service (DaaS), 324, 350
Data assets, 199, 210, 324, 463, 607–610
Database backup, 380
Databases, 390, 464–467, 544
Data catalog, 470, 502, 506, 512, 515
Data cleaning, 62, 255–256
Data coercion, 170, 180, 182–183
Data completeness, 166–167
Data context, 198, 203, 207, 208
Data documentation, 198, 199, 220–222
Data engineering tasks, ml, 254
Data exploration, 66–68, 255
Data extraction and loading
    Avro, 59
    CSV, 56
    feather, 58
    features, 56
    HDF5, 57
    ORC, 59
    Parquet, 58, 59
    pickle, 60
Data formats, 56, 62, 102, 167, 230, 326
Data frames, 42, 69, 98, 99, 120, 150, 207, 241–243
DataFrameSchema, 180–183, 194
Data governance, 369, 470
Data integration, 256, 284, 416, 469, 506, 577, 616
Data lakes, 455, 468, 506, 530
Data lifecycle management, 471, 474

Data locality, 250, 251
Data manipulation, 41, 42, 93, 103, 133, 161
Data munging, 41, 243
Data pipelines, 29, 62, 277, 370, 371, 601, 602
Data preprocessing techniques
    cross validation, 266, 267
    MinMaxScaler(), 264
    one hot encoding, 265
    RobustScaler(), 263
Data Processing, 2, 59, 94, 161, 225, 280, 380–381
Dataproc service, 554, 561
Data products, 324, 369, 574
Data projects, 361
Data protection, 470–471
Data quality reports, 433
Data range, 167
Data reshaping, 75
    crosstab(), 82
    melt() method, 81
    pivot() function, 76
    pivot table, 75
    stack() method, 78
    unstack(), 80
Data sources, 198, 199, 210, 255
Data splitting, 257
Data transformation, 65
    aggregations and group-by, 113
    basic operations
        arithmetic operations, 110
        df.select() method, 111
        df.with_columns() method, 111, 112
        expressions, 109
        with_columns() method, 111
    context

INDEX

filter, 106
group-by context, 106–109
selection, 104, 105
dataset, 66, 68
machine learning pipelines, 65
set of products, 67
String operations, 113
Data uniqueness, 166
Data validation
advantages, 165
definition, 165
disadvantages, 165
machine learning models, 164
Pandera (*see* Pandera)
principles (*see* Principles, data validation)
Pydantic library (*see* Pydantic)
RDBMSs, 163
specifications and rules, 164
Data validation workflow
copay_paid value, 215
data context, 208
data pipeline, 207
data source, 210
expectation suite, 210
get_context() method, 208
JSON document, 215
validator, 211–213
Data warehouses, 467, 468, 573
Data wrangling, 41, 256
data structures
data frame, 43, 44
series, 42
spreadsheets, 42
indexing, 44
Pandas, 42
Debian-based operating system, 374
Debugging, 425, 433

Decorators, 6, 192–195, 400
Delta Lake, 455
Deployment, 254, 260, 418, 445, 568, 575
df.select() method, 111
df.with_columns() method, 111
Diagonal acyclic graph (DAGs), 389
create variables, 404
DAG runs, 397
example, 395
function, 396
list, 398
macros, 407
nodes, 389
operator, 397, 398
params, 406, 407
Python context manager, 396
sensor, 399
task flow, 400, 401
tasks, 397
templates, 407
variables, 403–405
view, 395
workflow control
branching, 408
BranchPythonOperator, 408
ShortCircuitOperator, 410
triggers, 411, 413
typical scenario, 408
workflow management and orchestration system, 395
Xcom, 401–402
Dict, 7
Directed acyclic graph, 231, 232, 234, 236, 383
Disaster recovery, 460, 534
Distributed computing, 252, 276–278, 454, 460, 468, 569
Django, 323, 330, 347

625

DNS, *see* Domain Name System (DNS)
Docker, 311, 470, 521, 608
Documentation, 136, 198, 220, 221, 432, 595
Document databases, 465, 584
Domain Name System (DNS), 453

E

Eager evaluation, 93–96, 102, 247
Edges, 231, 232, 236, 389, 466
Elasticity, 279, 459
Email notification, 381
Encryption, 280, 404, 456, 574
Endpoints, 292, 324, 327, 567, 569
Engineering data pipelines, 2, 362
Enumerate functions, 8–10
ETL, *see* Extract, transform, and load (ETL)
Event processing, 278
Executor, 390–392
Expectations, 199, 203–207, 223, 257
Expectation store, 203, 208, 210, 222, 223
Expressions, 2, 11, 103, 123, 126, 242
Extract, transform, and load (ETL), 321, 365–367

F

Factorize function, 158, 159
factorize() method, 85–86
FastAPI, 323
    advantages, 330
    APIRouter objects, 334
    browser, 334, 335
    core concepts, 332
    create, 352, 353
    Curl command, 341
    database integration
        commands, 343
        create database, 344
        database user, 344, 345
        db1.py, 345, 347
        MySQL, 344
        mysql prompt, 344
        table, 345
    dependency injection, 342, 343
    documentation link, 358
    executing, 334
    filters, 336
    GET request documentation, 341
    GET request illustration, 342, 354
    get_student() function, 334
    HTTP error, 335
    middleware, 354, 355
    ML API endpoint
        middleware, 356, 357
        pickle file model, 355, 356
        POST request, 358
    Open API standards documentation, 340
    pass variables, 333
    prediction, 358, 359
    Pydantic integration, 337–339
    query parameters, 335, 337
    response_model attribute, 339
    RESTful data API, 350
        database connection, SQLAlchemy, 350, 351
        data validation class, 351
        define class, 351
        new student/transaction, 351, 352
    setup/installation, 331
    students and course data, 333
Fault tolerance, 247, 280, 282, 459, 460
Feature engineering, 133, 157, 158, 256–257, 562

INDEX

Field function, 173
Filter context, 106
Firewalls, 453, 578
Flask, 330
Flow runs, 418, 421–424, 431, 432, 438, 447
Flows, 397, 415, 417, 421, 424, 432, 442

G

GCP, *see* Google Cloud Platform (GCP)
Generator, 8, 14, 396
get_student() function, 334, 335
GIL, *see* Global interpreter lock (GIL)
Git, 16
    branching, 24
    cloning, 24
    code database, 16
    features, 30
    forking, 25
    GitHub, 17, 18
    pull request, 26
    Python code, 22
    repository, 21, 23
    and Secure Shell, 17
    SSH daemon process, 18
    tracking features, 22
GitHub account's settings, 19
GitHub server, 20, 23, 24
.gitignore file, 29, 30, 493
Global interpreter lock (GIL), 225–227
Google BigQuery, 552–554
Google Bigtable, 466, 548, 549, 551
Google Cloud, 461, 470, 532, 533
Google Cloud CLI, 535, 536
Google Cloud console page, 533
Google Cloud Platform (GCP), 469, 535, 536
    and AWS, 532

cloud computing services, 532
Cloud SQL, 541–544, 547
Cloud Storage, 534
compute engine (*see* Google Compute Engine)
Google BigQuery, 552–554
Google Bigtable, 548, 549, 551
Google Dataproc, 554–559
Google Vertex AI Workbench service, 559–561
new Google account/log, 533
organizational node, 532
Vertex AI (*see* Google's Vertex AI)
Google Cloud SDK, 535, 536
Google Cloud SQL, 541–547
Google Colab, 138–140, 147
Google Colab notebooks, 562
Google Compute Engine
    accelerator optimized, 537
    access VM instance through SSH, 541
    API enabled, 538
    compute optimized, 537
    create virtual machine, 537, 538
    enable Google Compute Engine API, 538
    general-purpose engine, 537
    Linux virtual machine, 538
    memory optimized, 537
    provision new virtual machine, 539
    storage optimized, 537
    types, 537
    VM instances list, 540
Google Dataproc, 554–559
Google Dataproc's Spark cluster, 555
Google IAM, 534
Google's Cloud Storage, 534
    traffic split, 567

INDEX

Google's Vertex AI
    AutoML feature, 562
    Colab notebooks, 562
    deployment dashboard, 568, 569
    deploy to endpoint, 568
    endpoint, model, ml, 566, 567
    import new model, 565, 566
    machine learning model, 562
    machine learning platform, 562
    model registry, 564
    online prediction, 567
    traffic split, 567
    and Vertex AI Workbench, 559–562
Government cloud, 458
GPU-accelerated environments, 161
GPU programming
    CPU instructions, 135
    Kernels, 135, 136
    memory management, 136
Gradient-boosted trees, 261
Graph databases, 466, 584
Graphical user interface (GUI), 25, 383, 387, 474, 502, 533
GraphQL, *see* Graph query language (GraphQL)
Graph query language (GraphQL), 325
Great Expectations
    checkpoints, 216
    components, 198, 199
    data documentation, 220, 221
    data validation libraries, 223
    definition, 198
    Expectations store, 222, 223
    functionality and environment, 207
    setup and installation
        CLI, 201
        data source, 203
        data sources, 202
        project, 198, 200
        project structure creation, 201, 202
        Python packages, 200
        relational databases, 203
        SQLAlchemy package, 200
        stores, 198
        virtual environments, 200
    writing expectations, 203–207
Grid search, 269, 270
Group-by context, 104, 106–109
Group-by function, 107, 152, 153
groupby() method, 89
GUI, *see* Graphical user interface (GUI)

H

Hadoop, 59, 231, 278, 454–455, 468
Hadoop Distributed File System (HDFS), 455
Hadoop ecosystem, 451, 454, 455
HDFS, *see* Hadoop Distributed File System (HDFS)
Hierarchical Data Format, 57
High availability, 459, 555, 587
Hive tool, 455
Horizontal scaling, 458, 459
Host, 135, 136, 287, 326, 458
h5py, 238
HTTP, *see* HyperText Transfer Protocol (HTTP)
HTTP methods, 325, 328, 329, 340
HTTP status code, 329, 330
Hybrid cloud, 457
Hyperparameters, 257, 259, 260, 269, 270, 562, 611
Hyperparameter tuning, 259
    grid search, 269
    incremental search, 270

random search, 270
HyperText Transfer Protocol (HTTP), 325, 328, 538, 602
Hypervisor software, 454
Hypothesis testing, 178, 186, 187

I

IAM Identity Center, 481–483, 486, 488
Identity and access management, 463, 483, 488, 523
Incremental search, 270–272
Indexing, 44
    loc and .iloc methods, 48
    multi-indexing, 49
    in Pandas data frame, 45
    parameter, 45, 46
    query, 48
    rows and columns, 48
    time delta, 52
Infrastructure as a service, 462, 577
Inner join, 69, 73–74, 117, 150
Internet, 452, 453, 457
Internet protocol (IP) address, 356, 453
isna() method, 63–64

J

JavaScript Object Notation (JSON), 57, 59, 102–103, 144, 222, 301, 465, 550
Jinja, 407
"joblib"-based algorithms, 261
JSON, *see* JavaScript Object Notation (JSON)
JSON file formats, 144
JSON schemas, 170, 174–176

K

Keras, 262
Kernels, 135, 136, 161
Key hashing, 283
Key-value stores, 466
Kubernetes cluster, 391

L

Lambdas, 5
Lazy evaluation, 93–97, 100, 189, 232, 247
Lazy frame, 100–102, 128
Left anti join, 151–152
Left join, 69–71, 114–116, 148
Left semi join, 151
Link artifact, 433, 434
Linux distribution, 137, 538
List comprehension, 11, 12
Logging, 283, 313, 375, 416, 574

M

Machine learning (ML), 355, 606
    classifier model, 614
    compute instance
        adding, 611, 612
        create, 613
        illustration, 613, 614
        naming, 612
    data assets, 609, 611
    data engineering tasks, 254
    diabetes, 615
    function_app.py file, 615
    GPUs, 612
    home page, 608
    jobs, 611
    model code, 614, 615

INDEX

Machine learning (ML) (*cont.*)
    model deployment, 616
    storage account, 608
    workspace, 607, 608
Machine learning data pipeline workflow
    data clean, 255, 256
    data exploration, 255
    data integration, 256
    data source, 255
    data splitting, 257
    data wrangling, 256
    deployment, 260
    feature engineering, 256
    feature selection, 257
    final test, 259
    hyperparameter tuning, 259
    model evaluation, 259
    model selection, 258
    monitoring, 260
    retraining, 260
    training, 258
Macros, 407–408, 413
Markdown artifacts, 434–436
mean() function, 239
melt() method, 81–82
Memory management, 134, 136, 161
Merge() method, 69, 70
Microsoft, 451, 461, 572
Microsoft Azure, 572
    authentication, 594
    blob storage, 574
        container, 576
        create, 574
        deployment, 574, 575
        sessions, 576
        uploading data, 575, 576
    Cosmos DB, 584
        APIs, 584
        categories, 585
        choosing API, 585
        cluster configuration, 588
        configuration, 587
        database sharding, 588
        firewall rules, 588, 589
        hardware resources, 586
        instance, 585
        JSON document, 589, 590
        MongoDB cluster, 587
        provisioned throughput, 584
        resource group, 586
        serverless model, 584
        units, 584
    data factory, 593, 594
        blob storage, 596–598
        console, 596
        copy data tool, 597, 598
        creation, 594
        data integration, 596
        data pipeline, 600
        data system connection, 596
        destination data system, 599, 600
        home page, 595
        linked service, 596, 597
        monitoring/rerunning, 601, 602
        source data system, 598, 599
        validation process, 601
    management group, 572, 573
    ML (*see* Machine learning (ML))
    resource group, 573
    resources, 574
    SQL, 577
        backup storage redundancy, 582
        database, 578, 579
        database server, 580
        data warehouse, 591
        eDTU, 580

elastic pool, 579, 580
    hardware resources, 581
    logical server, 581
    Python application, 582, 583
    SaaS, 578
    server, 577
    serverless pools, 591
  storage classes, 577
  subscriptions, 573
  synapse analytics, 591–593
Microsoft's Cosmos DB, 466
MinMaxScaler(), 264–265
Missing values
  data cleaning and transportation, 62
  data entry, 61
  methods and treatments
    isna() method, 63
    notna(), 64
  NA, 63
  NaN, 62
  NaT, 63
  None, 62
ML, *see* Machine learning (ML)
Multi-cloud, 458
Multi-indexing, 49
  arrays, 49
  columns, 50
  data frame, 50, 51
Multiprocessing, 94, 227
Multi-thread scheduler, 234
myVariable, 169

N

NA, *see* Not available (NA)
NAN, *see* Not a number (NAN)
NaT, *see* Not a time (NaT)
ndJSON, 102, 103

Networking concepts
  DNS, 453
  Firewalls, 453
  IP address, 453
  Ports, 453
  virtualization, 454
  Virtual Private Cloud, 453
Nodes, 231–232, 249, 389, 454, 466
NoSQL database systems
  column-oriented databases, 465, 466
  document databases, 465
  graph databases, 466
  key-value stores, 466
  Schema on Write *vs.* Schema on Read, 464
  time series databases, 467
  vector databases, 467
NoSQL data store, 547
Not a number (NAN), 62, 73, 136
Not a time (NaT), 63
Not available (NA), 63
Notifications, 381, 416, 418
NumPy arrays, 94, 234, 236
NumPy's data type object, 43
NVIDIA drivers, 138
NVIDIA GPU, 135–137
NVIDIA offerings, 140

O

Object relational mapping (ORM), 347
  Alembic, 350
  SQLAlchemy, 347
    engine, 348
    query API, 349, 350
    session, 348, 349
Object storage, 443, 463, 464, 474, 489, 506, 530

INDEX

Observability, 361, 371, 417
One hot encoding, 265–266
OpenWeather API, 324
Operational expenditure, 452
Optimized Row Columnar (ORC), 59
ORC, *see* Optimized Row Columnar (ORC)
Organizations, 165, 230, 253, 361, 454, 532
ORM, *see* Object relational mapping (ORM)
Outer join, 69, 72–73, 116, 150

P

Pandas, 42, 49, 55, 69
    and NumPy, 42
    objects, 91
Pandera
    Checks, 183–186
    data coercion, 182, 183
    data frame schema, 180–182
    declarative schema definition, 178
    definition, 178
    installation, 179, 180
    lazy validation, 179, 189–192
    statistical validation, 178, 186–188
Pandera decorators
    @check_input, 193
    @check_output, 193
    DataFrameSchema, 194, 195
    data validators, 194
    decoratortest, 193
    hello_world() function, 193
    validation condition, 195
    validation parameters, 193
Parallelism, 100, 229, 236, 247, 612
Parallel processing
    concepts, 227
    core, 227

GIL, 226
history, 226
identify CPU Cores, 227, 228
large programming task into several smaller tasks, 226
multiprocessing library, 228
process, 227
Python, 226
thread, 227
Params, 406–407
Parquet, 58, 59, 103, 144, 504, 591, 609
Partitioning, 89, 226, 232, 257
Path parameters, 327, 332–337
Peer-to-peer computing, 452
Pickle files, 564
Pickling, 232–233, 564
Pig Latin, 455
Pivot_table() method, 76–78
Platform as a service, 462, 471, 577, 578
Polars
    advanced methods
        dataset, 120
        missing values identification, 121
        pivot function, 122
        unique values identification, 122
    combing objects (*see* Combing multiple Polars objects)
    data extraction and loading
        CSV file, 102
        JSON, 102, 103
        Parquet, 103
    data structures, 94, 97–103
    data transformation (*see* Data transformation, Polars)
    definition, 93
    lazy evaluation and eager evaluation, 94
    lazy *vs.* eager evaluation, 95, 96

INDEX

multi-threading and parallelization, 94
objects, 130
Python data analysis ecosystem, 94
Rust, 94
syntax, 94
Polars CLI
    SQL query, 128–130
    structure, 128
Polars/SQL interaction
    copay_paid, 124
    data type conversions, 125
    random dataset, 124
    random_health_data, 126
    SQL context object, 126
    SQL queries, 123, 127
Ports, 453
Postgres database syntax, 31
POST method, 339, 528, 529
Preceding data frame, 83
Prefect, 415
    async, 416
    backend Prefect database, 420
    caching, 416
    development, 421 (*see also* Prefect development)
    future, 440
    installation, 417, 418
    logging, 416
    notifications, 416
    observability, 417
    open source workflow orchestration system, 420
    Prefect Cloud, 417
    Prefect Core, 417
    Python functions, 416
    retries, 416
    scheduling, 416
    second-generation, 416
    secrets management system, 420
    server, 418–420
    set up, 417
    shallow, 416
    user interface, web browser, 420
Prefect development
    artifacts (*see* Artifacts)
    blocks, 442–444
    flow run output, 431
    flow runs, 421–423
    flows, 421
    interface, 424
    log trace, prefect flow, 424
    persisting results, 428–432
    results, 427, 428
    state change hooks, 441, 442
    states, 438–440
    task runners, 446, 447
    tasks, 424–427
    variables, 444–446
Principles, data validation
    data accuracy, 166
    data completeness, 166, 167
    data consistency, 167
    data format, 167
    data range, 167
    data uniqueness, 166
    referential integrity, 168
Private cloud, 457
Protocol Buffers (Protobuf), 284, 301
Public cloud, 453, 457
Pull requests, 25, 27, 28
Pydantic
    applications, 168
    constrained type, 176
    data schemas declaring, 168
    definition, 168
    field function, 173

Pydantic (*cont.*)
  installation, 170
  JSON schemas, 174, 176
  Pydantic models, 171–173
  type annotations, 168–170
  validation and serialization logic, 170
  validators, 176–178
Pydantic V2, 170
Python, 225, 226
Python 3, 138, 140
Python codes, 22, 23, 95, 193, 227, 400
Python decorators, 6, 192, 416
Python library, 42, 94, 136, 227, 230, 307, 330, 350
Python object, 42–44, 60, 196, 234, 298, 564
Python programming language, 2, 62, 135
  concepts, 2
  f string, 2, 3
  function arguments
    args parameter, 4
    kwargs, 5
  functions, 3, 4
  preceding code, 4
Python script, 22, 161, 292, 301, 345, 375, 421
Python type annotations, 168–170
PyTorch, 261–262, 519, 520

Q

Query parameters, 332, 333

R

Random module, 13
  choice function, 14
  getrandbits() method, 13
  randint(), 13
  range, 13
  sample method, 15
  seed() method, 16
  shuffle() method, 15
Real-time data pipelines, 277, 279, 322
Redshift, 467, 497–502
Referential integrity, 168
Relational database service (RDS), *see* Amazon RDS
Remote computing, 452
Representational state transfer (REST), 325, 328, 329
REST, *see* Representational state transfer (REST)
RESTful services, 328
RobustScaler(), 263–264
Rust, 93, 94, 96, 170

S

Scalability, 350, 458–459
Scheduler, 234, 249, 391
Scheduling, 250, 369, 377, 416
SchemaError, 189
scikit-learn, 160, 230, 260–263, 565
Selection context, 104, 105, 111
Semi join, 118, 119
Sensors, 61, 279, 399–400
Sequential execution, 446
SequentialTaskRunner, 446
Serialization, 170, 232–233, 298, 305
Series, 41–43, 97, 98, 143, 254, 389
Serverless computing, 455, 459, 616
ShortCircuitOperator, 408, 410
Similarity search, 467
SimpleImputer, 272, 273
Simple object access protocol (SOAP), 325
Single sign-on (SSO), 481, 482

Single-thread scheduler, 234
Skorch library, 261
Software as a Service (SaaS), 462, 577, 578
SOAP, *see* Simple object access protocol (SOAP)
Spark, 451, 455, 468, 531, 559, 561, 609
"spark.jars" file, 561
SQL, *see* Structured query language (SQL)
SQLAlchemy, 200, 347–348, 350, 390
SQL Server, 493, 542, 578
SSH connection, 20
SSO, *see* Single sign-on (SSO)
Stacking, 258
Stack() method, 78–79
Starlette, 331–332
State change hooks, 441, 442, 449
State object, 439, 440
Streaming data, 278, 284–286, 309, 321, 322
Stream processing
    Kafka Streams, 307
    kSQL, 307
    ksqlDB, 307, 309
        API key and secret, 313
        apt repository, 311
        auto.offset.reset, 318
        CLI instructions, 313
        command line interface, 311, 313, 314
        Confluent Kafka, 309
        Confluent platform, 312
        creation, 309, 310
        environment variable and path, 312
        help command, 312
        insert data, 319
        Java, 311
        naming/sizing, 310
        prerequisites, 311
        public key, 311
        query, 319, 320
    Stateful processing, 308
    Stateless processing, 308
    topics
        create schema, 316–318
        creation, 314, 315
        naming/partitions, 315, 316
Structured query language (SQL), 1
    queries, 31, 38, 39
    self-joins, 33
    tables, 32
    temporary tables, 36
    view
        materialized, 35
        standard, 34
    window functions, 37
sum() method, 64, 109
Systemd, 381

T

Table artifact, 436–438
Task graphs, 236, 245
Task runners, 446–448
Tasks, 231, 397, 424–427
T.compute(), 237
Templates, 407, 483
TensorFlow, 160, 262, 519, 560, 565
Time delta indexes, 52
    single value, 54
    subset, 55
Time series databases, 467
"to_hdf5" method, 239
transform() function, 153, 154
Type annotations refresher, 168–170
Type hinting, 7
Typing module, 7–8

INDEX

U

Uniform resource identifier (URI), 326, 327, 565
Uniform resource locator (URL), 326, 327, 329, 332, 335–337
unstack() method, 80–81
Unsupervised learning, 258, 261
URI, *see* Uniform resource identifier (URI)
URL, *see* Uniform resource locator (URL)
User-defined function, 154–156

V

Validate method, 189
Validators, 176–178, 194, 215, 222
Variables, 2–4, 62, 81, 156–158, 444, 445
Vector databases, 467, 584
Vertex AI Workbench, 559–562, 569
Vertical scaling, 459
Virtualization, 454, 457
Virtual Private Cloud, 453–454
Virtual Private Network, 454

W

wait() function, 248
Webhooks, 325

Web server, 57, 283, 331, 390, 404, 420
Window function, 37, 133, 159
Workflow configuration, 368
　example, 368
　organization, 368
　parameters, 368, 369
　product/solution, 368
Workflow orchestration
　automations, 370
　centralized administration and management of process, 369
　data pipelines, 370, 371
　define workflow, 369
　error handling and routing, 370
　ETL data pipeline workflow, 365–367
　example, 362, 364
　integration with external systems, 370
　scheduling, 369
　tasks/activities, 362
　utilization, resources, 369
　workflow configuration, 368, 369
World Wide Web, 452

X, Y, Z

Xcom, 383, 401–402
XGBoost, 160, 230, 261, 262

GPSR Compliance

The European Union's (EU) General Product Safety Regulation (GPSR) is a set of rules that requires consumer products to be safe and our obligations to ensure this.

If you have any concerns about our products, you can contact us on

ProductSafety@springernature.com

In case Publisher is established outside the EU, the EU authorized representative is:

Springer Nature Customer Service Center GmbH
Europaplatz 3
69115 Heidelberg, Germany